J773

£125 est

SA

D0271302

Understanding
the Process
of Aging

OXIDATIVE STRESS AND DISEASE

Series Editors

LESTER PACKER, PH.D.
University of California
Berkeley, California

ENRIQUE CADENAS, M.D., PH.D.
University of Southern California School of Pharmacy
Los Angeles, California

1. Oxidative Stress in Cancer, AIDS, and Neurodegenerative Diseases, *edited by Luc Montagnier, René Olivier, and Catherine Pasquier*
2. Understanding the Process of Aging: The Roles of Mitochondria, Free Radicals, and Antioxidants, *edited by Enrique Cadenas and Lester Packer*

Additional Volumes in Preparation

Redox Regulation of Cell Signaling and Its Clinical Application, *edited by Lester Packer and Junji Yodoi*

Related Volumes

Vitamin E in Health and Disease: Biochemistry and Clinical Applications, *edited by Lester Packer and Jürgen Fuchs*

Vitamin A in Health and Disease, *edited by Rune Blomhoff*

Free Radicals and Oxidation Phenomena in Biological Systems, *edited by Marcel Roberfroid and Pedro Buc Calderon*

Biothiols in Health and Disease, *edited by Lester Packer and Enrique Cadenas*

Handbook of Antioxidants, *edited by Enrique Cadenas and Lester Packer*

Handbook of Synthetic Antioxidants, *edited by Lester Packer and Enrique Cadenas*

Vitamin C in Health and Disease, *edited by Lester Packer and Jürgen Fuchs*

Lipoic Acid in Health and Disease, *edited by Jürgen Fuchs, Lester Packer, and Guido Zimmer*

Flavonoids in Health and Disease, *edited by Catherine Rice-Evans and Lester Packer*

Understanding the Process of Aging

The Roles of Mitochondria, Free Radicals, and Antioxidants

edited by

ENRIQUE CADENAS

University of Southern California School of Pharmacy
Los Angleles, California

LESTER PACKER

University of California
Berkeley, California

MARCEL DEKKER, INC.　　　　　NEW YORK · BASEL

ISBN: 0-8247-1723-6

This book is printed on acid-free paper.

Headquarters
Marcel Dekker, Inc.
270 Madison Avenue, New York, NY 10016
tel: 212-696-9000; fax: 212-685-4540

Eastern Hemisphere Distribution
Marcel Dekker AG
Hutgasse 4, Postfach 812, CH-4001 Basel, Switzerland
tel: 44-61-261-8482; fax: 44-61-261-8896

World Wide Web
http://www.dekker.com

The publisher offers discounts on this book when ordered in bulk quantities. For more information, write to Special Sales/Professional Marketing at the headquarters address above.

Copyright © 1999 by Marcel Dekker, Inc. All Rights Reserved.

Neither this book nor any part may be reproduced or transmitted in any form or by any means, electronic or mechanical, including photocopying, microfilming, and recording, or by any information storage and retrieval system, without permission in writing from the publisher.

Current printing (last digit):
10 9 8 7 6 5 4 3 2 1

PRINTED IN THE UNITED STATES OF AMERICA

Series Introduction

Oxygen is a dangerous friend. Overwhelming evidence indicates that oxidative stress can lead to cell and tissue injury. However, the same free radicals that are generated during oxidative stress are produced during normal metabolism and thus are involved in both human health and disease.

Free radicals are molecules with an odd number of electrons. The odd, or unpaired, electron is highly reactive as it seeks to pair with another free electron.

Free radicals are generated during oxidative metabolism and energy production in the body.

Free radicals are involved in:

Enzyme-catalyzed reactions

Electron transport in mitochondria

Signal transduction and gene expression

Activation of nuclear transcription factors

Oxidative damage to molecules, cells, and tissues

Antimicrobicidal action of neutrophils and macrophages

Aging and disease

Normal metabolism is dependent upon oxygen, a free radical. Through evolution, oxygen was chosen as the terminal electron acceptor for respiration. The two unpaired electrons of oxygen spin in the same direction; thus, oxygen is a biradical, but is not a very dangerous free radical. Other oxygen-derived free radical species, such as superoxide or hydroxyl radicals, formed during metabolism or by ionizing radiation are stronger oxidants and are therefore more dangerous.

In addition to research on the biological effects of these reactive oxygen species, research on reactive nitrogen species has been gathering momentum. NO, or nitrogen monoxide (nitric oxide), is a free radical generated by NO synthase (NOS). This enzyme modulates physiological responses such as vasodilation or signaling in the brain. However, during inflammation, synthesis of NOS (iNOS) is induced. This iNOS can result in the overproduction of NO, causing damage. More worrisome, however, excess NO can react with superoxide to produce the very toxic product peroxynitrite. Oxidation of lipids, proteins, and DNA can result, thereby increasing the likelihood of tissue injury.

Both reactive oxygen and nitrogen species are involved in normal cell regulation in which oxidants and redox status are important in signal transduction. Oxidative stress is increasingly seen as a major upstream component in the signaling cascade involved in inflammatory responses, stimulating adhesion molecule and chemotractant production. Hydrogen peroxide, which breaks down to produce hydroxyl radicals, can also activate NFκB, a transcription factor involved in stimulating inflammatory responses. Excess production of these reactive species is toxic, exerting cytostatic effects, causing membrane damage, and activating pathways of cell death (apoptosis and/or necrosis).

Virtually all diseases thus far examined involve free radicals. In most cases, free radicals are secondary to the disease process, but in some instances free radicals are causal. Thus, there is a delicate balance between oxidants and antioxidants in health and disease. Their proper balance is essential for ensuring healthy aging.

The term *oxidative stress* indicates that the antioxidant status of cells and tissues is altered by exposure to oxidants. The redox status is thus dependent upon the degree to which a cell's components are in the oxidized state. In general, the reducing environment inside cells helps to prevent oxidative damage. In this reducing environment, disulfide bonds (S—S) do not spontaneously form because sulfhydryl groups kept in the reduced state (SH) prevent protein misfolding or aggregation. This reducing environment is maintained by oxidative metabolism and by the action of antioxidant enzymes and substances, such as glutathione, thioredoxin, vitamins E and C, and enzymes such as superoxide dismutase (SOD), catalase, and the selenium-dependent glutathione and thioredoxin hydroperoxidases, which serve to remove reaction oxygen species.

Changes in the redox status and depletion of antioxidants occur during oxidative stress. The thiol redox status is a useful index of oxidative stress mainly because metabolism and NADPH-dependent enzymes maintain cell glutathione (GSH) almost completely in its reduced state. Oxidized glutathione (glutathione disulfide, GSSG) accumulates under conditions of oxidant exposure, and this changes the ratio of oxidized to reduced glutathione; and increased ratio indicates oxidative stress. Many tissues contain large amounts of glutathione 2–4 mM in erythrocytes or neural tissues and up to 8 mM in hepatic tissues. Reactive oxygen and nitrogen species can directly react with glutathione to lower the levels of this substance, the cell's primary preventative antioxidant.

Current hypotheses favor the idea that lowering oxidative stress can have a clinical benefit. Free radicals can be overproduced or the natural antioxidant system defenses weakened, first resulting in oxidative stress, and then leading to oxidative injury and disease. Examples of this process include heart disease and cancer. Oxidation of human low-density lipoproteins is considered the first step in the progression and eventual development of atherosclerosis, leading to cardiovascular disease. Oxidative DNA damage initiates carcinogenesis.

Compelling support for the involvement of free radicals in disease development comes from epidemiological studies showing that an enhanced antioxidant status is associated with reduced risk of several diseases. Vitamin E and prevention of cardiovascular disease is a notable example. Elevated antioxidant status is also associated with decreased incidence of cataracts and cancer, and some recent reports have suggested an inverse correlation between antioxidant status and occurrence of rheumatoid arthritis and diabetes mellitus. Indeed, the number of indications in which antioxidants may be useful in the prevention and/or the treatment of disease is increasing.

Oxidative stress, rather than being the primary cause of disease, is more often a secondary complication in many disorders. Oxidative stress diseases include inflammatory bowel disease, retinal ischemia, cardiovascular disease and restenosis, AIDS, ARDS, and neurodegenerative diseases such as stroke, Parkinson's disease, and Alzheimer's disease. Such indications may prove amenable to antioxidant treatment because there is a clear involvement of oxidative injury in these disorders.

In this new series of books, the importance of oxidative stress in diseases associated with organ systems of the body will be highlighted by exploring the scientific evidence and the medical applications of this knowledge. The series will also highlight the major natural antioxidant enzymes and antioxidant substances such as vitamins E, A, and C, flavonoids, polyphenols, carotenoids, lipoic acid, and other nutrients present in food and beverages.

Oxidative stress is an underlying factor in health and disease. More and more evidence is accumulating that a proper balance between oxidants and antioxidants is involved in maintaining health and longevity, and that altering this balance in favor of oxidants may result in pathological responses causing functional disorders and disease. This series is intended for researchers in the basic biomedical sciences and clinicians. The potential for healthy aging and disease prevention necessitates gaining further knowledge about how oxidants and antioxidants affect biological systems.

Since the beginning of the twentieth century, the human lifespan has undergone dramatic decreases in mortality and the result has been to shift the human survival curve versus chronological age to the right, considerably increasing average lifespan. The main factors leading to this dramatic improvement have been better hygiene, better nutrition, and medical advances especially in decreasing infectious diseases. Further improvements in human lifespan now will depend more critically upon intrinsic factors, a balance between metabolism and genetic factors, and environmental influences. Oxidants and antioxidants must be maintained in a delicate balance in order to reduce degenerative diseases of aging. In 1984, Dehnam Harman, who originated the free radical theory on aging, stated "Very few individuals, if any, reach their potential maximum lifespan; they die instead prematurely of a wide variety of diseases—the vast majority being from free radical diseases." The truth of this statement seems more evident in these days in that cell signaling and gene expression have been found to exhibit steps sensitive to oxidants, redox changes, and antioxidants. Hence, including this volume, *Understanding the Process of Aging: The Roles of Mitochondria, Free Radicals, and Antioxidants* seems appropriate.

Lester Packer
Enrique Cadenas

Preface

The idea that free radicals may play a causal role in the aging process by producing cumulative oxidative damage was advanced in 1956 by Denham Harman. With time, the free radical theory of aging has incorporated other ideas, such as the concept of imbalance between pro- and antioxidants as the basis for oxidative molecular damage, thus evolving into the oxidative stress hypothesis of aging. Early on, Harman also suggested that the mitochondrion was particularly vulnerable to oxidative stress, a concept redefined by Miquel and his colleagues in the 1980s, who proposed that senescence was a by-product of free radical attack on the mitochondrial genome of postmitotic cells.

Mitochondria have a prominent place in the oxidative stress hypothesis of aging for several reasons.

1. These organelles are the most abundant and important subcellular site of oxyradical production in mammalian organs and the rate of oxidant generation increases with age.
2. Mitochondrial DNA is in close proximity to the sites of oxyradical generation and is not protected by histones; consequently, oxidatively modified bases in mitochondrial DNA are more prevalent than in nuclear DNA.
3. Mitochondria are susceptible to accruing informational errors through oxidative damage to their DNA (and RNAs), which encodes about 10% of the respiratory chain proteins.
4. Mitochondrial dysfunction elicited by oxyradicals has profound effects for the cell: this dysfunction can be amplified into cellular death by triggering apoptotic pathways.

Thus the mitochondrial DNA hypothesis of aging incorporates most of the basic developments since the original free radical hypothesis was formulated over 40 years ago.

This book is devoted largely to this current view of aging, to the role of mitochondria in cellular aging, apoptosis, and neurodegeneration, and to the evaluation of mutations of mitochondrial DNA in human aging and age-related diseases. Chapters on mitochondrial production of oxygen-centered radicals, its pathophysiological regulation by nitric oxide, and implications for mitochondrial function provide an unprecedented insight into the mechanistic basis underlying the cellular alterations inherent in the aging process.

Each chapter summarizes the current knowledge in the field, provides an assessment of the role of mitochondria in aging and age-related diseases, and poses new questions pertaining to this intense and fascinating area of research.

Enrique Cadenas
Lester Packer

Contents

Contributors

Bruce N. Ames, Ph.D.   Department of Molecular and Cell Biology, University of California, Berkeley, California

Miguel A. Asensi, Ph.D.   Department of Physiology, Faculty of Medicine, University of Valencia, Valencia, Spain

Angelo Azzi, M.D.   Institute for Biochemistry and Molecular Biology, University of Bern, Bern, Switzerland

Bernard M. Babior, M.D., Ph.D.   Department of Molecular and Experimental Medicine, The Scripps Research Institute, La Jolla, California

Andreas Becker   Fachberiche Chemie, Philipps-Universität, Marburg, Germany

Kenneth B. Beckman, Ph.D.   Department of Molecular and Cell Biology, University of California, Berkeley, California

Alberto Boveris, Ph.D.   Department of Physical Chemistry, School of Pharmacy and Biochemistry, University of Buenos Aires, Buenos Aires, Argentina

Ulf T. Brunk, M.D., Ph.D.   Division of Pathology II, Department of Neuroscience and Locomotion, University of Linköping, Linköping, Sweden

Enrique Cadenas, M.D., Ph.D.   Department of Molecular Pharmacology and Toxicology, University of Southern California School of Pharmacy, Los Angeles, California

Catherine Freitag Clarke, Ph.D.   Department of Chemistry and Biochemistry, University of California, Los Angeles, California

xi

Trudy Cornwell, Ph.D.   Department of Pathology, University of Alabama at Birmingham, Birmingham, Alabama

Gino A. Cortopassi, Ph.D.   Department of Molecular Biosciences, University of California, Davis, California

Lidia E. Costa, Ph.D.   Department of Physical Chemistry, School of Pharmacy and Biochemistry, University of Buenos Aires, Buenos Aires, Argentina

Victor M. Darley-Usmar, Ph.D.   Department of Pathology, University of Alabama at Birmingham, Birmingham, Alabama

Juan M. Esteve, Ph.D.   Department of Physiology, Faculty of Medicine, University of Valencia, Valencia, Spain

Steve Esworthy, Ph.D.   Department of Medical Oncology and Experimental Therapeutics, City of Hope National Medical Center, Duarte, California

Robert A. Floyd   Free Radical Biology and Aging Research Program, Oklahoma Medical Research Foundation, Oklahoma City, Oklahoma

Henry Jay Forman, Ph.D.   Department of Molecular Pharmacology and Toxicology, University of Southern California, Los Angeles, California

José García de la Asunción, Ph.D.   Department of Physiology, Faculty of Medicine, University of Valencia, Valencia, Spain

Roberta A. Gottlieb, M.D.   Department of Molecular and Experimental Medicine, The Scripps Research Institute, La Jolla, California

David M. Guidot, M.D.   Atlanta VAMC and Department of Medicine, Emory University, Atlanta, Georgia

Kenneth Hensley, Ph.D.   Free Radical Biology and Aging Research Program, Oklahoma Medical Research Foundation, Oklahoma City, Oklahoma

Marjan Huizing, Ph.D.   Department of Pediatrics, University Hospital, Nijmegen, The Netherlands

Yoko Inai, Ph.D.   Division of Molecular and Cell Biology, Institute of Medical Science, Kurashiki Medical Center, Kurashiki, Japan

Masayasu Inoue, Ph.D. Department of Biochemistry, Osaka City University Medical School, Osaka, Japan

Bernhard Kadenbach, Ph.D. Fachbereich Chemie, Philipps-Universität, Marburg, Germany

Tomoko Kanno, Ph.D. Division of Molecular and Cell Biology, Institute of Medical Science, Kurashiki Medical Center, Kurashiki, Japan

Linda K. Kwong, Ph.D. Department of Biological Sciences, Southern Methodist University, Dallas, Texas

Achim Lass, Ph.D. Department of Biological Sciences, Southern Methodist University, Dallas, Texas

Sandra Leist, Ph.D. Fachbereich Chemie, Philipps-Universität, Marburg, Germany

Lindsay Maidt, Ph.D. Free Radical Biology and Aging Research Program, Oklahoma Medical Research Foundation, Oklahoma City, Oklahoma

Joe M. McCord, Ph.D. Webb-Waring Antioxidant Research Institute and the University of Colorado Health Sciences Center, Denver, Colorado

Akitane Mori, Ph.D. Department of Neuroscience, Okayama University Medical School, Okayama, Japan

Jörg Napiwotzki, Ph.D. Fachbereich Chemie, Philipps-Universität, Marburg, Germany

Manabu Nishikawa, Ph.D. Department of Biochemistry, Osaka City University Medical School, Osaka, Japan

Takayuki Ozawa, M.D., Ph.D. Faculty of Medicine, University of Nagoya, Nagoya, Japan

Lester Packer, Ph.D. Department of Molecular and Cell Biology, University of California, Berkeley, California

Federico V. Pallardó, Ph.D. Department of Physiology, Faculty of Medicine, University of Valencia, Valencia, Spain

Fernando Palmieri, Ph.D.   Department of Pharmaco-Biology, University of Bari, Bari, Italy

Rakesh P. Patel, Ph.D.   Department of Pathology, University of Alabama at Birmingham, Birmingham, Alabama

Quentin N. Pye, Ph.D.   Free Radical Biology and Aging Research Program, Oklahoma Medical Research Foundation, Oklahoma City, Oklahoma

Annette Reith, Ph.D.   Fachbereich Chemie, Philipps-Universität, Marburg, Germany

Christoph Richter, Ph.D.   Eidgenössische Technische Hochschule, Zürich, Switzerland

K. A. Robinson, Ph.D.   Free Radical Biology and Aging Research Program, Oklahoma Medical Research Foundation, Oklahoma City, Oklahoma

Wim Ruitenbeek, Ph.D.   Department of Pediatrics, University Hospital, Nijmegen, The Netherlands

Juan Sastre, Ph.D.   Department of Physiology, Faculty of Medicine, University of Valencia, Valencia, Spain

Eisuke F. Sato, Ph.D.   Department of Biochemistry, Osaka City University Medical School, Osaka, Japan

Jeffery R. Schultz, Ph.D.   Department of Chemistry, Pacific Lutheran University, Tacoma, Washington

J. A. M. Smeitink, Ph.D.   Department of Pediatrics, University Hospital, Nijmegen, The Netherlands

Rajindar S. Sohal, Ph.D.   Department of Biological Sciences, Southern Methodist University, Dallas, Texas

C. A. Stewart, Ph.D.   Free Radical Biology and Aging Research Program, Oklahoma Medical Research Foundation, Oklahoma City, Oklahoma

Yoshiki Takehara, Ph.D.   Division of Molecular and Cell Biology, Institute of Medical Science, Kurashiki Medical Center, Kurashiki, Japan

Alexei Terman, Ph.D. Division of Pathology II, Department of Neuroscience and Locomotion, University of Linköping, Linköping, Sweden, and Institute of Gerontology, AMS of Ukraine, Kiev, Ukraine

J. M. F. Trijbels, Ph.D. Department of Pediatrics, University Hospital, Nijmegen, The Netherlands

Kozo Utsumi, Ph.D. Institute of Medical Science, Kurashiki Medical Center, Kurashiki, Japan

L. P. van den Heuvel, Ph.D. Department of Pediatrics, University Hospital, Nijmegen, The Netherlands

José Viña-Ribes, M.D., Ph.D. Department of Biochemistry and Molecular Biology, Faculty of Medicine, University of Valencia, Valencia, Spain

Patrick B. Walter, Ph.D. Department of Molecular and Cell Biology, University of California, Berkeley, California

Alice Wong Department of Molecular Biosciences, University of California, Davis, California

Munchisa Yabuki, Ph.D. Division of Molecular and Cell Biology, Institute of Medical Science, Kurashiki Medical Center, Kurashiki, Japan

Liang-Jun Yan Department of Biological Sciences, Southern Methodist University, Dallas, Texas

1

The Mitochondrial Production of Oxygen Radicals and Cellular Aging

Alberto Boveris and Lidia E. Costa
School of Pharmacy and Biochemistry, University of Buenos Aires, Buenos Aires, Argentina

Enrique Cadenas
University of Southern California School of Pharmacy, Los Angeles, California

I. PRODUCTION OF SUPEROXIDE RADICAL AND HYDROGEN PEROXIDE IN MAMMALIAN ORGANS

A series of organelles isolated from mammalian organs and other aerobic eukaryotic cells have been described as sources of superoxide radical anion (O_2^-) and hydrogen peroxide (H_2O_2). These two molecules, the products of the univalent and bivalent reduction of oxygen, are physiologically produced as a characteristic of aerobic metabolism and constitute normal intracellular metabolites. Superoxide radical is produced from one-electron transfer to the oxygen molecule, which originates a chemical species with an odd number of electrons, in other words an oxygen free radical, in short, an oxyradical. Membranes isolated from mitochondria, endoplasmic reticulum, and plasma membrane have been recognized as able to catalyze the univalent reduction of oxygen to O_2^- (1). Hydrogen peroxide is generated both as the product of O_2^- dismutation and as the product of two-electron transfer from flavin enzymes to the oxygen molecule. In mammalian cells most of the H_2O_2-producing enzymes are located in the peroxisomes (1,2).

Some enzymes, such as the cytosolic flavoprotein xanthine oxidase of rat liver, intestine, and heart (3,4) and the membrane-bound NADPH-cytochrome P-450 reductase of the endoplasmic reticulum (1,5,6), do both modes, univalent and bivalent, and reduce oxygen to O_2^- and H_2O_2. These two products of the partial reduction of oxygen, O_2^- and H_2O_2, are able in aerobic systems to react as coreactants that initiate a chain reaction of oxyradicals. The oxyradical chain reaction of aerobic systems constitutes a normal metabolic pathway whose rate follows mostly the rate of O_2^- and H_2O_2 production. Consequently, both production rates must be jointly considered in relation to the biological effects of the oxyradical chain reaction in mammalian organs and tissues (1,7).

Regarding oxyradical production in mammalian organs, it is convenient to first consider the mitochondrial production of O_2^- and H_2O_2, since mitochondria, which are present in all aerobic cells, appear as the quantitatively more important, and in many cases the unique reported source of O_2^- and H_2O_2. On the other hand, it seems convenient to take liver as a model organ since most of the data on O_2^- and H_2O_2 production in organelles has been obtained with mitochondria, peroxisomes, and microsomes isolated from rat liver, which in addition, exhibits a high xanthine oxidase activity. It is obvious that other organs lack the well-developed endoplasmic reticulum or the peroxisomes that are characteristic of the liver. Most of the organs show microperoxisomes which seem less effective than peroxisomes in terms of H_2O_2 production but are active organelles in providing catalase activity to the cell (1,2). The lung has a metabolically active endoplasmic reticulum that has been reported to produce both O_2^- and H_2O_2 (8).

The consideration of the physiological rate of the mitochondrial production of O_2^- and H_2O_2 is preceded by the estimation of the mitochondrial metabolic state in the organ under physiological conditions due to the very different rates of O_2^- and H_2O_2 production in the different mitochondrial metabolic states. The resting mitochondrial state 4, characterized by a relatively slow rate of respiration with high substrate supply and no availability of ADP, has a relatively high rate of O_2^- and H_2O_2 production, about 1 nmol H_2O_2/min/mg protein, due to the high reduction of the components of the respiratory chain (flavin semiquinone of the NADH dehydrogenase and ubisemiquinone) which by autoxidation yield O_2^-. In contrast, the active mitochondrial state 3, with a high rate of oxygen uptake and plain availability of respiratory substrate and ADP, shows a relatively slow rate of O_2^- and H_2O_2 production, about 0.1 nmol H_2O_2/min/mg protein, due to the highly oxidized states of the components of the mitochondrial respiratory chain. In the anoxic state 5, with no availability of O_2, there is no respiration and, of course, no partial reduction of oxygen to O_2^- or H_2O_2 (1,9,10).

The rates of oxygen uptake by isolated mitochondria in metabolic states 4 and 3, as well as the amount of mitochondrial protein per gram of organ and the oxygen uptake rates of the perfused organs, for rat liver and heart are given in Table 1. In the assumption that the whole oxygen uptake is accounted for by the

Table 1 Rates of Oxygen Uptake in Isolated Mitochondria and in Perfused Rat Liver and Heart[a]

	Oxygen uptake (nmol O$_2$/min/mg/protein)[b]		Content of mitochondria[c] (mg protein/g organ)	Oxygen uptake of perfused organ (μmol O$_2$/min/g organ)	Fraction of mitochondria (%)[d]	
	State 4	State 5			State 4	State 3
Rat liver	10	88	35	1.36[e]	64	36
Rat heart	28	135	53	3.05[f]	72	28

[a]Temperature: 30°C

[b]The rates of state 4 and 3 oxygen uptake were estimated for a substrate supply under physiological conditions such as (4 × rate with 5 mM malate and 5 mM glutamate as substrates) + (rate with 5 mM succinate as substrate)/5. Original data from Refs. 11–13.

[c]From Ref. 11.

[d]Fractions of mitochondria in state 4 (x) and in state 3 ($1 - x$) were calculated as (x) (state 4 oxygen uptake rate × content of mitochondria) + $(1 - x)$ (state 3 oxygen uptake rate × content of mitochondria) = oxygen uptake rate of the perfused organ.

[e]Calculated from Ref. 12.

[f]Calculated from Ref. 14; isolated beating heart, 240 beats/min.

Table 2   Rates of Mitochondrial Hydrogen Peroxide Production in Rat Liver and Heart[a]

	Production of H_2O_2 (nmol H_2O_2/min/g organ)[b]		Organ production of H_2O_2 (nmol H_2O_2/min/g organ)[c]			Organ H_2O_2 production/ organ O_2 uptake (%)[d]
	St. 4	St. 3	St. 4 mitos.	St. 3 mitos.	Total	
Rat liver	0.98	0.15	5.7	0.49	6.2	0.46
Rat heart	1.24	0.10	12.3	0.38	12.7	0.42

[a]Temperature: 30°C.
[b]From Refs. 9 and 10, rates at 220 μM O_2.
[c]Calculated for an organ steady-state concentration of 25 μM O_2 as $V_{25\mu M} = 0.26 \times V_{220\,\mu M}$ (Ref. 10); amount of mitochondrial protein/g organ form Table 1.
[d]Organ O_2 uptake form Table 1.
Abbreviations: st = state; mitos =mitochondria; prod. = production.

sum of the oxygen uptakes of the fraction of mitochondria respiring in state 4 and of the fraction of mitochondria respiring in state 3, the percentages of mitochondria in states 4 and 3 under physiological conditions are estimated as 64% and 36%, respectively, for rat liver and as 72% and 28%, respectively, for rat heart (Table 1).

The mitochondrial production of O_2^- and H_2O_2 is described in Table 2 in terms of H_2O_2 production, considering that, in mitochondria, O_2^- is the stoichiometric precursor of H_2O_2 (10,15). The H_2O_2 production rates measured in isolated mitochondria were calculated as organ physiological rates of mitochondrial H_2O_2 generation after considering the amount of mitochondria per gram of organ and the oxygen-dependence of the rate of mitochondrial O_2^- production (10). Finally, it can be estimated that under physiological conditions the mitochondrial production of H_2O_2 amounts to 0.46% and 0.42% of the rate of oxygen uptake of the perfused organ for rat liver and rat heart, respectively. Similarly, taking into account that O_2^- is the stoichiometric precursor of H_2O_2, it can be stated that the mitochondrial production of O_2^- amounts to 0.92% and 0.84% of the rate of oxygen uptake of the perfused organ for rat liver and rat heart, respectively. In the reactions of O_2^- and H_2O_2 utilization, the superoxide dismutase reaction gives back one O_2 molecule per two O_2^- molecules dismutated and the catalatic reaction of catalase gives back one O_2 molecule per two H_2O_2 molecules dismutated, whereas the glutathione peroxidase reaction reduces H_2O_2 with GSH as the hydrogen donor and does not give O_2 back. Assuming a two-thirds H_2O_2 utilization via the catalatic reaction of catalase and one-third through the glutathione peroxidase reaction, it follows that from each mole of O_2 primarily reduced to O_2^-, 0.67 mol will be returned as O_2 and 0.33 mol will be reduced to water (H_2O).

Accordingly, the pathway of the univalent reduction of oxygen contributes, under physiological conditions, to about 0.31% and 0.28% of the total mitochondrial oxygen uptake, used for energy and ATP production, in perfused rat liver and heart, respectively.

A series of organelles and subcellular structures of rat liver have been recognized as effective sources of O_2^- and H_2O_2. Production rates and steady-state concentrations, estimated for physiological conditions after considering substrate and oxygen availability, are given in Table 3. Taking into account that O_2^- is not permeable through the inner mitochondrial membrane (1,10), two compartments—mitochondria and cytosol—were considered for the estimation of the steady-state concentrations of O_2^- in the hepatocyte. To the contrary, considering the high diffusibility of H_2O_2 through biological membranes (1), mitochondria and cytosol were considered as a unique common compartment in the hepatocyte. The high concentration of catalase in the peroxisome and the corresponding sink effect (1,16) define a peroxisomal compartment with a low H_2O_2 steady-state concentration (17). No production of O_2^- has been reported in isolated peroxisomes (1,2). It is then apparent that O_2^- is kept at steady-state concentrations of 0.02 nM in the cytosol by the action of the Cu-Zn-superoxide dismutase and of 0.1 nM in the mitochondrial matrix by the action of mitochondrial Mn-superoxide dismutase (18). Similarly, peroxisomal catalase and mitochondrial and cytosolic glutathione peroxidases keep a mitochondrial H_2O_2 steady-state concentration of 0.05 µM and a cytosolic H_2O_2 steady-state concentration of 0.1 µM with a simultaneous steady-state level of about 0.01 µM H_2O_2 in the peroxisomes (17,18).

II. THE OXYRADICAL CHAIN REACTION IN MAMMALIAN ORGANS

The continuous generation of O_2^- and H_2O_2 by the organelles and enzymes of aerobic cells under physiological conditions sustains an also continuous chain reaction in which most of the reactants are oxyradicals or related chemical species (1). This free radical chain reaction is also known as the free radical reaction of lipoperoxidation, since lipids, fatty acids, and phospholipids are reactants and lipoperoxides and hydroperoxides are the stable products of the reaction. The oxyradical chain reaction of lipoperoxidation in aerobic cells and tissues is initiated by the production of O_2^-. This initiation reaction is followed by a rather specific propagation reaction in which O_2^- reduces Fe^{3+} bound to ferritin to free Fe^{2+}, which in turn cleaves H_2O_2 homolytically to yield hydroxyl radical (HO·). The production of HO· is considered to be the first classical free radical propagation reaction, which leads successively to alkyl (R·) and peroxyl (ROO·) radical

Table 3 Superoxide Radical and Hydrogen Peroxide Production Rates in Subcellular Fractions and Steady-State Concentrations in Rat Liver Under Estimated Physiological Conditions[a]

	Production rates (nmol/min/mg protein)		Amount of subcellular fraction (mg protein/g liver)	Steady-state concentrations	
	O_2^-	H_2O_2		O_2 (nM)	H_2O_2 (μM)
Mitochondria			35.3[b]		
Intramitochondria	0.36	0.18[c]		0.1[d]	0.05[e]
Cytosol	0	0.18[c]		—	0.1[e]
Peroxisomes			10.2[f]		
Intraperoxisomal	0	2.5[g]		—	0.01[h]
Cytosol	0	0.12[i]		0.02	0.1[e]
Microsomes			25.4[j]		
Cytosol	0.25[k]	0.43[k]		0.02[d]	0.1[e]
Nuclear membrane			2.3[l]		
Cytosol	0.15[l]	0.20[l]		0.02[d]	0.1[e]
Xanthine oxidase			4.0[m]		
Cytosol	0.002[m]	0.004[m]		0.02[d]	0.1[e]

[a]Temperature: 30°C

[b]From Ref. 11.

[c]From Table 2, considering O_2 as the stoichiometric precursor of H_2O_2 both produced in the mitochondrial matrix and H_2O_2 diffusing to the cytosol (Ref. 10).

[d]From Ref. 18.

[e]Measured in this laboratory and considered uniform in the mitochondrial and cytsolic compartments.

[f]From Ref. 19.

[g]Calculated from Ref. 19 for a steady-state concentration of 25 μM O_2 as $V_{25\,\mu M} = 0.12 \times V_{22\,\mu M}$ (Refs. 1, 19–20).

[h]Measured; from Ref. 17.

[i]Calculated assuming a 5% diffusion to the cytosol (Refs. 1, 16).

[j]Calculated from Refs. 17 and 19.

[k]Recalculated from Refs. 5 and 19 for a liver steady-state concentration of 25 μm O_2 as $V_{25\,\mu M} = 0.12 \times V_{220\,\mu M}$ (Ref. 20).

[l]Calculated from Refs. 21 and 22.

[m]Calculated from Refs. 2, 19, and 23.

production, the latter after O_2 addition. In this abbreviated notation R stands for the carbon chain residue of the fatty acid or phospholipid with an allylic carbon that undergoes hydrogen abstraction and oxygen addition. Peroxyl radicals with the $-COO\cdot$ group in a primary or secondary carbon, through the termination Russell reaction, produce excited states [compounds with an excited carbonyl group ($=CO^*$) or singlet molecular oxygen ($^1O_2$)]. Singlet oxygen is a chemiluminescent species that gives off one photon upon a bimolecular collision or encounter of two singlet oxygen molecules and can be utilized as a by-product indicator of the rate of the free radical reaction of lipoperoxidation (24–26). The determination of chemiluminescence is a suitable assay for the whole free radical reactions of lipoperoxidation. It has been used in isolated cells such as hepatocytes (27), protozoa (28), and polymorphonuclear leukocytes during phagocytosis (29) and in perfused and in situ organs (5,24–26). The calculated steady-state concentrations of O_2^- and the calculated and measured steady-state concentrations of H_2O_2 for liver and liver mitochondria (Table 3) can be used to estimate the steady-state concentrations of $HO\cdot$, $R\cdot$, $ROO\cdot$, and $^1O_2$ using published reaction constants and the multiple steady-state approach (18). The physiological steady-state concentration of 0.1 fM $^1O_2$ is calculated and obtained from the direct measurement of organ chemiluminescence. Table 4 shows the results of such an exercise and lists the estimated steady-state concentrations of oxygen species and oxyradicals, collectively termed reactive oxygen species, that constitute and sustain the oxyradical chain reaction of lipoperoxidation. The species are ordered kinetically according to their occurrence in the chain reaction. There is an inverse relationship between chemical reactivity and steady-state level, the more reactive species are kept at lower concentrations and vice versa.

Table 4 Steady-State Concentrations of Reactive Oxygen Species in Rat Liver Mitochondria and Rat Liver[a]

| | | Steady-state concentration | |
| | | Liver | Liver |
Chemical species	Type	mitochondria	(organ)
Superoxide (O_2^-)	Oxygen free radical	0.1 nM	0.02 μM[b]
Hydrogen peroxide (H_2O_2)	Reactive oxygen species	0.05 μM	0.1 μM
Hydroxyl radical ($HO\cdot$)	Oxygen free radical	0.06 aM	0.06 aM
Alkyl radical ($R\cdot$)	Carbon free redical	6 aM	6 aM
Peroxyl radical ($ROO\cdot$)	Carbon and oxygen free radical	2 nM	2 nM
Singlet oxygen (O_2^1)	Reacitve oxygen species	0.1 fM	0.1 fM

[a]The reactive oxygen species are ordered according to their kinetic occurrence in free redical chain reaction of lipoperoxidation. Data from Ref. 18. For details see text.
[b]Cytosolic concentration.

III. MITOCHONDRIAL HYDROXYL RADICAL PRODUCTION AND DAMAGE TO mtDNA

A widely discussed theory of cellular aging, first outlined by Szilard (30) and Orgel (31), considers the process in terms of the accumulation of molecular damage and informational errors in the informational (DNA and RNA) molecules. The damage may well be produced by some of the reactive oxygen species, such as HO· and ROO· which are able to abstract hydrogen from the purinic and pyrimidinic bases of nucleic acids or as $^1O_2$ which is able to react with DNA and RNA bases yielding additional products. Although most of the damage produced by the reactive oxygen species will be dealt with by the specialized systems of DNA repair, it is accepted that random alterations may accumulate and oxidative damage to mitochondrial and nuclear DNA is known to occur in vivo (32,33).

The "free radical hypothesis" (34,35) becomes more attractive when applied to mitochondria in comparison to the whole cell: (1) mitochondria are the relatively more important subcellular site of oxyradical (O_2^- and H_2O_2) production in mammalian organs; (2) O_2^- steady-state concentration in the mitochondrial matrix is about 5 to 10 times higher than in the cytosolic and nuclear spaces; (3) mtDNA is in close proximity to the sites of oxyradical generation; (4) mtDNA is not protected by the histones associated with nuclear DNA; (5) mtDNA normally has a level of oxidatively modified bases which is 10 to 20 times higher than the one in nuclear DNA; (6) some peptides that constitute dehydrogenases and cytochromes are coded in mtDNA; and (7) accumulation of faulty synthetized proteins might compromise energy transduction.

One of the key points in the consideration of the free radical hypothesis of mitochondrial aging is the evidence that the mitochondrial steady states of O_2^- and H_2O_2 under normal or physiological conditions are able to produce damage to mtDNA. This key point can be divided in three separate issues that can be experimentally answered: (1) the production of HO· or other strong oxidant by isolated mitochondrial preparations; (2) the reaction of HO· or other strong oxidant with mtDNA to chemically modify the bases; and (3) the existence of a detectable amount of modified mtDNA bases in aged mitochondria or tissues. The first two questions will be dealt with here and the latter one will be dealt with in the next section.

Production of HO·, as detected by spin trapping with 5,5′-dimethyl-1-pyrroline-N-oxide (DMPO), was reported in beef heart submitochondrial particles supplemented with NADH or succinate in the presence of antimycin (36). The effect of inhibitors, radical scavengers, and metal chelators on radical-DMPO adduct formation are consistent with the occurrence of active redox pools of Cu and Fe in submitochondrial particles, which elicit HO· formation through an O_2^--driven Fenton reaction (36). The attack of HO· on the desoxyguanosine base of DNA yields 8-HO-desoxyguanosine (8-HO-dG) which is conveniently detected by high-performance liquid chromatography (HPLC) (32,33). The rates of production of H_2O_2 and the amount of 8-HO-desoxyguanosine detected in mtDNA in rat liver are qualitatively related, as shown in Table 5. The rates of HO· and H_2O_2 generation are understood as being linearly related, but these production rates are not supposed to be strictly correlated with 8-HO-dG formation since (1) the actual site of HO· formation in mitochondria is not known, although its formation depends on the steady-state reduction of the inner membrane; (2) highly reactive HO· is able to diffuse several molecular diameters from its site of formation before it reacts; (3) part of mtDNA seems attached to the inner membrane at the point where mtDNA duplication starts; and (4) formation of 8-HO-dG could result from site-specific damage, entailing H_2O_2 cleavage by redox active metals (Cu, Fe) bound to mtDNA. Thus the stereo- and regioselectivity of HO· formation in the mtDNA vicinity determine the chemical structure of the formed products (36).

IV. MITOCHONDRIAL DAMAGE DURING AGING

Another of the key points in the consideration of the free radical hypothesis of mitochondrial aging is the evidence that aged mitochondria or mitochondria isolated from aged animals show some loss of the capacity for energy production. The concept of a decreased capacity of energy production in aged organs or aged individuals is immediately associated with the decreased basal metabolic rate observed by Rubner in 1883 (37). However, the marked changes in the basal metabolic rate when it is expressed as kilojoules per day per kilogram of body weight or as kilojoules per day per square meter of body surface observed in humans are almost no changes when basal metabolic rate is expressed as kilojoules per day per kilogram of vital organ (heart, brain, liver, or kidneys) (37,38). The changes in body composition that occur with age and an almost constant energy expenditure in vital organs and tissues in adult humans (values in kilojoules per day per kilogram of organ: heart, 1840; kidneys, 1840; brain, 1000; liver, 830; skeletal muscle, 55; adipose tissue, 20) apparently account for a constant basal metabolic rate when taking organ energy expenditure into consideration (37,38). Since in physiological conditions most of the mitochondria are in the resting state 4 (Table 1), the data do not invalidate the hypothesis of a decreased maximal capacity of energy production in aged organs. The hypothesis will then be that basal metabolic rate is related to mitochondrial resting and controlled state 4 and that the active state 3, as the maximal physiological rate of oxygen uptake and energy (ATP) production, will be the target of the mitochondrial aging process. However, state 4 respiration has been reported increased in

Table 5 Formation of 8-HO-Desoxiguanosine and Production of H_2O_2 in Rat Liver Mitochondria[a]

Substrate	[8-OH-dg] (pmol/μ mtDNA)	[8-OH-dG]/10^3[dG]	H_2O_2 production (nmol/min/mg protein)
Endogenous	0.10	0.12	0.16
Succinate	0.29	0.36	0.50
Succinate plus antimycin	0.34	0.42	1.20

[a]Data from Ref. 36.

mitochondria isolated from aged mice, which would indicate a partial uncoupling in aged mitochondria (39).

An approach to the effect of aging in the eventual maximal energy output in the organs was followed through the comparison of the mitochondrial content of the tissue. Morphometric differences in mitochondrial volume density and mitochondrial membrane surface have been reported to correlate with decreased metabolic activity in homeotherms versus poikilotherms (at similar body mass and temperature) (37). Similarly, available literature indicates decreased mitochondrial volume density and mitochondrial membrane surface in aging (37,40,41), strongly suggesting that mitochondria in aged cells, decreased in size and number, will progressively provide less energy for cellular demands. Mitochondria also appear as the unique cellular organelle that shows morphological alterations with age (42). Recently, by using forward angle light scattering in a cell sorter, two mitochondrial populations, with small and large size, were reported for rat liver and brain; the large size mitochondria markedly increased in aged animals (43). The existence of the two mitochondrial populations, the young and undamaged organelles and the old and injured mitochondria, certainly help explain earlier reports concerning changes in mitochondrial size, shape, and number (37).

Mitochondria isolated from aged animals also show a decreased electron transfer capability. The activity of complexes I (NADH-dehydrogenase) and IV (cytochrome oxidase) were reported decreased in the heart of aged rats (44) and cytochrome oxidase activity, the common marker of oxidative metabolism, was found to decrease with age in skeletal muscle (45) and brain (46,47). The hypothesis of decreased maximal energy production during aging was recently tested in a study in which a decreased active state 3 respiration, the maximal physiological rate of oxygen uptake and ATP synthesis, was found decreased by about 45–50% in liver and heart mitochondria isolated from senescence-accelerated mice (48). The observation appears to have been confirmed at the University of Sao Paulo, Sao Paulo, Brazil (V. Junqueira, personal communication), where liver mitochondria isolated from 6- to 24-month-old rats show a decline of about 35% in the active state 3 oxygen uptake. Concerning the coupling of exergonic electron transfer and endergonic ADP phosphorylation in mitochondria isolated from aged animals, recent information in the senescence-accelerated mouse indicate normal ADP:O ratios and unchanged acceptor respiratory controls (48). However, decreased mitochondrial membrane potential (43) and increased state 4 respiration (39) have been reported in mitochondria from old animals which appears to indicate that aged mitochondria have an increased membrane permeability to H^+ and therefore an increased H^+ backleakage that decreases the efficiency of mitochondrial energy transduction.

In addition to the decrease in electron transfer and oxygen uptake, liver mitochondria isolated from old animals show increases in TBARS (47) and

GSSG (43) content. Brain and liver mitochondria isolated from old rats show decreased membrane potentials as well as increased peroxide and 8-HO-dG contents (43) and an increase in mtDNA deletions (49). Human skin fibroblasts obtained from donors 0 to 97 years of age show age-dependent decreases in cytochrome oxidase and overall polypeptide synthesis. However, there was not a decrease in the copies of mtDNA nor an increase in the deletion mutated mtDNA, suggesting that the age-related phenotype was nuclear (50). Apparently a normal nucleus-mitochondria cross talk would be impaired upon aging. Mitochondria isolated from old mice and flies show increased state 4 respiration and H_2O_2 production, as a clear indication of mitochondrial dysfunction (35,39,51). The mitochondrial functional changes correlate positively with increased protein carbonyl and TBARS content and negatively with life expectancy (35,39). It has also been reported that in birds, which have a high aerobic metabolism, a low rate of oxyradical production by mitochondria is associated to a relatively long life span (52). The combination of decreased state 3 respiration and increased production of O_2^- and H_2O_2, which seems associated with aging, has been observed in isolated mitochondria in some physiopathological situations, such as liver mitochondria after ischemia-reperfusion (23) and skeletal muscle mitochondria after septic shock (53). It is apparent that in both ischemia-reperfusion and in septic shock high levels of NO are to be expected, and, consequently, then an inhibition of electron transfer in the cytochrome b–cytochrome c segment of the respiratory chain and an increased rate of O_2^- production (54). Increased rates of production and steady-state concentrations of O_2^- and NO will inevitably lead to an increased steady-state level of peroxynitrite ($ONOO^-$) and thus to mitochondrial damage (55). Therefore, it seems that both direct functional damage with a multiplicative effect with respect to the generation of reactive oxygen species and indirect functional damage through mtDNA damage or impairment of the nucleus-mitochondia cross talk occurs with aging.

The cells appear to defend themselves against mitochondrial damage with constant mitochondrial turnover ($t_{1/2}$ of about 2–4 weeks), removing those damaged mitochondria that have increased rates of O_2^- and H_2O_2 production (32). The likely process would be a sort of mitochondrial apoptosis triggered by highly increased intramitochondrial steady states of the reactive oxygen species. An increased mitochondrial fragility and a release to the cytosol of complexes I, II, III, and IV has been observed in liver and muscle from aged rats (A. Navarro, personal communication).

V.   THE FREE RADICAL THEORY OF AGING

The biological importance of oxygen free radicals was advanced by Gerschman et al. (56) after recognizing that (1) oxygen free radicals are the common mechanism of oxygen and radiation toxicity; (2) an increase in prooxidants or a decrease in antioxidants will equally lead to cell damage; and (3) oxygen toxicity is a continuous phenomenon. The concepts were incorporated into a general free radical theory of aging by Harman (57). Support for this theory was provided by the increases in life span of rats and mice obtained by supplementation of the diet with antioxidants such as 2-mercaptoethylamine, butylated hydroxytoluene (58,59) and 2-ethyl-6-methylhydroxypyridine (60). A potent antioxidant mixture, *Ginkgo biloba* extract Egb 761, given to young rats prevented the age-associated changes in mitochondrial morphology and function (42). Additional support for the free radical theory of aging came from the linear relationship observed by Cutler (61) after plotting the ratio of superoxide dismutase activity (protection against oxyradicals) over basal metabolic rate (aggression due to the metabolic generation of oxyradicals) against maximal life span in a series of mammals and primates. In summary, the free radical theory of aging and its subsidiary, the free radical hypothesis of mitochondrial aging, have reached maturity and are ready to suggest and promote new experiments.

REFERENCES

1. Chance B, Sies H, Boveris A. Hydroperoxide metabolism in mammalian organs. Physiol Rev 1979; 59:527–605.
2. Singh I. Mammalian peroxisomes: antioxidant enzymes and oxidative stress. In: Davies KJA, Ursini F, eds. The Oxygen Paradox Padova: CLEUP University Press, 1995: 209–222.
3. Fridovich I. Quantitative aspects of the production of superoxide anion radical by xanthine oxidase. J Biol Chem 1970; 245:4053–4057.
4. Brass CA, Narciso J, Gollan JL. Enhanced activity of the free radical producing enzyme xanthine oxidase in hypoxic rat liver. J Clin Invest 1991; 87:424–431.
5. Boveris A, Fraga CG, Varsavsky AI, Koch OR. Increased chemiluminescence and superoxide production in the liver of chronically ethanol-treated rats. Arch Biochem Biophys 1983; 227:534–541.
6. Puntarulo S, Cederbaum AI. Inhibition of the oxidation of hydroxyl radicals scavenging agents after alkaline phosphatase treatment of rat liver microsomes. Biochem Biophys Acta 1991; 1074:12–18.
7. Fridovich I. Superoxide dismutases. Annu Rev Biochem 1975; 44:147–159.
8. Turrens JF, Freeman BA, Crapo JD. Hyperoxia increases hydrogen peroxide release from lung mitochondria and microsomes. Arch Biochem Biophys 1982; 217:411–421.
9. Boveris A, Chance B. The mitochondrial generation of hydrogen peroxide. Biochem J 1973; 134:707–716.
10. Boveris A, Cadenas E. Production of superoxide radicals and hydrogen peroxide in

mitochondria. In: Oberley L, ed. Superoxide Dismutase. Boca Raton: CRC Press, 1982; 15–30

11. Costa LE, Boveris A, Koch OR, Taquini AC. Liver and heart mitochondria in rats submitted to chronic hypobaric hypoxia. Am J Physiol 1988; 255:C123–C129.

12. Videla LA, Villena MI, Donoso G, Giulivi C, Boveris A. Changes in oxygen consumption induced by t-butyl hydroperoxide in perfused rat liver. Effect of free radical scavengers. Biochem J 1984; 223:879–883.

13. Costa LE, Mendez G, Boveris A. Oxygen dependence of mitochondrial function measured by high resolution respirometry in long term hypoxic rats. Am J Physiol 1997; 273:C852–C858.

14. Poderoso JJ, Carreras MC, Peralta J, et al. Nitric oxide regulates oxygen uptake and promotes hydrogen peroxide release by the isolated beating rat heart. Am J Physiol 1998; 274:C112–C119.

15. Boveris A, Cadenas E. Mitochondrial production of superoxide anions and their relationship to the antimycin-insensitive respiration. FEBS Lett 1975; 54:311–314.

16. Poole B. Diffusion effects in the metabolism of hydrogen peroxide by rat liver peroxisomes. J Theor Biol 1975; 51:149–167.

17. Oshino N, Chance B, Sies H, Bucher T. The role of H_2O_2 generation in perfused rat liver and the reaction of catalase compound I and hydrogen donors. Arch Biochem Biophys 1973; 154:117–131.

18. Boveris A, Cadenas E. Cellular sources and steady-state levels of reactive oxygen species. In: Biadasz Clerch L, Massaro JL, eds. Oxygen, Gene Expression and Cellular Function. New York: Marcel Dekker, 1997: 1–25.

19. Boveris A, Oshino N, Chance B. The cellular production of hydrogen peroxide. Biochem J 1972; 128:617–630.

20. Chance B, Boveris A. Hyperoxia and hydroperoxide metabolism. In: Robin ED, ed. Extrapulmonary Manifestations of Respiratory Disease. New York: Marcel Dekker, 1978: 185–237.

21. Bartoli GM, Galeotti T, Azzi A. Production of superoxide anion and hydrogen peroxide in rat liver and Ehrlich ascites cell nuclei. Biochim Biophys Acta 1977; 497: 622–626.

22. Puntarulo S, Cederbaum AI. Effect of phenobarbital and 3-methylcholanthrene treatment on NADPH- and NADH-dependent production of reactive oxygen intermediates by rat liver nuclei. Biochim Biophys Acta 1992; 1116:17–23.

23. Gonzalez-Flecha B, Cutrin JC, Boveris A. Time course and mechanism of oxidative stress and tissue damage in rat liver subjected to in vivo ischemia-reperfusion. J Clin Invest 1993; 91:456–464.

24. Boveris A., Cadenas E, Reiter R, Filipkowski M, Nakase Y, Chance B. Organ chemiluminescence: non-invasive assay for oxidative radical reactions. Proc Natl Acad Sci USA 1980; 77:347–351.

25. Cadenas E, Sies H. Low level chemiluminescence as indicator of singlet molecular oxygen in biological systems. Methods Enzymol 1984; 105:221–231.

26. Cadenas E, Boveris A, Chance B. Chemiluminescence of biological systems. In: Pryor WB, ed. Free Radicals in Biological Systems, vol. 6. New York: Academic Press, 1984:41–87.

27. Turrens JF, Giulivi C, Boveris A. Increased spontaneous chemiluminescence from

liver homogenates and isolated hepatocytes upon inhibition of O_2^- and H_2O_2 utiliza tion. Free Radic Biol Med 1986; 2:135–140.

28. Lloyd D, Boveris A, Reiter R, Filipkowski M, Chance B. Chemiluminescence of *Acanthamoeba castellanii.* Biochem J 1979; 184:149–156.

29. Kakinuma K, Cadenas E, Boveris A, Chance B. Low level chemiluminescence of intact polymorphonuclear leukocytes. FEBS Lett 1979; 102:38–42.

30. Szilard L. On the nature of the aging process. Proc Natl Acad Sci USA 1959; 45: 30–45.

31. Orgel L. The maintenance of accuracy of protein synthesis and its relevance to aging. Proc Natl Acad Sci USA 1963; 49:519–525.

32. Ames BN, Shigenaga MK, Hagen TM. Oxidants, antioxidants, and the degenerative diseases of aging. Proc Natl Acad Sci USA 1993; 90:7915–7922.

33. Park JW, Ames BN. Methylguanine adducts in DNA are normally present at high levels and increase on aging: analysis by HPLC with electrochemical detection. Proc Natl Acad Sci USA 1988; 85:7467–7470.

34. Miquel J, Fleming J. Theoretical and experimental support for an "oxygen radical injury" hypothesis of cell aging. In: Johnson JE, Waldorf R, Harman D, Miquel J, eds. Free Radicals, Aging and Degenerative Diseases, vol 8. New York: Alan R. Liss, 1986; 51–74.

35. Sohal RS. The free radical hypothesis of aging. An appraisal of the current status. Aging Clin Exp Res 1993; 5:3–17.

36. Giulivi C, Boveris A, Cadenas E. Hydroxyl radical generation during mitochondrial electron transfer and the formation of 8-hydroxydesoxyguanosine in mitochondrial DNA. Arch Biochem Biophys 1995; 316:909–916.

37. McCarter RJM. Energy utilization. In: Masoro EJ, ed. Handbook of Physiology, 11: Aging. Oxford: Oxford University Press, 1995: 95–118.

38. Elia M. Organ and tissue contribution to metabolic rate. In: Kinney JM, Tucker HN, eds. Energy Metabolism. New York: Raven Press, 1991: 61–79.

39. Sohal RS, Ku HH, Agarwal S, Forster MJ, Lal H. Oxidative damage, mitochondrial oxidant generation and antioxidant defenses during aging and in response to food restriction in the mouse. Mech Ageing Develop 1994; 74:121–133.

40. Burns EM, Kruckerberg TW, Comerford LE, Muschmann MT. Thinning of capillary walls and declining number of endothelial mitochondria in the cerebral cortex. J Gerontol 1979; 34:642–650.

41. Herbener GHA. A morphometric study of age dependent changes in the mitochon- drial population of mouse liver and heart. J Gerontol 1976; 31:8–16.

42. Tauchi H, Sato T. Age changes in size and number of mitochondria of human hepatic cells. J Gerontol 1968; 23:454–463.

43. Sastre J, Millan A, Garcia de la Asuncion J, et al. A Gingko biloba extract (Egb 761) prevents mitochondrial aging by protecting against oxidative stress. Free Radic Biol Med 1998; 24:298–304.

44. Hayakawa M, Sugiyama S, Hattori K, Takasawa M, Osawa T. Age associated dam- age in mitochondrial DNA in human hearts. Mol Cell Biochem 1993; 119:95–103.

45. Trounce I, Byrne E, Marzuki S. Decline in skeletal muscle mitochondrial respiratory chain function. Possible factor in ageing. Lancet 1989; ii:637–639.

46. Benzi G, Pastoris O, Marzatico RF, Villa RF, Curti D. The mitochondrial electron

transfer alteration as a factor involved in brain aging. Neurobiol Aging 1992; 13: 361–368.

47. Martinez M, Ferrandiz ML, de Juan E, Miquel J. Age related changes in glutathione and lipid peroxide content in mouse synaptic mitochondria: relationship to cytochrome oxidase decline. Neurosci Lett 1994; 170:121–124.

48. Nakahara H, Kanno T, Inai Y, Mitochondrial dysfunction in the senescence accelerated mouse. Free Radic Biol Med 1998; 24:85–92.

49. Kang CM, Kristal BS, Yu BP. Age-related mitochondrial DNA deletions: effect of dietary restriction. Free Radic Biol Med 1998; 24:148–154.

50. Hayashi JI, Ohta S, Kagawa Y, et al. Nuclear but not mitochondrial genome in human age-related mitochondrial dysfunction. Functional integrity of mitochondrial DNA from aged subjects. J Biol Chem 1994; 269:6878–6883.

51. Sohal RS, Arnold LA, Sohal BH. Age-related changes in antioxidant enzymes and prooxidant generation in rat tissues with reference to parameters in two insect species. Free Radic Biol Med 1990; 9:495–500.

52. Barja G, Cadenas S, Rojas C, Perez-Campo R, Lopez-Torres M. Low mitochondrial free radical production per unit of O_2 consumption can explain the simultaneous presence of high longevity and high aerobic metabolic rate in birds. Free Radic Res 1994; 21:317–328.

53. Llesuy S, Evelson P, Gonzalez-Flecha B, et al. Oxidative stress in muscle and liver of rats with septic syndrome. Free Radic Biol Med 1994; 16:445–451.

54. Poderoso JJ, Carreras MC, Lisdero C, et al. Nitric oxide inhibits electron transfer and increases superoxide radical production in rat heart mitochondria and submitochondrial particles. Arch Biochem Biophys 1996; 328:85–92.

55. Radi R, Rodriguez M, Castro L, Telleri R. Inhibition of mitochondrial electron transport by peroxynitrite. Arch Biochem Biophys 1994; 308:89–95.

56. Gerschman R, Gilbert DL, Nye SW, Dwyer P, Fenn WO. Oxygen poisoning and x-irradiation: a mechanism in common. Science 1954; 19:623–629.

57. Harman D. Aging: a theory based on free radical and radiation chemistry. J Gerontol 1956; 11:298–300.

58. Harman D. Prolongation of the normal life-span and inhibition of spontaneous cancer by antioxidants. J Gerontol 1961; 16:247–254.

59. Harman D. Prolongation of life: role of free radical reactions in aging. J Am Geriatr Soc 1969; 17:721–735.

60. Emanuel NM. Free radicals and the action of inhibitors of radical processes under pathological states and aging in living organisms and in man. Q Rev Biophys 1976; 9:283–308.

61. Cutler R. Anti-oxidants, aging and longevity. In: Pryor WA, eds. Free Radicals in Biology, vol. 6. Orlando, FL: Academic Press, 1994: 381–395.

2

Walking a Tightrope: The Balance Between Mitochondrial Generation and Scavenging of Superoxide Anion

David M. Guidot
Atlanta VAMC and Emory University, Atlanta, Georgia

Joe M. McCord
Webb-Waring Antioxidant Research Institute and the University of Colorado Health Sciences Center, Denver, Colorado

I. INTRODUCTION

All aerobic organisms possess superoxide dismutases (SODs) (1). All eukaryotes have a mitochondrial form of SOD with manganese at the catalytic site and a cytosolic form with Cu and Zn at the catalytic site. In addition, mammals have an extracellular form that also contains Cu and Zn (2). The toxicity of superoxide was initially inferred by the nearly ubiquitous distribution of the enzymes (1), but was not really demonstrated until it was determined that the elimination of the Fe- and Mn-containing SODs of *Escherichia coli* resulted in an organism with marked sensitivity to growth in oxygen, and that this phenotype could be reversed if the mutant organism were supplied with the ability to produce the human Cu,Zn-SOD (3). In mammals, the picture is somewhat more complicated, but great insight has been provided recently with the production of knockout mice deficient in each of the three mammalian SODs. The results have been rather surprising. Mice unable to produce the cytosolic Cu,Zn-SOD appear quite normal and do not even show increased sensitivity to hyperoxia (4). Mice unable to produce the extracellular SOD also appear quite normal, but do show some increased sensitivity to hyperoxia (5). Mice unable to produce mitochondrial SOD have a serious cardiomyopathy and neurodegeneration, and die within days of their birth (6,7). Thus, in the mammalian organism, which has much more compartmentalization than simple prokaryotes, we see that the mitochondrial protection against superoxide radical is supremely important for development, function, and survival of the organism. In sharp contrast, loss of protection for the cytosolic or extracellular compartments causes hardly any observable deficit, at least under normal (nonstressed) conditions.

What are the sources of O_2^- in aerobic organisms? Studies have demonstrated O_2^- production by the mitochondrial electron transport chain, cytochrome P-450 system, oxidation of hemoglobin and catecholamines, xanthine oxidase, prostaglandin metabolism, and numerous other biological reactions in vitro (8). However, the degree to which each of these reactions contributes to O_2^- production in vivo is unknown. Because mitochondrial respiration accounts for more than 95% of all oxygen consumption in aerobic cells, we hypothesized that the electron transport chain is a predominant source of O_2^- production, particularly during states of "oxidative stress" such as exposure to hyperoxia.

II. RESPIRING MITOCHONDRIA PRODUCE SUPEROXIDE ANION IN VIVO

Evidence that mitochondria produce significant amounts of O_2^- is largely inferential and is primarily based on experiments using chemical inhibitors of electron transport. However, disruption of electron transport in mammalian cells is lethal because it eliminates all but glycolytic ATP generation. Therefore, biochemical inhibitors of electron transport, such as rotenone and cyanide, can only be used in isolated mitochondrial preparations. There is extensive evidence that O_2^- is produced by electron transport when electron flow is interrupted with these inhibitors (8,9). However, these observations do not unequivocally mean that the electron transport chain forms significant amounts of O_2^- in vivo. Resolution of this uncertainty has been further hampered because oxygen radicals are extremely short-lived species whose measurement is difficult in vivo.

We chose to address these dilemmas by studying mitochondrial respiration and O_2^- generation in yeast, specifically, the well-characterized *Saccharomyces cerevisiae* (10). We took this approach for several reasons. First, yeast are eukaryotic cells with an intracellular organization and biochemistry that is similar to higher eukaryotes, including mammals. Second, yeast are facultative aerobes that can grow in the absence of functional electron transport, a feature that permits disruption of the respiratory chain at discrete sites. The impact of these specific disruptions on O_2^- generation can then be determined in vivo. Third, yeast can be manipulated genetically to create strains deficient in putative sources of O_2^- generation and/or O_2^- scavenging.

We reasoned that a deficiency in the mitochondrial form of superoxide dismutase could serve as a "bioassay" for mitochondrial O_2^- generation in vivo. Several naturally occurring *S. cerevisiae* mutants that lacked the cytosolic Cu,Zn-containing superoxide dismutase were identified and found to be sensitive to oxygen. However, no comparable mutations in the mitochondrial enzyme were identified. Fortuitously, the gene for the Mn-containing superoxide dismutase in *S. cerevisiae* was cloned and subsequently deleted by gene disruption, and the resulting mutant that lacked the mitochondrial form of superoxide dismutase was sensitive to growth in oxygen (11). This suggested that the source of O_2^- in those experiments was the mitochondria, as those cells had the cytosolic form of the enzyme that accounts for approximately 85% of the total superoxide dismutase in yeast. In order to test the hypothesis that mitochondrial respiration produced O_2^- in vivo, we capitalized on this recent work and made our own *S. cerevisiae* constructs that were deficient in the Mn-containing superoxide dismutase, and combined this gene disruption with concomitant disruptions in electron transport. Studies on isolated mitochondria and submitochondrial particles in vitro provided evidence that O_2^- production in the electron transport chain occurred at sites proximal to cytochrome-*c* oxidase (12–14), with most of the evidence favoring the ubisemiquinone site as the principal source. We predicted that a proximal disruption of electron transport, but not a distal disruption of cytochrome-*c* oxidase, would eliminate mitochondrial O_2^- production and enable the Mn-SOD-deficient yeast to grow normally in oxygen. A schematic of the putative dynamics between O_2^- production by the respiratory chain and O_2^- scavenging by Mn-SOD is depicted in Figure 1.

First, we constructed a Mn-SOD-deficient strain of *S. cerevisiae* and determined that it grew normally in room air in enriched medium but it failed to grow in hyperoxia (>95% oxygen) (10). We next constructed Mn-SOD-deficient yeast that lacked either all of the components of electron transport (the Rho0 state) or lacked only the cytochrome-*c* oxidase component (COX(−)). We then determined that Mn-SOD-deficient yeast that lacked all of the components of electron transport grew normally in hyperoxia, whereas the selective disruption in cytochrome-*c* oxidase did not restore growth in hyperoxia (10). These results are illustrated in Figure 2. These experiments provided strong evidence that respiring mitochondria produce significant amounts of O_2^- in vivo.

We can draw several other conclusions from this study. Mn-SOD-deficient yeast grew normally in rich medium under normoxic conditions. This observation implies that, at least under conditions of relatively little biochemical and/or bioenergetic stress, the electron transport chain is extremely "tight" and O_2^- is not produced in toxic amounts. The percentage of electrons that leak from the respiratory chain under normal conditions may be only 1–2% (8). Alternatively,

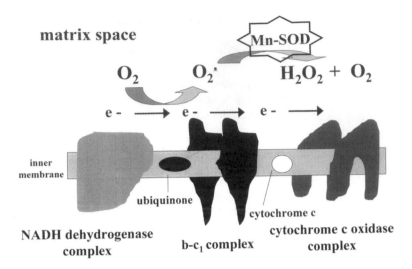

Figure 1 Schematic representation of superoxide anion (O_2^-) production by the electron transport chain within the mitochondrial inner membrane, and its scavenging within the mitochondrial matrix by manganous superoxide dismutase (Mn-SOD).

other nonenzymatic scavengers of O_2^- may exist within mitochondria that can detoxify smaller amounts of O_2^-. However, in hyperoxia Mn-SOD had no biochemical replacement within the mitochondria and was essential for cell growth. In other states of biological stress, such as ischemia or sepsis, uncoupling of electron transport may occur and O_2^- production most likely increases even when O_2 tension is normal or low. Under these conditions, mitochondrial respiration is impaired, ATP production is inhibited (15), and cellular injury can be induced both by the subsequent O_2^- production and by the accompanying depletion of energy stores. In these scenarios, Mn-SOD may be overwhelmed, resulting in mitochondrial injury from O_2^-.

Our results also localized the site of O_2^- production by the electron transport chain in vivo to sites which are proximal to cytochrome-c oxidase, in accordance with studies performed in vitro (12,14,16). Selective disruption of this complex failed to restore normal growth in hyperoxia to the Mn-SOD-deficient yeast. Partial restoration of growth in hyperoxia in this strain may reflect the effect that cytochrome-c oxidase deletion has on the redox state of proximal components in electron transport, specifically coenzyme Q (CoQ). In the normal state, the majority of CoQ exists as the free radical ubisemiquinone which can donate an electron to O_2 to form O_2^-. However, when electron flow from cytochrome-c_1 to cytochrome-c oxidase is inhibited, CoQ exists primarily in the reduced ubi-

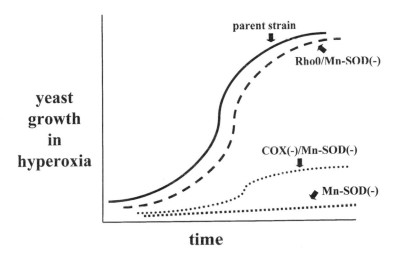

yeast growth in hyperoxia

time

Figure 2 Schematic representation of the growth characteristics of yeast (*Saccharomyces cerevisiae*) in hyperoxia (>95% O_2). All of the strains are isogenic with the exceptions that the Mn-SOD(−) strain is deficient in Mn-SOD, the Rho0/Mn-SOD(−) strain is deficient in all components of electron transport as well as Mn-SOD, and the COX(−)/Mn-SOD(−) strain is deficient in the terminal complex of electron transport (i.e., cytochrome-*c* oxidase) as well as Mn-SOD. The complete absence of electron transport allowed the Mn-SOD(−) strain to grow normally in hyperoxia, thereby providing evidence that electron transport is an important source of superoxide anion production in hyperoxia in vivo. This figure is modified from the actual growth curves in Ref. 10.

quinol form which will not readily produce O_2^- (17). Thus, deletion of cytochrome-*c* oxidase may reduce O_2^- generation from the proximal CoQ site. This interpretation is consistent with studies that show no evidence for O_2^- release from cytochrome-*c* oxidase in vitro.

The biological consequences of excess mitochondrial O_2^- production are likely substantial. A variety of disease states, including ischemia/reperfusion injury and hyperoxia, are associated with increased mitochondrial oxidant generation, mitochondrial oxidative injury, and/or ATP depletion (18–21).

These results supported prior assumptions regarding the site of O_2^- production within the mitochondrial electron transport chain and the relevant importance of cellular O_2^- generation. Nonetheless, many questions remained regarding the intracellular balance of O_2^- and other reactive oxygen species. For example, sensitive targets for O_2^- and other reactive oxygen species, as well as the other biologically relevant sources for their production, remained largely ill defined or altogether unknown.

We therefore initiated new experiments that were designed to identify specific intracellular targets of O_2^- generated by mitochondrial respiration. We reasoned that our genetic approach in yeast would allow us to test the hypothesis that excessive production of O_2^- by mitochondria, particularly in hyperoxia, could damage cytosolic as well as mitochondrial targets. Although studies on isolated mitochondria had failed to identify leaks of O_2^- into the extramitochondrial space, there had previously been no way to test this in vivo. We predicted that yeast deficient in the cytosolic form of superoxide dismutase and known to be sensitive to oxygen would suffer less oxidative stress if mitochondrial respiration, and therefore O_2^- generation, was eliminated. To our subsequent surprise, and scientific delight, we could not have been more wrong.

III. RESPIRING MITOCHONDRIA SCAVENGE EXTRAMITOCHONDRIAL SUPEROXIDE ANION: THE "VACUUM CLEANER" EFFECT

Our yeast model provided further evidence that mitochondrial respiration produces significant amounts of O_2^- in vivo. These findings were consistent with the long-standing assumption that electron transport, albeit critical for normal cellular function, imposes a substantial oxidative stress on the cell. The critical role that the superoxide dismutases, particularly Mn-SOD, play in the antioxidant defense strategy of the cell has been characterized in diverse models of acute and chronic oxidative injury. The prevailing view was that the mitochondrial and cytosolic forms of the enzyme only scavenged O_2^- within their respective intracellular compartments. We speculated that mitochondrial respiration, particularly during severe stress such as hyperoxia or sepsis, when O_2^- by the electron transport chain can increase dramatically, could damage cytosolic targets by leaking O_2^- into this space. To test this hypothesis, we constructed yeast mutants that were deficient in Cu,Zn-SOD with and without concomitant disruptions in electron transport. We predicted that yeast deficient in Cu,Zn-SOD would be sensitive to oxygen, and that eliminating mitochondrial respiration would decrease this oxygen sensitivity.

Surprisingly, yeast deficient in both Cu,Zn-SOD and electron transport grew poorly in oxygen compared to yeast deficient only in Cu,Zn-SOD, yeast deficient only in electron transport, or the parent strain (Fig. 3). As neither defect individually had a substantial effect on growth in oxygen, the oxygen sensitivity in yeast with the combined disruptions was unexpected. Furthermore, since yeast deficient in both electron transport and Cu,Zn-SOD grew normally in nitrogen (not shown), the growth inhibition in oxygen could reasonably be ascribed to oxidant stress and not an unidentified metabolic defect caused by the combined genetic disruptions. Perhaps even more perplexing, yeast that were deficient in both Mn-SOD and Cu,Zn-SOD were not sensitive to room air oxygen concentra-

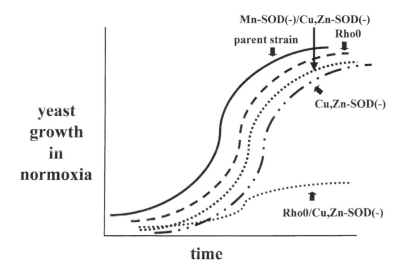

yeast
growth
in
normoxia

time

Figure 3 Schematic representation of the growth characteristics of yeast (*Saccharomyces cerevisiae*) in normoxia (21% O_2). All of the strains are isogenic with the exceptions that the Cu,Zn-SOD($-$) strain is deficient in Cu,Zn-SOD, the Rho0 strain is deficient in all of the components of electron transport, the Rho0/Cu,Zn-SOD($-$) strain is deficient in all components of electron transport as well as Cu,Zn-SOD, and the Mn-SOD($-$)/ Cu,Zn-SOD($-$) strain is deficient in both forms of SOD. The absence of electron transport impaired growth of the Cu,Zn-SOD($-$) strain in normoxia, even though absence of both Mn-SOD and Cu,Zn-SOD did not significantly impair growth under the same conditions. All of the strains had identical growth curves in nitrogen (not shown). This figure is modified from the actual growth curves in Ref. 46.

tions and grew far better than the yeast that lacked Cu,Zn-SOD and electron transport but possessed Mn-SOD (Fig. 3).

How could we interpret this unexpected result? One possibility was that in the absence of electron transport, cytosolic O_2^- production increased as ATP production by glycolysis increased. However, glycolysis should not generate O_2^-, and one would predict that O_2^- production in yeast that lacked both forms of SOD but still had intact mitochondrial respiration would be greater than in the yeast that lacked only the Cu,Zn-SOD and could not generate O_2^- via mitochondrial respiration. Was there a previously unidentified superoxide dismutase in the cytosol that assumed greater responsibility for detoxifying O_2^- when mitochondrial respiration and ATP production are unavailable? We could not identify even a trace of superoxide dismutase activity, at least as reflected by the ability to inhibit cytochrome-*c* reduction by xanthine oxidase-/xanthine-generated O_2^-, that was not accounted for by the Mn-SOD and the Cu,Zn-SOD in any of our yeast constructs under any conditions. In particular, yeast that had disruptions in both the Mn-SOD and Cu,Zn-SOD genes had no SOD activity when total cell extracts were tested.

When we considered our findings from a different perspective, another possibility arose. In the yeast that lacked Cu,Zn-SOD activity and therefore were vulnerable to O_2^- damage of cytosolic targets, it was more advantageous to have mitochondrial respiration than to have mitochondrial SOD. That is to say, mitochondrial respiration decreased the cytosolic O_2^- burden more than did Mn-SOD, and therefore appeared to act as an additional SOD in vivo. We speculated that mitochondria scavenged cytosolic O_2^- that diffused across the outer mitochondrial membrane, and somewhat facetiously termed this phenomenon the "vacuum cleaner effect."

To test this new hypothesis, we isolated mitochondria from yeast and measured their capacity to scavenge O_2^- in vitro. We employed a traditional SOD assay in which xanthine oxidase/xanthine is used to generate O_2^-, and SOD activity is determined by measuring the inhibition of cytochrome-c reduction in the reaction mixture. We predicted that mitochondrial respiration, induced by adding respiratory substrates to the isolated mitochondria, would increase the "apparent" SOD activity of the mitochondria above the baseline SOD activity that was attributable to the Mn-SOD enzyme (Fig. 4).

We determined that isolated yeast mitochondria had a respiration-dependent increase in O_2^- scavenging. Specifically, the addition of pyruvate and malate to the isolated mitochondria significantly increased O_2^- scavenging, and this effect was completely eliminated if either antimycin A (which inhibits mitochondrial respiration) or FCCP (carbonyl cyanide p-(trifluoromethoxy) phenyl hydrazone, which uncouples electron transport) were added with the pyruvate and malate (Fig. 5). These results argued that mitochondrial respiration somehow scavenged O_2^- quite efficiently and, apparently, nonenzymatically. Further evidence that enzymatic scavenging, at least by the mitochondrial form of SOD, did not account for this effect was revealed when we determined that mitochondria isolated from yeast deficient in Mn-SOD had the same respiration-dependent capacity to scavenge O_2^- as mitochondria that contained Mn-SOD (Fig. 6).

We extended these studies to examine if mammalian mitochondria could scavenge extramitochondrial O_2^-. We isolated rat liver mitochondria and determined that, as expected, mammalian mitochondria also have a respiration-dependent capacity to scavenge O_2^-. This was not surprising, at least not by this time in our work, because yeast and mammalian mitochondria are remarkably similar. In particular, the essential structure and function of the electron transport chain and the inner membrane are virtually identical in all eukaryotic aerobic organisms.

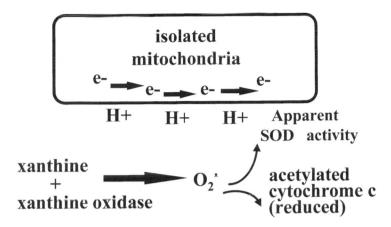

Figure 4 Scheme for the modified assay of SOD activity used to determine the respiration-dependent (i.e., nonenzymatic) scavenging of extramitochondrial superoxide anion by isolated mitochondria in vitro. A standard SOD assay is employed wherein respiring mitochondria are assayed for their ability to inhibit cytochrome-c reduction by O_2^- that is generated by xanthine/xanthine oxidase. The "apparent SOD activity" is calculated by determining how much more O_2^- is scavenged during respiration (induced by adding respiratory substrates) than when the mitochondria are resting (before adding respiratory substrates).

We had started with the surprising result that mitochondrial respiration increased oxidative stress in yeast that were deficient in Mn-SOD, but decreased oxidative stress in yeast that were deficient in Cu,Zn-SOD. This initially paradoxical finding in vivo was supported and explained by the isolated mitochondrial studies. Both yeast and mammalian mitochondria had a significant increase in superoxide scavenging during respiration compared to the resting state. Antimycin, which inhibits electron transport and thus prevents proton gradient generation, and FCCP, which increases electron transport but dissipates the proton gradient, both eliminated this increase. Furthermore, the addition of ADP and inorganic phosphate, which increases oxygen consumption and electron transport but does not increase the transmembrane potential, did not increase respiration-dependent scavenging by yeast mitochondria. These data argue that superoxide scavenging requires a polarized inner mitochondrial membrane and not electron transport per se.

Perhaps even more importantly, respiration-dependent superoxide scavenging did not require Mn-SOD. Mitochondria isolated from yeast that lacked Mn-SOD had the same respiration-dependent capacity to scavenge superoxide as mi-

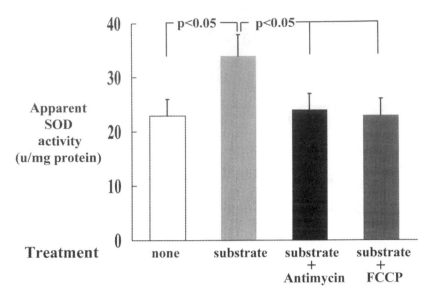

Figure 5 The apparent SOD activity of isolated yeast mitochondria. The total activity is the enzymatic activity accounted for by Mn-SOD + respiration-dependent (nonenzymatic) activity. Adapted from Ref. 46.

tochondria with Mn-SOD. The increased scavenging induced by pyruvate and malate is inconsistent with an enzymatic mechanism (such as a previously un-identified superoxide dismutase) and argues for a nonenzymatic mechanism. We speculate that the inner mitochondrial membrane, which during respiration trans-locates protons to its outer surface, enhances the spontaneous dismutation of ex-tramitochondrial O_2^- that diffuses into the intermembrane space by presenting a localized proton-rich (i.e., acidic) environment. The second-order rate constant for the spontaneous dismutation of superoxide anion is about 10^8/M/s at pH 4.8, but is more than two orders of magnitude slower at cytosolic pH (22). As O_2^- rapidly dismutes, a concentration gradient of O_2^- between the cytosol and the inner membrane is created which would draw in O_2^- in a manner analogous to how O_2 diffuses down the concentration gradient maintained by mitochondrial O_2 consumption. Our proposed scheme for mitochondrial scavenging of extra-mitochondrial O_2^- is shown in Figure 7.

While the primary mechanism of mitochondrial scavenging appears to be nonenzymatic, enhanced enzymatic dismutation may occur in vivo. At the inner mitochondrial membrane, O_2^- will protonate to form the uncharged hydroperoxyl radical ($HO_2\cdot$) that could diffuse into the mitochondrial matrix (its solubility is similar to O_2) and be enzymatically dismuted by Mn-SOD. Dismutation con-

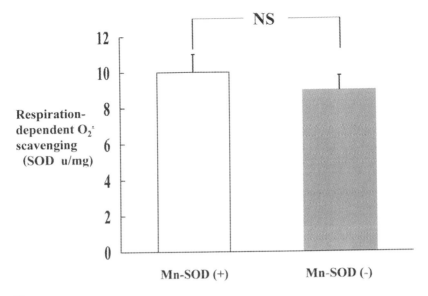

Figure 6 The respiration-dependent (nonenzymatic) O_2^- scavenging of mitochondria isolated from yeast that have Mn-SOD(+) and from yeast that lack the enzyme (Mn-SOD(−)). Adapted from data presented in Ref. 46.

sumes two molecules of O_2^- and yields one molecule of H_2O_2, which can also cause oxidative damage. However, H_2O_2 is rapidly reduced to H_2O by catalase and/or other cellular peroxidases. Yeast have a unique cytochrome-c peroxidase which scavenges mitochondrial H_2O_2 (23), while a specific mitochondrial catalase has been identified in rat heart and kidney mitochondria (24,25). Thus mitochondria are equipped to reduce H_2O_2 generated during O_2^- scavenging. It is important to stress that we do not really know how mitochondrial respiration scavenges O_2^-, and much work remains in order to elucidate the precise mechanisms underlying this previously unidentified phenomenon. There is precedent for superoxide anion being attracted to a polarized region, in that the surface of the Cu,Zn-SOD molecule has a charge distribution that is proposed to ''guide'' superoxide anion to the catalytic site (26). For now we can state that mitochondrial scavenging of O_2^- occurs, and at least in yeast that lack Cu,Zn-SOD, mitochondrial respiration is an important antioxidant in vivo. Therefore, rather than viewing mitochondria as only generating O_2^- in vivo, we now must consider that mitochondrial production and scavenging of O_2^- are actually in balance, and that the net result in the normal state may actually favor O_2^- scavenging. The next important question is what role does mitochondrial respiration play in antioxidant defenses during oxidative stress?

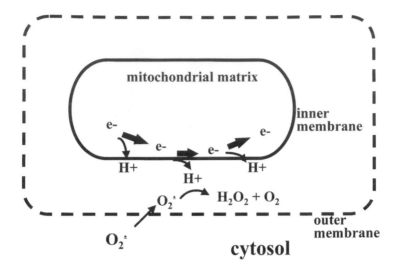

Figure 7   Proposed mechanism for mitochondrial scavenging of extramitochondrial su-peroxide anion (O_2^-). Cytosolic diffusing through the outer mitochondrial membrane will encounter the proton-rich inner membrane of the respiring mitochondria. In this region, spontaneous dismutation of O_2^- to H_2O_2 and O_2 is accelerated. Protonated superoxide anion, the hydroperoxyl radical (HO_2) could also diffuse across the inner membrane and undergo enzymatic dismutation by Mn-SOD within the matrix. H_2O_2 generated by either mechanism can be detoxified by catalase or other peroxidases within the mitochondria. The net consumption of O_2^- creates a concentration gradient for O_2^- which favors diffu-sion from the cytosolic to the mitochondrial space. Adapted from Ref. 46.

IV. RESPIRATION-DEPENDENT SCAVENGING OF SUPEROXIDE IN LUNG MITOCHONDRIA IS AFFECTED BY OXIDATIVE STRESS

We extended our previous observations and examined mitochondrial scavenging of superoxide in rat lungs under different conditions of oxidative stress. We spec-ulated that respiration-dependent scavenging of cytosolic superoxide could be an important component of the antioxidant defense system that had heretofore not been examined in animal models. Acute oxidative cellular injury results from diverse biological insults including ischemia-reperfusion, infections, radiation, and hyperoxia. We have had a long-standing interest in acute lung injury and therefore focused our attention on lung mitochondrial function. The acute respira-tory distress syndrome (ARDS) is a devastating form of edematous lung injury in which oxidant-antioxidant imbalance is a central feature (27). Supportive care of ARDS patients frequently requires high concentrations of inspired oxygen which may worsen oxidative stress. Although the precise mechanisms by which hyperoxia damages the lungs are unknown, considerable experimental evidence implicates toxic species of reactive oxygen as proximal mediators of tissue injury (28,29).

Multiple treatments, including systemic or intratracheal delivery of endotoxin, tumor necrosis factor (TNF), or interleukin-1 (IL-1), induce tolerance to subsequent hyperoxia exposure in animal models (30–33). Although acquired tolerance to hyperoxia remains poorly understood, these treatments all increase lung enzyme activity and/or messenger RNA (m-RNA) levels of Mn-SOD. As mitochondria are a significant source of superoxide anion generation in hyperoxia in vivo (10), induction of Mn-SOD could protect against hyperoxic stress. However, there are multiple cytosolic sources of superoxide anion that could also contribute to oxidative stress in hyperoxia (8). In addition, while some studies suggest that the cytosolic antioxidant Cu,Zn-SOD is modestly induced by sublethal hyperoxia (34) or endotoxin treatment (35), there is considerable evidence that Cu,Zn-SOD is not induced by endotoxin (31,36) or the combination of TNF and IL-1 (32). Furthermore, although pulmonary endothelial cells increase Mn-SOD activity ninefold following endotoxin treatment in vitro, Cu,Zn-SOD activity is decreased (37). Since the pulmonary endothelium appears to be a critical target in hyperoxia-induced lung disease, these studies raise questions as to the mechanisms of endotoxin-induced tolerance to hyperoxia.

With this background, we hypothesized that mitochondrial respiration-dependent scavenging of extramitochondrial O_2^- could explain some of these questions. We first determined the effect of hyperoxia on lung mitochondrial scavenging of O_2^-. Lung mitochondria were isolated from rats exposed to varying periods of hyperoxia (>95% O_2) and their superoxide scavenging capacity determined in vitro. Hyperoxia dramatically decreased respiration-dependent superoxide scavenging in a time-dependent manner (Fig. 8). Importantly, this antioxidant capacity was reduced by more than 50% at a time when antioxidant enzyme activities remained at control levels (38), suggesting that loss of mitochondrial scavenging of cytosolic superoxide may be a critical early step in the development of hyperoxic lung injury. Therefore, hyperoxia exposure, which increases mitochondrial generation of O_2^-, tips the balance in favor of O_2^- production over scavenging, resulting in a toxic accumulation of O_2^- within the cell and ultimately cell injury or death.

We next determined the effect of endotoxin pretreatment, which induces tolerance to hyperoxia, on lung mitochondrial scavenging. Rats were pretreated with endotoxin (5 µg intratracheally), and 24 h later lung mitochondrial superoxide scavenging was determined. In addition, some rats were pretreated with endotoxin and 24 h later were exposed to hyperoxia for 48 h and compared to untreated rats exposed to hyperoxia. Endotoxin pretreatment increased lung mitochondrial superoxide scavenging compared to lung mitochondria from untreated rats. More

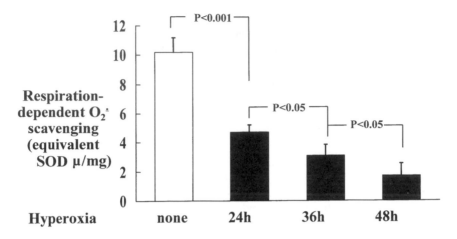

Figure 8   Respiration-dependent (nonenzymatic) scavenging of extramitochondrial O_2^- by lung mitochondria isolated from untreated rats and from rats exposed to hyperoxia (>95% O_2) for 24, 36, or 48 h. Adapted from Ref. 47.

importantly, lung mitochondria from endotoxin-pretreated rats exposed to hyperoxia had the same superoxide scavenging capacity as untreated, air-breathing rats. Therefore, endotoxin pretreatment maintained mitochondrial superoxide scavenging in hyperoxia exposure. In fact, endotoxin treatment more than doubled the respiration-dependent scavenging of lung mitochondria in rats exposed to hyperoxia compared to lung mitochondria from hyperoxia-exposed rats not given endotoxin. These results are shown in Figure 9.

In contrast, we determined that Mn-SOD activity, when normalized for the amount of protein in the crude mitochondrial fraction, was not increased by endotoxin (Fig. 10). While many previous studies have shown that endotoxin pretreatment increases lung Mn-SOD m-RNA (31,32,39) and lung Mn-SOD protein (32), we found no increase in SOD activity of our mitochondrial preparations when activity was normalized to mitochondrial protein. A potential explanation is that increased Mn-SOD in response to endotoxin may accompany increases in other mitochondrial proteins and even increases in mitochondrial number or volume per cell. Mn-SOD may increase per unit of tissue, as has been reported, but not increase per mitochondria, which would be consistent with our findings. Therefore, endotoxin pretreatment appears to increase O_2^- scavenging in hyperoxia such that, even if O_2^- production is increased, the net result still favors O_2^- scavenging and the cell is protected from the otherwise harmful effects of hyperoxia.

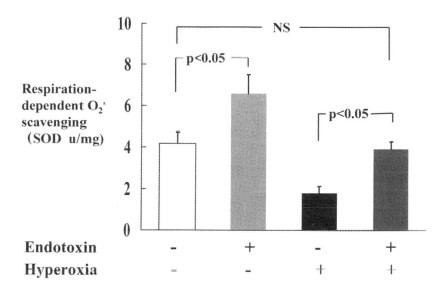

Figure 9 Respiration-dependent (nonenzymatic) scavenging of extramitochondrial O_2^- by lung mitochondria isolated from untreated rats and from rats treated with endotoxin (5 μg intratracheally 36 h earlier) with or without exposure to hyperoxia (>95% O_2) for 24 h. Adapted from Ref. 47.

IV. MITOCHONDRIAL BIOGENESIS AS A RESPONSE TO OXIDATIVE STRESS

Our studies suggest an important antioxidant function for respiring mitochondria. As noted above, the cellular response to any stress, including oxidative insults, is critically dependent on mitochondrial respiration. The increased energy demands alone depend on mitochondrial ATP production, and the myriad other functions performed by these organelles, including scavenging of O_2^-, are ultimately mediated by the polarized inner membrane. However, there is surprisingly little evidence that cells increase mitochondria in response to oxidative stress. An important finding is that the number of mitochondria in alveolar epithelial cells doubles after 1 week of sublethal hyperoxia exposure (85% O_2) in rats (40). What about the extensive literature on Mn-SOD induction in diverse animal models of oxidative stress and, in particular, acquired tolerance to oxidative stress? Mn-SOD is a mitochondrial protein, and its induction may reflect in part a more generalized increase in mitochondria. This makes sense, at least teleologically, in that an isolated increase in Mn-SOD would not directly augment mitochondrial respira-

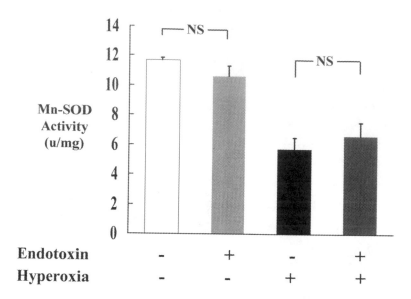

Figure 10 Mn-SOD activity of lung mitochondria isolated from untreated rats and from rats treated with endotoxin (5 μg intratracheally 36 h earlier) with or without exposure to hyperoxia (>95% O_2) for 24 h. Adapted from Ref. 47.

tion and increase ATP production. Indirectly, an increase in Mn-SOD would prevent oxidative damage to mitochondria, particularly the components of electron transport, and would thereby preserve respiration. However, the observation that alveolar epithelial cells increase the number of mitochondria per cell during exposure to sublethal hyperoxia (40) led us to consider that mitochondrial biogenesis might be a coordinated response in other cells as well.

Recently we have examined mitochondrial biogenesis in the livers of rats after endotoxin treatment, which increases Mn-SOD expression and activity in a variety of tissues and confers protection against diverse types of oxidative stress. Rats were treated with endotoxin (2 mg/kg IP) or saline and their hepatocytes were isolated 48 h later. The isolated hepatocytes were then stained with either rhodamine-123 or 10-N nonyl-acridine orange (NAO) and examined by confocal microscopy. Rhodamine-123 is concentrated within mitochondria in proportion to the net potential across the inner membrane, and the relative fluorescence can be used as an index of the mitochondrial membrane potential (41). 10-N nonyl-acridine orange is a fluorescent probe that binds to the inner mitochondrial membrane independently of the membrane potential, and therefore its

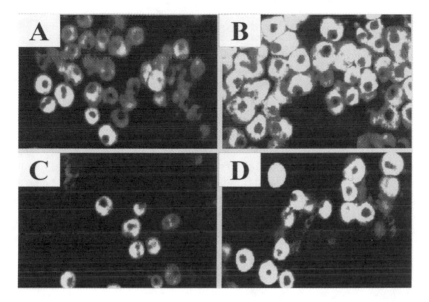

Figure 11 Confocal microscopy images of isolated hepatocytes stained with fluorescent dyes that concentrate within mitochondria. Shown are hepatocytes from untreated rats (panel A) and hepatocytes from endotoxin-treated rats (2 mg/kg intraperitoncally 36 h earlier) (panel B) incubated with rhodamine-123, which sequesters inside the mitochondrial matrix in proportion to the mitochondrial inner membrane potential, and hepatocytes from untreated rats (panel C) and hepatocytes from endotoxin-treated rats (panel D) incubated with 10-N nonyl-acridine orange (NAO), which sequesters within mitochondria in proportion to the inner mitochondrial membrane mass. (Unpublished data.)

relative fluorescence can be used as an index of mitochondrial inner membrane mass (42,43). We determined that endotoxin treatment increased both rhodamine-123 and NAO fluorescence in isolated hepatocytes (Fig. 11; unpublished data). We performed additional studies to verify that NAO fluorescence was indeed independent of the mitochondrial membrane potential (not shown), confirming the observations of other investigators (42,43). Therefore, endotoxin treatment, which is known to increase Mn-SOD activity and induce tolerance to oxidative stress in a variety of model systems, increases mitochondrial inner membrane mass as well as the transmembrane potential. This finding is consistent with the observed increase in mitochondrial number per alveolar epithelial cell in hyperoxia-exposed rats and suggests that mitochondrial biogenesis is a generalized response to oxidative stress.

V. SUMMARY

We started with a genetic approach in yeast in order to examine the long-standing biochemical question as to whether or not mitochondria generate superoxide anion in vivo. Surprisingly, we determined that mitochondrial respiration has a much more complex role in the control of intracellular levels of O_2^-. Mitochondrial respiration, via leak of electrons to oxygen, generates significant amounts of O_2^- even under normal conditions, and this production can increase greatly to toxic levels under various pathological conditions. However, respiring mitochondria scavenge cytosolic O_2^-, and this scavenging appears to be so efficient that the net result under normal conditions is that mitochondria actually behave as ''organellar SODs.'' Furthermore, mitochondrial scavenging of O_2^- is influenced by oxidative stress, and this previously unidentified antioxidant function appears to explain some of the observations regarding oxidant injury and tolerance in animal models.

If it is true that mitochondrial scavenging of O_2^- is dependent on the respiratory function of the mitochondria, regardless of whether our proposed mechanism in Figure 7 is correct, then this previously unidentified function changes our previous assumptions about mitochondria and their role in the response to oxidative injury. Perhaps this should not be all that surprising. After all, although electron transport evolved to generate ATP efficiently and allowed higher organisms to develop, the polarized inner membrane that results from electron transport serves other cellular needs as well. In addition to ATP production, the mitochondrial transmembrane potential is also utilized to pump calcium and other ions across the mitochondrial membrane (44), as well as proteins (45) that are required for metabolic pathways within the mitochondrial matrix (such as the Krebs cycle). Therefore, it appears reasonable that an overall increase in mitochondrial respiratory capacity, either by increasing the number of mitochondria per cell or the amount of inner membrane per mitochondria, would be a potent response to oxidative stress. Superoxide scavenging may be just one of a myriad of mitochondrial functions that are ultimately powered by the electron transport chain.

We propose that mitochondria, and not just individual components such as Mn-SOD, are a critical antioxidant response element. The studies that we have outlined in this chapter suggest a more complex balance between O_2^- production and scavenging within mitochondria than we had previously envisioned. It now appears that mitochondria actively influence O_2^- homeostasis both inside the matrix space and in the cytosolic space as well (Fig. 12). Furthermore, both enzymatic (i.e., Mn-SOD) and nonenzymatic (i.e., respiration-dependent) O_2^- scavenging can be increased as part of the overall cellular defense against oxidative injury.

Further investigations will undoubtedly turn up even more surprising results than those already uncovered. One could expect, and indeed hope, that our

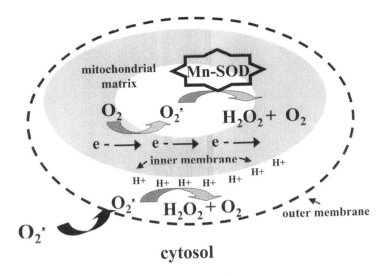

Figure 12 Proposed scheme for the overall role of mitochondria in superoxide anion (O_2^-) production and scavenging within the cell. Mitochondrial respiration produces O_2^- which under normal conditions is scavenged by Mn-SOD within the mitochondrial matrix. In addition, mitochondrial respiration scavenges cytosolic O_2^- that diffuses into the intermembrane space, thereby complementing the O_2^- scavenging by Cu,Zn-SOD. Therefore, mitochondria are directly involved in O_2^- homeostasis throughout the cell.

understanding of how mitochondrial respiration influences oxidative stress will be subject to frequent revision.

REFERENCES

1. McCord JM, Keele BB Jr, Fridovich I. An enzyme-based theory of obligate anaerobiosis: the physiological function of superoxide dismutase. Proc Natl Acad Sci USA 1971; 68:1024–1027.
2. Marklund SL. Human copper-containing superoxide dismutase of high molecular weight. Proc Natl Acad Sci USA 1982; 79:7634–7638.
3. Natvig DO, Imlay K, Touati D, Hallewell RA. Human copper-zinc superoxide dismutase complements superoxide dismutase-deficient *Escherichia coli* mutants. J Biol Chem 1987; 262:14697–14701.
4. Ho YS, Gargano M, Cao J. Mice lacking copper/zinc superoxide dismutase show no increased sensitivity to hyperoxia [abstract]. Am J Respir Crit Care Med 1997; 155:A17.
5. Carlsson LM, Jonsson J, Edlund T, Marklund SL. Mice lacking extracellular super-

oxide dismutase are more sensitive to hyperoxia. Proc Natl Acad Sci USA 1995; 92:6264–6268.

6.   Li Y, Huang TT, Carlson EJ, et al. Dilated cardiomyopathy and neonatal lethality in mice lacking manganese superoxide dismutase. Nat Genet 1995; 11:376–381.

7.   Lebovitz RM, Zhang H, Vogel H, et al. Neurodegeneration, myocardial injury, and perinatal death in mitochondrial superoxide dismutase-deficient mice. Proc Natl Acad Sci USA 1996; 93:9782–9787.

8.   Cross AR, Jones OTG. Enzymic mechanisms of superoxide production. Biochim Biophys Acta 1991; 1057:281–298.

9.   Boveris A, Cadenas E. Mitochondrial production of superoxide anion and its relationship to the antimycin-insensitive respiration. FEBS Lett 1975; 54:311–314.

10. Guidot DM, McCord JM, Wright RM, Repine JE. Absence of electron transport (Rho0 state) restores growth of a manganese-superoxide dismutase-deficient *Saccharomyces cerevisiae* in hyperoxia: evidence for electron transport as a major source of superoxide generation in vivo. J Biol Chem 1993; 268:26699–26703.

11. van Loon A, Pesold-Hurt B, Schatz G. A yeast mutant lacking mitochondrial manganese-superoxide dismutase is hypersensitive to oxygen. Proc Natl Acad Sci USA 1986; 83:3820–3824.

12. Boveris A, Cadenas E, Stoppani AO. Role of ubiquinone in the mitochondrial generation of hydrogen peroxide. Biochem J 1976; 156:435–444.

13. Cadenas E, Boveris A, Ragan CI, Stoppani AO. Production of superoxide radicals and hydrogen peroxide by NADH-ubiquinone reductase and ubiquinol-cytochrome c reductase from beef heart mitochondria. Arch Biochem Biophys 1977; 180:248–257.

14. Turrens JF, Boveris A. Generation of superoxide anion by the NADH dehydrogenase of bovine heart mitochondria. Biochem J 1980; 191:421–427.

15. Zhang Y, Marcillat O, Giulivi C, Ernster L, Davies KJA. The oxidative inactivation of mitochondrial electron transport chain components and ATPase. J Biol Chem 1990; 265:16330–16336.

16. Imlay JA, Fridovich I. Assay of metabolic superoxide production in *Escherichia coli*. J Biol Chem 1991; 266:6957–6965.

17. Turrens JF, Alexandre A, Lehninger AL. Ubisemiquinone is the electron donor for superoxide formation by complex III of heart mitochondria. Arch Biochem Biophys 1985; 237:408–414.

18. Turrens JF, Freeman BA, Levitt JG, Crapo JD. The effect of hyperoxia on superoxide production by lung submitochondrial particles. Arch Biochem Biophys 1982; 217:401–410.

19. Jennings RB, Ganote CE. Mitochondrial structure and function in acute myocardial ischemic injury. Circ Res 1976; 38(suppl 1):1.

20. Frederiks WM, Marx F, Myagkaya GL. A histochemical study of changes in mitochondrial enzyme activities of rat liver after ischemia in vitro. Virchows Arch 1986; 51:321–329.

21. Nishida T, Shibata H, Koseki M, et al. Peroxidative injury to the mitochondrial respiratory chain during reperfusion of hypothermic rat liver. Biochim Biophys Acta 1987; 890:82–88.

22. Klug D, Rabani J, Fridovich I. A direct demonstration of the catalytic action of superoxide dismutase through the use of pulse radiolysis. J Biol Chem 1972; 247: 4839–4842.

23. Yonetani T. Studies on cytochrome c peroxidase. II. Stoichiometry between enzyme, H_2O_2, and ferrocytochrome c and enzymic determination of extinction coefficients of cytochromes. J Biol Chem 1965; 240:4509–4514.

24. Radi R, Sims S, Cassina A, Turrens J. Roles of catalase and cytochrome c in hydroperoxide-dependent lipid peroxidation and chemiluminescence in rat heart and kidney mitochondria. Free Radic Biol Med 1993; 15:653–659.

25. Radi R, Turrens J, Chang L, Bush K, Crapo J, Freeman B. Detection of catalase in rat heart mitochondria. J Biol Chem 1991; 266:22028–22034.

26. Getzoff ED, Tainer JA, Weiner PK, Kollman PA, Richardson JS, Richardson DC. Electrostatic recognition between superoxide and copper, zinc superoxide dismutase. Nature 1983; 306:287–290.

27. Repine JE. Scientific perspectives on adult respiratory distress syndrome. Lancet 1992; 339:466–469.

28. Jamieson D. Oxygen toxicity and reactive oxygen metabolites in mammals. Free Radic Biol Med 1989; 7:87–108.

29. Cadenas E. Biochemistry of oxygen toxicity. Annu Rev Biochem 1989; 58:79–110.

30. White CW, Ghezzi P, Dinarello CA, Caldwell SA, McMurtry IF, Repine JE. Recombinant tumor necrosis factor/cachectin and interleukin 1 pretreatment decreases lung oxidized glutathione accumulation, lung injury, and mortality in rats exposed to hyperoxia. J Clin Invest 1987; 79:1868–1873.

31. Tang G, Berg JT, White JE, Lumb PD, Lee CY, Tsan M. Protection against oxygen toxicity by tracheal insufflation of endotoxin: role of Mn SOD and alveolar macrophages. Am J Physiol 1994; 266(10):L38–L45.

32. White CW, Lewis-Molock Y, Suzuki K, et al. Effects of cytokines and endotoxin on lung manganese superoxide dismutase expression and immunohistochemical distribution. Chest 1994; 105(3):85S–86S.

33. Tsan M, Lee CY, White JE. Interleukin 1 protects rats against oxygen toxicity. J Appl Physiol 1991; 71:688–697.

34. Freeman BA, Mason RJ, Williams MC, Crapo JD. Antioxidant enzyme activity in alveolar type II cells after exposure of rats to hyperoxia. Exp Lung Res 1986; 10: 203–222.

35. Iqbal J, Clerch LB, Hass MA, Frank L, Massaro D. Endotoxin increases lung Cu,Zn superoxide dismutase mRNA: O_2 raises enzyme synthesis. Am J Physiol 1989; 257: L61–L64.

36. Rahman I, Massaro D. Endotoxin treatment protects rats against ozone-induced lung edema: with evidence for the role of manganese superoxide dismutase. Toxicol Appl Pharmacol 1992; 113:13–18.

37. Shiki Y, Meyrick BO, Brigham KL, Burr IM. Endotoxin increases superoxide dismutase in cultured bovine pulmonary endothelial cells. Am J Physiol 1987; 252: 436–440.

38. Frank L, Bucher JR, Roberts RJ. Oxygen toxicity in neonatal and adult animals of various species. J Appl Physiol 1978; 45:699–704.

39. Visner GA, Dougall WC, Wilson JM, Burr IA, Nick HS. Regulation of manganese

superoxide dismutase by lipopolysaccharide, interleukin-1, and tumor necrosis factor. J Biol Chem 1990; 265:2856–2864.

40. Harris JB, Chang LY, Crapo JD. Rat lung alveolar type I epithelial cell injury and response to hyperoxia. Am J Respir Cell Mol Biol 1991; 4:115–125.

41. Johnson LV, Walsh ML, Bockus BJ, Chen LB. Monitoring of relative mitochondrial membrane potential in living cells by fluorescence microscopy. J Cell Biol 1981; 88:526–535.

42. Abderrahman M, Pettit JM, Ratinaud M, Julien R. 10-N nonyl-acridine orange: a fluorescent probe which stains mitochondria independently of their energetic state. Biochem Biophys Res Commun 1989; 164:185–190.

43. Maftah A, Petit JM, Julien R. Specific interaction of the new fluorescent dye 10-N nonyl-acridine orange with inner mitochondrial membrane: a lipid-mediated inhibition of oxidative phosphorylation. FEBS Lett 1990; 260:236–240.

44. Saris NE, Akerman EO. Uptake and release of bivalent cations in mitochondria. In: Sanadi DR, ed. Current Topics in Bioenergetics, vol. 10. San Diego: Academic Press, 1980:103–179.

45. Eilers M, Oppliger W, Schatz G. Both ATP and energized inner membrane are required to import a purified precursor protein into mitochondria. EMBO J 1987; 6: 1073–1077.

46. Guidot DM, Repine JE, Kitlowski AD, et al. Mitochondrial respiration scavenges extramitochondrial superoxide anion via a non-enzymatic mechanism: an additional role for electron transport. J Clin Invest 1995; 96:1131–1136.

47. Guidot DM. Endotoxin treatment increases lung mitochondrial scavenging of extramitochondrial superoxide in hyperoxia-exposed rats. Arch Biochem Biophys 1996; 326:266–270.

3

The Biochemistry of Nitric Oxide and Peroxynitrite: Implications for Mitochondrial Function

Rakesh P. Patel, Trudy Cornwell, and Victor M. Darley-Usmar
*University of Alabama at Birmingham,
Birmingham, Alabama*

I. INTRODUCTION

Mitochondrial metabolism is essential for aerobic organisms and as such is a potential target for the removal of either cancerous cells or parasites. The cell most frequently employed in this role is the macrophage and as part of its cytotoxic armamentarium it produces free radicals, notably nitric oxide (NO), and superoxide (O_2^-). A theme we will explore in this article is how the independent and cooperative interaction of these two free radicals plays a role in cell death and inflammation. An interesting aspect that only now is emerging is that the effects of NO on mitochondria may not be restricted to the pathological interactions of inflammatory cells but may also extend to the physiology of respiration.

The efficiency of aerobic metabolism is intimately related to the complete reduction of oxygen to water in the mitochondrion. An enzyme, cytochrome c oxidase, is designed specifically for this task by catalyzing a concerted four-electron transfer reaction, which also avoids the production and release of partially reduced oxygen intermediates. Oxygen is reduced to water by four electrons, which are supplied by various substrates and carried via the electron transport chain and cytochrome c to cytochrome c oxidase. Both mechanistic and structural aspects of this metabolic process are now well understood. The oxygen-binding site of this protein, a binuclear heme-copper ion center denoted as heme a_3 and Cu_B, shows high affinity for this substrate but, in common with many other redox centers, can also bind the signaling molecule, NO. The oxidizing species, often referred to as reactive oxygen species, most pertinent to mitochondrial metabolism are the partially reduced intermediates of oxygen and are O_2^-, hydrogen peroxide (H_2O_2), and the hydroxyl radical ($OH\cdot$). All these species have been implicated in the pathology of human disease and are used in host defense mechanisms (1). However, until recently a cytotoxic role for O_2^- had only been inferred from in vitro studies in which it was shown that (a) mitochondria, under conditions of mitochondrial blockade, produced O_2^- from a number of loci and (b) the mitochondrion contains manganese superoxide dismutase (Mn-SOD), an enzyme which catalyzes the dismutation of O_2^- to H_2O_2. This is circumstantial evidence for a role of O_2^- in mitochondria but falls somewhat short of compelling evidence for a cytotoxic role. For example, these data are also consistent with a cell signaling function for O_2^- and/or H_2O_2. Recently the gene for Mn-SOD has been knocked out, and this is a lethal mutation (2). This clearly implicates Mn-SOD as essential for mitochondrial well-being, although molecular mechanisms remain uncertain. Superoxide per se is relatively unreactive. However, the finding that the reaction between NO and O_2^-, two radicals that are formed in biological systems, results in the formation of the oxidant peroxynitrite ($ONOO^-$) (3) provides a mechanistic basis for these observations which we will discuss herein.

II. FREE RADICALS AND MITOCHONDRIA: GENERAL ASPECTS

Free radicals are molecules possessing unpaired electrons and arise as necessary intermediates in mitochondrial metabolism, for example, in the form of ubisemiquinone, as well as what we presume to be unwanted side products such as O_2^-. Recently, it has been recognized that free radicals can also act as mediators of cell signaling which reveals their role and places mechanisms for their formation in a more ambiguous light (4–7). Clearly, in this context we cannot view all sources of O_2^- as the perversity of otherwise perfect electron transfer systems.

Thus most biochemical reactions that result in free radical formation are well controlled and so are prevented from propagating themselves, a reaction process which amplifies oxidant damage. However, some radicals or oxidants, when produced outside these tightly controlled biochemical pathways, can induce specific modifications to biomolecules, alter function, and so contribute to the development of disease.

Free radical formation in biological systems is thus an essential part of normal physiology and a potential route to the pathophysiology associated with a number of diseases. As will be seen, this is well exemplified in the mitochondrion through the interactions of free radicals with transition metal centers in the organelle.

III. THE ORIGIN OF NO IN BIOLOGICAL SYSTEMS: NO SYNTHASES

Since the interaction of NO will be an important aspect of our discussion we will briefly provide background covering the biochemistry of the enzymes and NO. For a more detailed discussion of the many aspects of NO biology we refer the reader to the excellent reviews available on this topic (8–10). The free radical NO is synthesized by a family of enzymes termed NO synthases (NOS) which use L-arginine, NADPH, and oxygen as substrates (8–11). The elucidation of the biology of this important signaling molecule has lent many insights into the principles and actions of free radicals in a biological setting. Although free radicals can be extremely reactive, this is not a general rule and many are capable of displaying specificity. For example, an interesting aspect of NO biochemistry in biological systems is its interaction with metalloproteins containing iron or copper and its reactions with other free radicals. Both aspects are significant in understanding the effects of NO on mitochondrial function.

As mentioned earlier, NO is highly discriminating in its reactions, which are essentially restricted to high-affinity binding with transition metal centers in proteins or rapid reactions with other free radicals. It can freely cross the hydrophobic core of phospholipid membranes and this, coupled with a persistence of some 4–5 s in a biological setting, results in a sphere of influence which extends well beyond its site of formation (12). Given this finding, it is perhaps surprising to find that the location of NO synthase is not exclusively cytosolic, but it can be attached to intracellular structures such as the caveoli in the endothelium or dystrophin in skeletal muscle (13–16). The reasons for these associations are not entirely clear. They are likely to involve important elements in the regulation of the enzymes and perhaps play a role in ensuring specificity. In this respect the recent reports indicating that NOS localization in human skeletal muscle occurs in the proximity of mitochondria, and that an NOS is present in the mitochondrial matrix [mitochondrial NOS (mNOS)] are intriguing (17,18).

The characterization of the mNOS is not yet complete, but a great deal is known of the other isoforms, of which there are three subtypes with different characteristics of activation and with different concentrations of NO which can be formed on maximal activation. In a later section we will discuss in more detail the interaction of NO with cytochrome c oxidase and mitochondria.

IV. THE BIOLOGICAL EFFECTS OF NO AND CONCENTRATION

In discussing novel aspects of NO signaling it is important to place the finding in the context of what we already know about what appears to be a major signal transduction process: the activation of soluble guanylate cyclase. This endogenous "receptor" for NO contains a ferrous heme iron which when bound to the ligand stimulates the enzyme to synthesize cyclic guanosine monophosphate (cGMP) (19,20). The subsequent events in the signaling cascade are illustrated in Figure 1. In the example shown, the activation of protein kinase G–dependent pathways leads to a decrease in cytosolic Ca^{2+} and subsequent relaxation of vascular smooth muscle cells (VSMC) (21–23).

Recent studies have also shown that protein kinase G-Iα (PKG-Iα) plays an essential role in controlling the phenotypic modulation of vascular smooth muscle cells (24). The change in vascular smooth muscle cells from a contractile to a synthetic form is important in the response of the vasculature to inflammation.

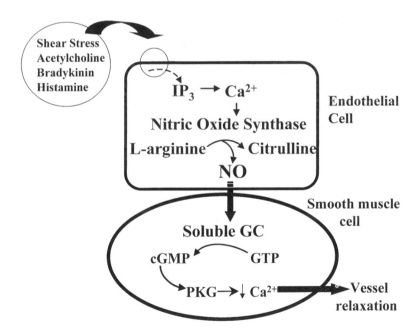

Figure 1 The NO-cGMP signal transduction pathway. Endothelial nitric oxide synthase (eNOS) can be activated physiologically through the action of both physical (e.g., shear stress) and agonist-dependent (e.g., acetylcholine) stimuli. The subsequent rise in intracellular calcium stimulates NO production from eNOS. NO then diffuses through to the underlying smooth muscle cell where it binds to and activates soluble guanylate cyclase (soluble GC), which catalyzes the hydrolysis of guanosine triphosphate (GTP) to cyclic guanosine monophosphate (cGMP). The latter activates protein kinase G (PKG) which in turn causes a decrease in the cytosolic calcium concentration and thus promotes smooth muscle cells and hence vessel relaxation.

The loss of PKG in cultured VSMCs is correlated with a reduced ability of cGMP to lower intracellular Ca^{2+}. This synthetic or "secretory" phenotype is responsible for the extensive deposition of extracellular matrix in the atherosclerotic lesion (24,25). The removal of PKG-Iα activity, one of the most sensitive pathways to NO in the cell, under conditions associated with remodeling presumably allows the functioning of other NO-dependent signaling pathways.

Are NO-dependent signaling pathways absent in cells which do not contain the PKG-1α cGMP signal transduction pathway? Probably not, since all NO-dependent responses do not show the saturation characteristics typical of many biological mediators that function through receptor-mediated interactions. For example, in cells where this pathway is absent or when binding to guanylate cyclase becomes maximal, other interactions come into play. Figure 2 illustrates this concept by showing the effect of NO on different biochemical processes as a function of concentration. The concentrations of NO required to elicit activation of PKA via PKG cannot be achieved by endothelial NOS but could result from activation of inducible NOS, or perhaps neuronal NOS. An example of this response is the NO-dependent inhibition of VSMC proliferation which results from PKA activation by cGMP (26).

The fate of NO in biological systems is still not entirely clear. It has been

NO concentration (M)

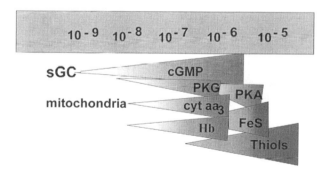

Figure 2 Concentration of NO and the activation of cell signaling pathways. The molar concentration of NO is shown increasing from left to right with the activation of soluble guanylate cyclase (sGC) to form cGMP which then proceeds to activate PKG or PKA. The interaction with mitochondria is shown at the level of cytochrome-*c* oxidase (cyt aa₃) and at higher concentrations other metalloproteins such as the iron sulfur centers (FeS). Nitric oxide can also rapidly bind to the ferrous heme of hemoglobin (Hb). Modulation of some ion channels may occur through mechanisms which are thiol dependent. At concentrations of NO above 1 μM reactions with oxygen generate reactive nitrogen species capable of modifying biomolecules (33).

suggested that the rapid reaction of NO with ferrous hemoglobin is a major pathway for its safe removal (12,27,28). Other possibilities include the reaction with free radicals or metabolism by mitochondria or other electron transfer systems (3,28–32). The reaction with oxygen requires two NO molecules for every oxygen atom and is likely to be too slow under physiological conditions to account for any significant degree of NO metabolism except in membranes (33).

V. NITRIC OXIDE AND MITOCHONDRIA

Some of the earliest studies examining the effects of NO on mitochondrial function were directed at dissecting the mechanisms through which macrophages could kill tumors. These studies describe an irreversible inhibition of complexes I and II of the respiratory chain in the mitochondria of cells coincubated with activated macrophages (34). At the time it had not been shown that NO was the mediating agent produced by these cells which resulted in these defects of mitochondrial metabolism. On realization that inducible NOS could indeed elicit such responses, these effects were characterized in more detail. A combination of spectroscopic and biochemical studies revealed that, under these conditions, the FeS centers of electron transfer proteins were targets for iron mobilization and accounted for the irreversible inhibition of electron transport in mitochondria by NO. In addition, other FeS centers in proteins such as aconitase were also demonstrated to be targets of inhibition (32). More recently, the effects of NO induced in neuronal cells was examined and similar results were reported, with the interesting addition that a reversible inhibition of the terminal member of the mitochondrial electron transport chain cytochrome c oxidase also occurred (35–37).

Most of these studies were performed with cell culture systems rather than isolated mitochondria. The recent availability of stable compounds which could be used to deliver NO at controlled rates to mitochondrial preparations allowed examination of the direct effects of NO on mitochondrial function (38–43). In this case the effects were surprising since it was immediately clear that NO had little or no direct effect on any mitochondrial electron transfer proteins other than a reversible inhibition of cytochrome c oxidase (42–44).

In retrospect, this should have been anticipated, since spectroscopic studies of the binding of NO to the enzyme had been used to gain insight into the molecular ligands chelating the redox centers (45–48). However, the concentrations used were extremely high and a physiological role for NO had not yet been envisaged. The recent crystal structure of eukaryotic cytochrome c oxidase has placed the four redox centers, two copper atoms and two heme a_3 prosthetic groups, in close proximity (49,50). Cytochrome a_3 and Cu_B together form the oxygen binding site in which the oxygen atom bridges between the two metals.

Studies looking directly at the interaction of NO and cytochrome c oxidase have shown that it is in the oxygen binding site of the enzyme that NO binds and in fact competes with oxygen for the cytochrome a_3 site (38,42–44,51). This explains numerous reports which have demonstrated that NO is a more potent inhibitor of mitochondrial respiration at lower oxygen tensions (51,52). NO-mediated inhibition of cytochrome c oxidase only occurs during enzyme turnover (42), and thus requires substrates shuttling electrons into the respiratory chain. Rapid kinetic studies indicate that NO does not affect electron transfer between cytochrome c and the enzyme nor electron transfer between the different redox centers within the enzyme (43,44). The final inhibited state of the enzyme is NO bound to the reduced binuclear center, that is, $Fe_{a3}^{2+}-NO-Cu_B^+$, the site where oxygen binds and is reduced to water. Infrared spectroscopic studies have shown that NO binds specifically to this site, with no evidence that binding occurs at either $heme_a$ or Cu_A (53). Thus the oxygen concentration is critical in determining the extent to which the enzyme can be inhibited by NO. However, simulation of the competitive nature of this inhibition, using the known rate constants for oxygen and NO binding to the binuclear center (O_2: $k = 1 \times 10^8$ M/s; NO: $k = 4 \times 10^7$ M/s) suggests that at oxygen concentrations up to 30 µM, which are in the physiological range, the rate of enzyme inhibition should be much slower than experimentally observed (42,44). This has led to the postulate that NO binds to an enzyme intermediate for which oxygen has a low affinity. A partially reduced binuclear center (i.e., $a_3^{3+}-Cu_B^+$) has been proposed as such an intermediate (42) (Fig. 3). More recent data indicate that NO directly reduces Cu_B^{2+} to Cu_B^+, forming NO^+ which then becomes hydrated to HNO_2 (54). By assessing the degree of NO-dependent enzyme inhibition at different oxygen tensions, a K_i value of approximately 0.1 µM for NO (in the absence of oxygen) was calculated (42), which is similar to values obtained for inhibition of respiration in whole mitochondria (41). These concentrations of NO can be formed in vivo and imply that this may be a relevant mechanism by which the activity of cytochrome c oxidase may be controlled. The physiological consequences of such regulation remain obscure.

Another intriguing aspect of this mode of inhibition is its reversibility, which has been observed in studies with purified cytochrome c oxidase and in mitochondrial particles (41,42–44,51,52). The reversibility may be accounted for by either the slow reaction of NO with oxygen to yield NO_2 (44) or by a direct enzyme-dependent reduction to N_2O (55). This inhibition is still reversible under the conditions used to prepare mitochondria, and thus also explains why earlier studies failed to detect effects on cytochrome c oxidase in mitochondria isolated from cells previously exposed to NO. Since no effects on complexes I and II were reported, a more complex explanation for the NO-dependent inhibition of these enzymes must be invoked. This will be discussed in more detail in section VI.

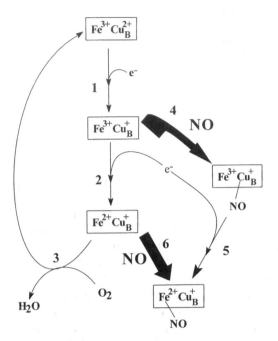

Figure 3 Interaction of NO with binuclear center in cytochrome c oxidase. NO inhibits mitochondrial respiration by interacting with the binuclear center of cytochrome c oxidase shown prior to step 1 in its oxidized state ($Fe^{3+}Cu_B^{2+}$). Inhibition is only seen during enzyme turnover, that is, when electrons are supplied to the binuclear center (step 1). Addition of another electron (step 2) forms the fully reduced enzyme ($Fe^{2+}Cu_B^+$), which in turn reduces O_2 to H_2O (step 3). Points of interaction with NO are shown with the broad arrows. Nitric oxide has been postulated to interact with a partially reduced binuclear center ($Fe^{3+}Cu_B^+$; step 4) in which copper exists in the cuprous oxidation state and for which O_2 has a low affinity (42,54). Addition of an electron to this intermediate (step 5) facilitates binding of NO to the ferrous iron, thus forming the final inhibited state of the enzyme ($Fe^2-NO-Cu_B^+$). This species can also be formed through a reaction between NO and the fully reduced enzyme (step 6), a process which directly competes with reduction of O_2. To explain the reversibility of inhibition, NO has been postulated to either autooxidize to form nitrogen dioxide (44) or be reduced to N_2O (55).

Direct reaction of NO with thiols on key metabolic proteins and ion channels have been reported (56). The biochemical basis of these interactions is now being investigated and may reveal alternative routes for NO signaling. Mitochondria contain a number of proteins with thiol groups which are essential for their activity (57). At the present juncture no reports of NO-dependent thiol modification of a respiratory enzyme has been reported. However, the conjunction of both

NO and metal centers which could promote S-nitrosothiol formation make this a realistic prospect.

VI.   FUNCTION OF THE INTERACTION BETWEEN NO AND MITOCHONDRIA

Can the inhibition of cytochrome c oxidase by NO have a pathological or physiological function? This is an important issue since an admittedly teleological point of view is that if an isoform of NOS is to be found in the mitochondrion it is likely to have an organelle-specific function. Perhaps more compelling is the finding that administration of NOS inhibitors to animal models enhances oxygen consumption via respiration (58). These observations indicate that NO affects basal rates of oxygen consumption, and thus may be a modulator of cellular respiration in vivo (59). Under some circumstances, as will be discussed later, this interaction could contribute to the cytotoxic effects of NO. This is not a simple issue to address since, as we discussed previously, NO binding to a reduced iron center is an intrinsic chemical property of the molecule. One approach to this problem is to define, on a hypothetical basis, some of the possibilities and address these in turn. These include modulation of respiration, oxygen sensing, and regulation of apoptosis and these are discussed in more detail in other chapters in this volume.

VII.   PEROXYNITRITE AND MITOCHONDRIA

In addition to reaction with transition metal centers a facile reaction of NO is with other free radicals. This aspect has been explored in the most detail in the context of the rapid reaction with O_2^- to form the oxidant known as peroxynitrite ($ONOO^-$). In fact, one of the most abundant free radicals in aerobic biological systems is O_2^-. Since the discovery of SODs it has been argued that their location in a particular site in the cell indicates that this free radical is being produced in this location. In this respect it is interesting that mitochondria contain a SOD different from that in the cytosol which has manganese in its active site. Various sites of superoxide formation have been identified in mitochondria including complexes I, II, and III. However, it is important to note that O_2^- formation from mitochondria cannot be detected except as the product of its dismutation, hydrogen peroxide. Only when submitochondrial particles are prepared free of Mn-SOD can O_2^- itself be detected. If O_2^- were produced in the intermembrane space it is likely that it would react with cytochrome c. Since in this reaction O_2^- is a reductant, the product is oxygen. These mechanisms ensure that O_2^- is kept at a low concentration. The only reaction that we currently know of which

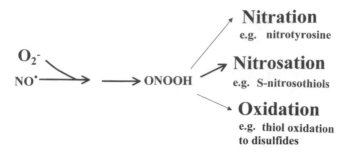

Figure 4 Reactions that peroxynitrite may mediate. Peroxynitrite can promote nitration, nitrosation, and oxidation of biological molecules including proteins and lipids (60–63). Such modifications can significantly alter the biological function of the affected molecules and are involved in the irreversible inhibition of mitochondrial respiration mediated by peroxynitrite.

could compete with the SOD catalyzed dismutation of O_2^- is that with NO which proceeds at a rate close to the diffusion limit.

It is plausible that NO-mediated inhibition of cytochrome c oxidase results in O_2^- formation in other segments of the respiratory chain which contributes to the formation of $ONOO^-$. This could also result in release of Ca^{2+} from mitochondria and may explain the damage to FeS centers which occurs on extended exposure to NO. Peroxynitrite mediates three types of reactions which could be relevant to mitochondrial function (Fig. 4). Oxidation reactions mediated by $ONOO^-$ can be either one- or two-electron transfers. Nitration or nitrosation of lipids, proteins, carbohydrates, or thiols results in the formation of stable compounds (60–63). Some of these, such as nitrated aromatic amino acids, are also immunogenic (64). Indeed, this fact has been used as one way to demonstrate $ONOO^-$ formation in vivo. Three-nitro-tyrosine (nTyr), the product of the addition of a nitro group (NO_2) to the *ortho* position of the hydroxyl group of tyrosine can also be detected by other analytical techniques such as HPLC or mass spectrometry (65,66). The nitration reaction could proceed through two intermediates; nitrogen dioxide or the nitronium ion and in the latter case metalloenzymes, notably SOD, can enhance this reaction (67). Alternative routes of formation include the reaction of nitrite with hypochlorite derived from the enzymatic activity of myeloperoxidase, but these are unlikely to occur within the matrix of the mitochondrion (68). Does the reaction of NO and O_2^- occur in the mitochondrion? In support of this possibility, nitration of Mn-SOD has been demonstrated to occur in organ rejection and is associated with a loss in enzyme function (66). A scenario can be envisaged therefore where $ONOO^-$-dependent damage to Mn-

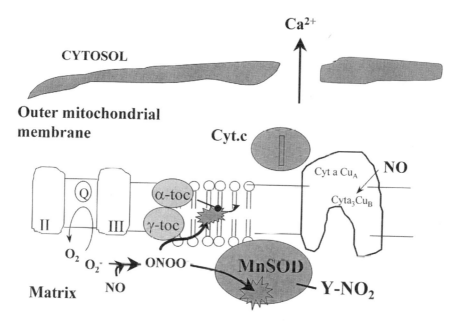

Figure 5 Inhibition of mitochondrial respiration by NO. This figure depicts NO interactions within mitochondria leading to inhibition of cytochrome c oxidase and formation of peroxynitrite ($ONOO^-$). NO binds rapidly with ferrous heme iron of the heme a_3-Cu_B complex of cytochrome c oxidase resulting in inhibition of mitochondrial respiration and hence of proton translocation. The subsequent depolarization of the membrane causes Ca^{2+} efflux into the cytosol. Furthermore, the inhibition of cytochrome c oxidase can enhance O_2^- formation via the reaction of electron carriers [e.g., ubiquinone (Q)] with oxygen. Superoxide can then react with NO to form $ONOO^-$, which can exacerbate injury by inhibiting oxidative phosphorylation through inhibition of other mitochondrial enzymes. Peroxynitrite can also initiate lipid peroxidation reactions which would contribute to mitochondrial dysfunction. This can be inhibited by scavenging of peroxynitrite and of lipid peroxyl radicals by α-tocopherol (α-toc) and γ-tocopherol (γ-toc). The nitration of Mn-SOD by $ONOO^-$ in the mitochondrial matrix will result in increased superoxide and establish a vicious cycle culminating in destruction of the organelle.

SOD leads to a higher concentration of O_2^- which in turns leads to formation of more $ONOO^-$ (Fig. 5).

Oxidation of thiols by $ONOO^-$ has been linked to calcium release via the mitochondrial transition pore (69,70). Peroxynitrite also irreversibly inhibits mitochondrial respiration by inactivating complexes I, II, and V. The mechanisms by which this occurs remain unknown, although lipid peroxidation reactions and membrane thiol cross-linking have been implicated (40,71,72). The similarity in the pattern of inhibition of respiration observed on addition of ONOO⁻ to isolated mitochondria and addition of NO to intact cells (40) suggests that in the latter case ONOO⁻ is being generated within the mitochondrion in close proximity to components of the electron transport chain (73). This would bypass the protective effect of endogenous antioxidant molecules such as glutathione, which have been shown to efficiently block the deleterious effects of ONOO⁻ in mitochondrial respiration (74).

In addition to mitochondrial electron transfer, the organelle contains other essential metabolic pathways which utilize electron transport pathways resulting in the intermediate formation of protein-bound free radicals. These are proteins involved in processes such as DNA synthesis, for example, ribonucleotide reductase. Since mitochondria contain proteins encoded for by both the nuclear and mitochondrial genomes, it may be particularly sensitive to NO during replication of the organelle.

VIII. CONCLUSIONS

The important role for NO in many physiological processes at the cell or organ level has now been recognized and many of the molecular mechanisms defined in some detail. However, although the effects of this signaling molecule on mitochondrial function have been known for some time, only recently has evidence been presented supporting a physiological function for this interaction. This has stimulated renewed interest in this area, and as a result many new insights are now emerging. For example, if substantiated, the discovery of an mNOS is both exciting and perplexing since we have very little idea of what the function might be. As with other aspects of NO biochemistry, the mitochondrial response will depend on many factors including the concentrations of NO, oxygen, and prevailing activity of the electron transport chain. However, due to the rapid reaction of NO with metal ions and with other free radicals, NO can also be a mediator of cytotoxic processes, primarily through inhibition of mitochondrial respiration and formation of the potent oxidant ONOO⁻. While this complex spectrum of interactions certainly presents some unique challenges to the experimentalist, it will no doubt continue to stimulate a great deal of interest and activity in the future.

REFERENCES

1. Nussler AK, Billiar TR. Inflammation, immunoregulation, and inducible nitric oxide synthase. J Leuk Biol 1993; 54:171–178.

2. Li Y, Huang T-T, Carlson EJ, et al. Dilated cardiomyopathy and lethality in mutant mice lacking manganese superoxide dismutase. Nat Genet 1995; 11:376–381.
3. Beckman JS, Beckman TW, Chen J, Marshall PA, Freeman BA. Apparent hydroxyl radical production by peroxynitrite: implications for endothelial injury from nitric oxide and superoxide. Proc Natl Acad Sci USA 1990; 87:1620–1624.
4. Monteiro HP, Stern A. Redox modulation of tyrosine phosphorylation-dependent signal transduction pathways. Free Radic Biol Med 1996; 21:323–334.
5. Darley-Usmar VM, Wiseman H, Halliwell B. Nitric oxide and oxygen radicals: a question of balance. FEBS Lett 1995; 369:131–135.
6. Flohe L, Brigelius-Flohe R, Saliou C, Traber MG, Packer L. Redox regulation of NFκB. Free Radic Biol Med 1997; 22:1115–1126.
7. Suzuki YJ, Forman HJ, Sevanian A. Oxidants as stimulators of signal transduction. Free Radic Biol Med 1997; 22:269–285.
8. Knowles RG, Moncada S. Nitric oxide as a signal in blood vessels. Trends Biochem Sci 1993; 17:399–402.
9. Marletta MA. Nitric oxide synthase: aspects concerning structure and catalysis. Cell 1994; 78:927–930.
10. Moncada S, Palmer R, Higgs EA. Nitric oxide: physiology, pathophysiology and pharmacology. Pharmacol Rev 1991; 43:109–142.
11. Garvey EP, Furfine ES, Sherman PA. Purification and inhibitor screening of human nitric oxide synthase isozymes. Method Enzymol 1996; 268:339–349.
12. Lancaster JR. Stimulation of the diffusion and reaction of endogenously produced nitric oxide. Proc Natl Acad Sci 1994; 91:8137–8141.
13. Garcia-Cardena G, Fan R, Stern DF, Liu J, Sessa WC. Endothelial nitric oxide synthase is regulated by tyrosine phosphorylation and interacts with caveolin-1. J Biol Chem 271:27237–27240.
14. Feron O, Belhassen L, Kobzik L, et al. Endothelial nitric oxide synthase targeting to caveolae. Specific interactions with caveolin isoforms in cardiac myocytes and endothelial cells. J Biol Chem 1996; 271:22810–22814.
15. Garcia-Cardena G, Oh P, Liu J, Schnitzer JE, Sessa WC. Targeting of nitric oxide synthase to endothelial cell caveolae via palmitoylation: implications for nitric oxide signaling. Proc Natl Acad Sci USA 1996; 93:6448–6453.
16. Grozdanovic Z, Gosztonyi G, Gossrau R. Nitric oxide synthase I (NOS-I) is deficient in the sarcolemma of striated muscle fibers in patients with Duchenne muscular dystrophy, suggesting an association with dystrophin. Acta Histochem 1996; 98:61–69.
17. Bates TE, Loesch A, Burnstock G, Clark JB. Immunocytochemical evidence for a mitochondrially located nitric oxide synthase in brain and liver. Biochem Biophys Res Commun 1995; 213:896–900.
18. Bates TE, Loesch A, Burnstock G, Clark JB. Mitochondrial nitric oxide synthase: a ubiquitous regulator of oxidative phosphorylation. Biochem Biophys Res Commun 1996; 218:40–44.
19. Ignarro LJ, Degnan JN, Baricos WH, Kadowitz PJ, Wolin MS. Activation of purified guanylyl cyclase by NO requires heme. Biochim Biophys Acta 1982; 718:49–59.
20. Traylor TG, Sharma VS. Why NO? Biochemistry 1992; 31:2847–2849.

21. Lincoln TM, Cornwell TL. Intracellular cyclic GMP receptor proteins. FASEB J 1993; 7:328–338.

22. Butt E, Geiger J, Jarchau T, Lohmann SM, Walter U. The cGMP-dependent protein kinase—gene, protein, and function. Neurochem Res 1993; 18:27–42.

23. Francis SH, Noblett BD, Todd BW, Wells JN, Corbin JD. Relaxation of vascular and tracheal smooth muscle by cyclic nucleotide analogs that preferentially activate purified cGMP-dependent protein kinase. Mol Pharmacol 1988; 34:505–517.

24. Boerth NJ, Dey N, Cornwell TL, Lincoln TM. Cyclic GMP-dependent protein kinase regulates vascular smooth muscle cell phenotype. J Vasc Biol 1997; 34:245–259.

25. Campbell GR, Chamley-Campbell JH. Smooth muscle phenotypic modulation: role in atherogenesis. Med Hypotheses 1981; 7:729–735.

26. Cornwell TL, Arnold E, Boerth NJ, Lincoln TM. Inhibition of smooth muscle cells growth by nitric oxide and activation of cAMP-dependent protein kinase by cGMP. Am J Physiol 1994; 267:C1405–C1413.

27. Doyle MP, Hoekstra JW. Oxidation of nitrogen oxides by bound dioxygen in hemeproteins. J Inorg Biochem 1981; 14:351–358.

28. Sharma VS, Traylor TG, Gardiner R. Reaction of nitric oxide with heme proteins and model compounds of hemoglobin. Biochemistry 1987; 26:3837–3843.

29. Padmaja S, Huie RE. The reaction of nitric oxide with organic peroxyl radicals. Biochem Biophys Res Commun 1993; 195:539–544.

30. Richter C. Reactive oxygen and nitrogen species regulate mitochondrial calcium homeostasis and respiration. Biosci Rep 1997; 17:53–66.

31. Brown GC. Nitric oxide regulates mitochondrial respiration and cell functions by inhibiting cytochrome oxidase. FEBS Lett 1995; 369:136–139.

32. Henry Y, Lepoivre M, Drapier J-C, et al. EPR characterization of molecular targets for NO in mammalian cells and organelles. FASEB J 1993; 7:1124–1134.

33. Wink DA, Darbyshire JF, Nims RW, Saavedra JE, Ford PC. Reactions of the bioregulatory agent nitric oxide in oxygenated aqueous media: determination of the kinetics of oxidation and nitrosation by intermediates generated by the NO/O_2 reaction. Chem Res Toxicol 1993; 6:23–27.

34. Granger DL, Lehninger AL. Sites of inhibition of mitochondrial electron transport in macrophage-injured neoplastic cells. J Cell Biol 1982; 95:527–535.

35. Brown GC, Bolanos JP, Heales SJ, Clark JB. Nitric oxide produced by activated astrocytes rapidly and reversibly inhibits cellular respiration. Neurosci Lett 1995; 193:201–204.

36. Bolanos JP, Peuchen S, Heales SJ, Land JM, Clark JB. Nitric oxide-mediated inhibition of the mitochondrial respiratory chain in cultured astrocytes. J Neurochem 1994; 63:910–916.

37. Cooper CE, Brown GC. The interactions between nitric oxide and brain nerve terminals as studied by electron paramagnetic resonance. Biochem Biophys Res Commun 1995; 212:404–412.

38. Cleeter MW, Cooper JM, Darley-Usmar VM, Moncada S, Schapira AH. Reversible inhibition of cytochrome c oxidase, the terminal enzyme of the mitochondrial respiratory chain, by nitric oxide. Implications for neurodegenerative diseases. FEBS Lett 1994; 345:50–54.

39. Balakirev MY, Khramtsov VV, Zimmer G. Modulation of the mitochondrial permeability transition by nitric oxide. Eur J Biochem 1997; 246:710–718.

40. Cassina A, Radi R. Differential inhibitory action of nitric oxide and peroxynitrite on mitochondrial electron transport. Arch Biochem Biophys 1996; 328:309–316.

41. Poderoso JJ, Carreras MC, Lisdero C, et al. Nitric oxide inhibits electron transfer and increases superoxide radical production in rat heart mitochondria and submitochondrial particles. Arch Biochem Biophys 1996; 328:85–92.

42. Torres J, Darley-Usmar V, Wilson MT. Inhibition of cytochrome c oxidase in turnover by nitric oxide: mechanism and implications for control of respiration. Biochem J 1995; 312:169–173.

43. Torres J, Davies N, Darley-Usmar VM, Wilson MT. The inhibition of cytochrome c oxidase by nitric oxide using S-nitrosoglutathione. J Inorg Biochem 1997; 66:207–212.

44. Giuffre A, Sarti P, D'itri E, et al. On the mechanism of inhibition of cytochrome c oxidase by nitric oxide. J Biol Chem 1996; 271:33404–33408.

45. Stevens TH, Brudvig GW, Bocian DF, Chan SI. Structure of cytochrome a3-Cua3 couple in cytochrome c oxidase as revealed by nitric oxide binding studies. Proc Natl Acad Sci USA 1979; 76:3320–3324.

46. Boelens R, Rademaker H, Pel R, Wever R. EPR studies of the photodissociation reactions of cytochrome c oxidase-nitric oxide complexes. Biochim Biophys Acta 1982; 679:84–94.

47. Brudvig GW, Stevens TH, Chan SI. Reactions of nitric oxide with cytochrome c oxidase. Biochemistry 1980; 19:5275–5285.

48. Wever R, Boelens R, De Boer E, et al. The photoreactivity of the copper-NO complexes in cytochrome c oxidase and in other copper-containing proteins. J Inorg Biochem 1985; 23:227–232.

49. Tsukihara T, Aoyama H, Yamashita E, et al. The whole structure of the 13-subunit oxidized cytochrome c oxidase at 2.8 A. Science 1996; 272:1136–1144.

50. Iwata S, Ostermeier C, Ludwig B, Michel H. Structure at 2.8 A resolution of cytochrome c oxidase from *Paracoccus denitrificans*. Nature 1995; 376:660–669.

51. Brown GC, Cooper CE. Nanomolar concentrations of nitric oxide reversibly inhibit synaptosomal respiration by competing with oxygen at cytochrome c oxidase. FEBS Lett 1994; 356:295–298.

52. Takehara Y, Kanno T, Yoshioka T, Inoue M, Utsumi K. Oxygen dependent regulation of mitochondrial energy metabolism by nitric oxide. Arch Biochem Biophys 1995; 323:27–32.

53. Zhao X-J, Sampath V, Caughey WS. Infrared characterization of nitric oxide bonding to bovine heart cytochrome c oxidase and myoglobin. Biochem Biophys Res Commun 1994; 204:537–543.

54. Cooper CE, Torres J, Sharpe MA, Wilson MT. Nitric oxide ejects electrons from the binuclear centre of cytochrome c oxidase by reacting with oxidised copper: a general mechanism for the interaction of copper proteins with nitric oxide? FEBS Lett 1997; 414:281–284.

55. Zhao X-J, Sampath V, Caughey WS. Cytochrome c oxidase catalysis of the reduction of nitric oxide to nitrous oxide. Biochem Biophys Res Commun 1995; 212:1054–1060.

56. Stamler JS, Singel DJ, Loscalzo J. Biochemistry of nitric oxide and its redox-activated forms. Science 1992; 258:1898–1903.
57. Richter C, Gogradze V, Laffranchi R, et al. Oxidants in mitochondria: from physiology to diseases. Biochem Biophys Acta 1995; 1271:67–74.
58. Xie YW, Wolin MS. Role of nitric oxide and its interaction with superoxide in the suppression of cardiac muscle mitochondrial respiration. Involvement in response to hypoxia/reoxygenation. Circulation 1996; 94:2580–2586.
59. Shen W, Hintze TH, Wolin MS. Nitric oxide: an important signaling mechanism between vascular endothelium and parenchymal cell in the regulation of oxygen consumption. Circulation 1995; 92:3505–3512.
60. Moro MA, Darley-Usmar VM, Lizasoain I, The formation of nitric oxide donors from peroxynitrite. Br J Pharmacol 1995; 116:1999–2004.
61. Moro MA, Darley-Usmar VM, Goodwin DA, et al. Paradoxical fate and biological action of peroxynitrite on human platelets. Proc Natl Acad Sci USA 1994; 91:6702–6706.
62. Mayer B, Schrammel A, Klatt P, Koesling D, Schmidt K. Peroxynitrite-induced accumulation of cyclic GMP in endothelial cells and stimulation of purified soluble guanylyl cyclase. Dependence on glutathione and possible role of S-nitrosation. J Biol Chem 1995; 270:17355–17360.
63. O'Donnell VB, Chumley PH, Hogg N, et al. Nitric oxide inhibition of lipid peroxidation: kinetics of reaction with lipid peroxyl radicals and comparison with α-tocopherol. Biochemistry 1997; 36:15216–15223.
64. Beckmann JS, Ye YZ, Anderson PG, et al. Extensive nitration of protein tyrosines in human atherosclerosis detected by immunohistochemistry. Biol Chem Hoppe-Seyler 375:81–88.
65. Kaur H, Halliwell B. Evidence for nitric oxide-mediated oxidative damage in chronic inflammation. Nitrotyrosine in serum and synovial fluid from rheumatoid patients. FEBS Lett 1994; 350:9–12.
66. MacMillan-Crow LA, Crow JP, Kerby JD, Beckman JS, Thompson JA. Nitration and inactivation of manganese superoxide dismutase in chronic rejection of human renal allografts. Proc Natl Acad Sci USA 1996; 93:11853–11858.
67. Ischiropoulos H, Zhu L, Chen J, et al. Peroxynitrite-mediated tyrosine nitration catalyzed by superoxide dismutase. Arch Biochem Biophys 1992; 298:431–437.
68. Eiserich JP, Cross CE, Jones AD, Halliwell B, van der Vliet A. Formation of nitrating and chlorinating species by reaction of nitrite with hypochlorous acid. A novel mechanism for nitric oxide-mediated protein modification. J Biol Chem 1996; 271:19199–19208.
69. Schweizer M, Richter C. Peroxynitrite stimulates the pyridine nucleotide-linked Ca^{2+} release from intact rat liver mitochondria. Biochemistry 1996; 35:4524–4528.
70. Packer MA, Murphy MP. Peroxynitrite formed by simultaneous nitric oxide and superoxide generation causes cyclosporin-A-sensitive mitochondrial calcium efflux and depolarisation. Eur J Biochem 1995; 234:231–239.
71. Radi R, Rodriguez M, Castro L, Telleri R. Inhibition of mitochondrial electron transport by peroxynitrite. Arch Biochem Biophys 1994; 308:89–95.
72. Heales SJ, Bolanos JP, Land JM, Clark JB. Trolox protects mitochondrial complex IV from nitric oxide mediated damage in astrocytes. Brain Res 1994; 668:243–245.

73. Packer MA, Porteous CM, Murphy MP. Superoxide production by mitochondria in the presence of nitric oxide forms peroxynitrite. Biochem Mol Biol Int 1996; 40: 524–534.

74. Lizasoain I, Moro MA, Knowles RG, Darley-Usmar V, Moncada S. Nitric oxide and peroxynitrite exert distinct effects on mitochondrial respiration which are differentially blocked by glutathione or glucose. Biochem J 1996; 314:877–880.

4

Oxygen-Dependent Regulation of Biological Functions by Nitric Oxide

Kozo Utsumi, Yoshiki Takehara, Yoko Inai, Munehlsa Yabuki, and Tomoko Kanno
Kurashiki Medical Center, Kurashiki, Japan

Manabu Nishikawa, Eisuke F. Sato, and Masayasu Inoue
Osaka City University Medical School, Osaka, Japan

I. INTRODUCTION

Nitric oxide (NO), a short-lived gaseous radical synthesized by NO synthase (NOS), has been identified as an important regulatory molecule in neurotransmission, vasodilation, platelet aggregation, and in host defenses (1–3). Endogenously generated NO readily enters neighboring cells and reversibly affects the activities of various enzymes, such as protein kinase C and the calcium-dependent protease, calpain (4,5). Nitric oxide also inhibits mitochondrial functions in various cells (6–11) by interacting with cytochrome oxidase and ferrocytochrome-*c* (Cyt-*c*) to form nitrosyl ferrocytochrome complexes (12,13). NO has been shown to inhibit superoxide radical (O_2^-) production by neutrophils (14,15) through the formation of a nitrosyl complex with NADPH oxidase (16–18). Furthermore, NO de-energizes mitochondria (16,19), increases intracellular free calcium (20) thereby killing various cells either by an apoptotic or a nonapoptotic mechanism (21–26). Endothelial cells produce substantial amounts of NO, an essential mediator for regulating guanylate cyclase (GC) activity (1,27,28) and blood pressure (1,29).

   Because NO reacts with O_2 and O_2^-, its fate and functions might be affected by these reactants. It has been known for some time that the half-life of NO is extremely short under atmospheric conditions (several seconds) predominantly due to the rapid reaction with molecular oxygen ($k = 6 \times 10^6/M/s$) (1). Thus the biological activity of NO might be stronger at low oxygen tensions than at high tensions. Although effects of NO on various functions, such as enzymes, cells, and organs, have been studied in vitro, most of these experiments were carried out under atmospheric conditions where the oxygen concentration is significantly higher than in vivo.

Recent studies in this laboratory revealed that the duration of NO action depended greatly on the concentration of environmental oxygen. The biological effects of NO, such as modulation of mitochondrial functions (30–32), inhibition of ascites tumor cell respiration (33,34), inhibition of superoxide generation by neutrophils (35), enhanced DNA-fragmentation in HL-60 cells (36), and relaxation of aortic smooth muscles (37), were stronger and continued longer at physiologically low oxygen tensions (1–40 μM) than at air-saturated oxygen concentration (230 μM). The present work describes oxygen-dependent regulation of NO functions in various biological systems.

II. EFFECT OF NO AND OXYGEN CONCENTRATION

A. Effect of NO and Oxygen Concentration on Oxidative Phosphorylation of Mitochondria

The effect of NO on the respiration of mitochondria was examined under different oxygen tensions (Fig. 1). In the presence of a respiratory substrate and inorganic phosphate, ADP increased the respiration of rat liver mitochondria. The state 3 respiration was reversibly inhibited by NO in an oxygen-concentration-dependent manner. The duration of inhibition was shorter and the extent smaller at high oxygen concentrations compared with inhibition low oxygen tension (Fig. 1). This inhibition was reversed instantaneously by adding oxyhemoglobin (HbO$_2$) at any oxygen tension (30–32).

ATP synthesis also stopped when respiration was inhibited by NO, and the inhibition of ATP synthesis was also reversible. Mitochondria generate a membrane potential either by substrate oxidation or by ATP hydrolysis (19). The membrane potential of mitochondria was depolarized by NO in a reversible manner. The extent of depolarization depended on the concentration of NO and oxygen, depolarization occurred more markedly at low oxygen concentrations than at high tensions (Fig. 2).

Similar inhibition by NO of DNP-uncoupled respiration was observed in the presence of succinate, or α-ketoglutarate, or ascorbate-TMPD as substrates (Fig. 3) (32). The duration and extent of inhibition by NO was shorter and smaller at high oxygen concentrations than at low tensions. Kinetic analysis using specific

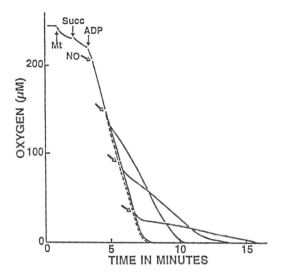

Figure 1 Oxygen dependent inhibition of mitochondrial respiration by NO. Oxidative phosphorylation was measured polarographically using a Clark-type oxygen electrode fitted to a 2 ml water-jacketed closed chamber at 25°C. Isolated mitochondria were suspended in a medium (0.5 mg protein/ml) consisting of 0.2 M sucrose, 10 mM KCl, 1 mM $MgCl_2$, 2 mM sodium phosphate, 10 mM Tris-HCl (pH 7.4), and 5 mM succinate at 25°C. State 3 respiration was initiated by adding 600 µM ADP. NO was added at different oxygen tensions (arrows) at a final concentration of 0.8 µM.

inhibitors revealed that the reversible inhibition of mitochondrial respiration and oxidative phosphorylation coincided with the reversible inhibition of cytochrome oxidase (30–32).

B. Effect of NO and Oxygen on the Respiration of Ascites Tumor Cells

NO also inhibited endogenous oxygen consumption of Ehrlich ascites tumor cells in a reversible manner; the inhibitory effect was more prolonged at low oxygen tensions than in air-saturated conditions (Fig. 4) (33). A similar effect of NO was also observed with ascites hepatoma (AH-130) cells (34). In the presence of various substrates, such as glutamate, glycerophosphate, succinate, or ascorbate + TMPD, digitonin-permeabilized tumor cells exhibited respiration by some NO-inhibitable mechanism (33,34).

Changes in the transmembrane potential are tightly coupled with cellular energy metabolism. In the presence of glucose, NO depolarized tumor cells tran-

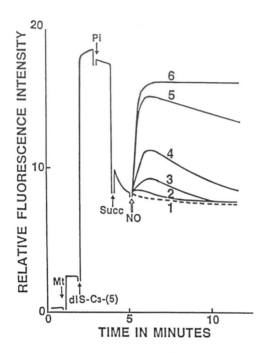

Figure 2 Oxygen dependency of NO-induced depolarization of mitochondria. The respiratory activity and membrane potential of rat liver mitochondria (0.1 mg protein/ml) were monitored simultaneously using a fluorospectrophotometer equipped with an oxygen electrode. The membrane potential was monitored by changes in the fluorescence intensity of cyanine dye diS-C$_3$-(5) (250 ng/ml) in a medium containing 0.2 M sucrose, 10 mM KCl, 1 mM MgCl$_2$, and 10 mM Tris-HCl (pH 7.4) at 25°C. The wavelength for excitation and emission was set at 622 and 670 nm, respectively. The reaction was started by adding 1.26 µM of NO. Oxygen tension at the time NO was added was 235 (2), 172 (3), 129 (4), 76 (5), and 6.6 µM, respectively. 1, control group in the absence of NO at air oxygen concentration (235 µM); Mt, 0.1 mg mitochondrial protein/ml; Pi, 2 mM phosphate buffer (pH 7.4); Succ, 5 mM sodium succinate.

siently in a dose- and oxygen-concentration-dependent manner. The steady-state level of the cellular membrane potential was further increased by adding glucose with a concomitant decrease in NO-dependent depolarization. The membrane potential of tumor cells was depolarized by a respiratory inhibitor, rotenone, and was polarized by glucose. In the presence of rotenone, the membrane potential of tumor cells was not depolarized by NO irrespective of the presence or absence of glucose (Fig. 5) (33). These results suggest that the energy metabolism of tumor cells greatly depends on NO and glycolysis. The glycolytic activity of

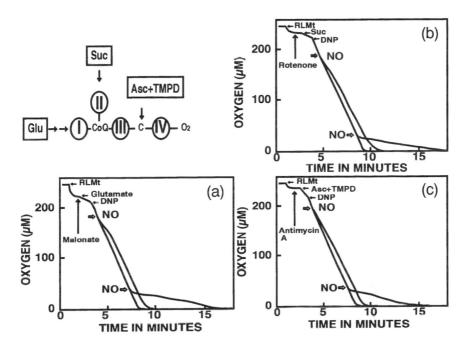

Figure 3 Effect of NO on the uncoupled respiration of mitochondria. Oxygen consumption was measured as in Figure 2 except for the presence of 20 µM DNP. Concentrations of added mitochondria in a, b, and c were 1, 0.5, and 0.25 mg/ml, respectively, and concentrations of NO were 1.2, 0.6, and 0.25 µM, respectively. (a) 1 mM malonate and 5 mM glutamate. (b) 0.1 µM rotenone and 5 mM succinate; (c) 1 nM antimycin A, 1 mM ascorbate, and 10 µM TMPD, respectively.

tumor cells was increased by NO-generating agent, 1-hydroxy-2-oxo-3,3-bis(2-aminoethyl)-1-triazene (NOC-18), as monitored by glucose consumption and lactate formation. The enhanced glycolysis was further increased by lowering oxygen tension. Thus NO regulates the energy metabolism of tumor cells especially under low oxygen tensions.

C. Effect of NO and Oxygen on Superoxide Generation by Neutrophils

The respiratory burst leading to O_2^- generation is one of the major responses of activated neutrophils. This response is mediated by NADPH oxidase, a membrane-bound flavoprotein-b-type cytochrome complex, which catalyses the reduction of molecular oxygen to O_2^- (38–40). Recently NO has been shown to inhibit

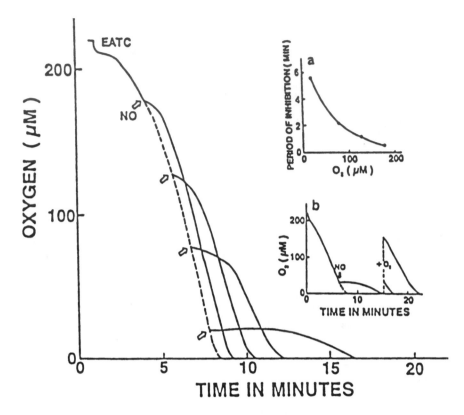

Figure 4 Oxygen-dependent inhibition of tumor cell respiration by NO. Oxygen consumption of Ehrlich ascites tumor cells (8×10^7 cells/ml) was measured in calcium-free Krebs Ringer phosphate (pH 7.4) at 37°C. NO was added at a final concentration of 3.7 µM. (a) Relationship between oxygen concentration and the duration of inhibition. (b) Reversibility of the inhibition.

O_2^- production by neutrophils as measured by the Cyt-c reduction method (14,41). However, NO also reacts with Cyt-c and forms a nitrosyl-Cyt-c complex (13). Thus the Cyt-c reduction method is not suitable for analyzing the effects of NO. To avoid such complex factors, the effect of NO on the respiratory burst of neutrophils was examined by measuring cellular oxygen consumption. Phorbol myristate acetate (PMA) stimulated oxygen consumption and O_2^- generation by neutrophils and both phenomena were inhibited reversibly by NO. The inhibitory effect of NO was suppressed by oxyhemoglobin (HbO₂), a NO scavenging agent. The inhibitory effect of NO increased with a concomitant decrease in oxygen tension in the medium. These results suggest that NO reversibly inhibits O_2^-

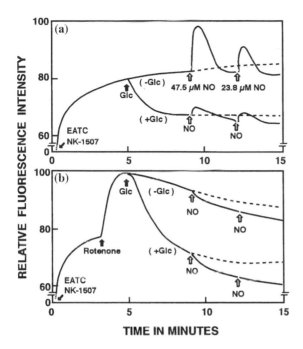

Figure 5 Transient depolarization of membrane potential of tumor cells by NO. Ehrlich ascites tumor cells (5 × 10⁵ cells/ml) were incubated in KRP containing 1 mM CaCl₂ and 50 ng/ml NK-1507 cyanine dye. Membrane potential was monitored at wavelengths of 570 and 530 nm for emission and excitation, respectively. The final concentrations of glucose, rotenone, and NO were 10 mM, 1 μM, and 23.8–47.5 μM, respectively. (a) Transient depolarization of membrane potential by NO in the presence or absence of glucose. (b) Effect of NO on the membrane potential in the presence of rotenone.

generation by neutrophils, especially under low oxygen tensions, thereby decreasing oxygen toxicity in hypoxic tissues (35).

D.　Effect of NO and Oxygen Concentration on Apoptosis of HL-60 Cells

Apoptosis is an energy-requiring active process which eliminates certain cells to maintain a dynamic balance between cell proliferation and death and is one of the important events in embryogenesis, metamorphosis, and hematopoiesis (42–44). Apoptosis usually involves fragmentation of nucleosomal DNA by endonuclease activity (45). Based on the observation that reactive oxygen species induce apoptosis by some antioxidant-inhibitable mechanism, oxidative stress has been postulated to play a role in the induction of apoptosis (46–48). Because, inhibitors of mitochondrial respiration and oxidative phosphorylation including NO inhibit cytochrome oxidase and depolarize the mitochondrial membrane potential, these compounds induce apoptosis in various cells (49). Cellular ATP is an important determinant for cell death either by apoptosis or necrosis (50). Because NO reversibly inhibits respiration and ATP synthesis in mitochondria and cells, the effects of NOC-18 and oxygen tension on the respiration, ATP synthesis, and apoptosis of HL-60 cells were tested. When respiration was inhibited by NOC-18, cellular ATP level decreased and DNA fragmentation was elicited. Both events were enhanced by decreasing oxygen tension and suppressed by adding NO-trapping agents, such as 2-(4-carboxyphenyl)-4,4,5,5-tetramethylimidazo-line-1-oxyl-3-oxide (carboxy-PTIO) and oxyhemoglobin (Fig. 6). The results suggest that NO plays a regulatory role in triggering apoptosis, particularly under low oxygen tensions (36).

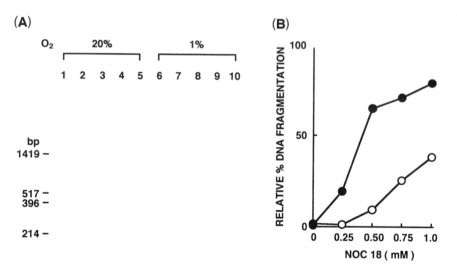

Figure 6 Effects of oxygen and NOC-18 on DNA fragmentation. HL-60 cells (2×10^5 cells/ml) were treated with various concentrations of NOC-18 for 4 h at 37°C under 20% or 1% oxygen tension. Total DNA was then extracted, electrophoresed on 2% agarose gel, and visualized by ethidium bromide staining. (A) The concentrations of NOC-18 were 0 (lanes 1 and 6), 0.25 (lanes 2 and 7), 0.50 (lanes 3 and 8), 0.75 (lanes 4 and 9), and 1.0 mM (lanes 5 and 10). The extent of DNA fragmentation was determined by measuring the image intensity of the lower bands and expressed as percent. (B) Oxygen tensions used were 20% (open circles) and 1% (closed circles).

E. Effects of NO and Oxygen Concentration on Smooth Muscle Relaxation

Hypoxia induces vasodilation of arteries (51) and adenosine has been considered as a possible mediator for hypoxia-dependent vasodilation (52). Hypoxia stimulates endothelial synthesis and the release of NO (53) which increases cGMP levels in arteries particularly during the early phase of hypoxia (54). However, low oxygen tension (28 mm $HgpO_2$) has been shown to inhibit NO synthase and the production of cGMP in bovine cerebellum (55). Furthermore, anoxia-dependent relaxation has been postulated to occur via some mechanism which is mediated by either prostaglandin, adenosine, endothelium-derived relaxing factor (EDRF), or by the opening of ATP-sensitive potassium channels (56,57). Exposure of aortic rings to anoxia for 5 min decreased the relaxation response to 59% (57–59). These observations suggested that endothelial cells might function as a sensor for local oxygen tension and the release of EDRF depending on pO_2. NO activates calcium-dependent and cGMP-dependent potassium channels (60). Both NO and NOC-18 induced the relaxation of rat aorta in a concentration-dependent manner. In vitro experiments for measuring vascular contraction and relaxation were generally carried out in a medium saturated with oxygen (600–800 μM). Such high concentrations of oxygen may affect the function of NO. To test this possibility, the effect of NO on aortic relaxation was examined under different oxygen concentrations. Although no detectable change in the degree of norepinephrine induced contraction in aortic specimens was induced simply by lowering the oxygen tension from 600 to 44 μM, NO-induced relaxation was markedly enhanced (Fig. 7). The extent of NOC-18-induced relaxation also depended on the concentration of oxygen. These results indicated that local oxygen concentration might play an important role in the regulation of NO-dependent vascular resistance of arteries.

F. Effects of NO and Oxygen Concentration on the Activity of Guanylate Cyclase

NO synthesized in endothelial cells readily enters in vascular smooth muscle cells and stimulates guanylate cyclase (GC) activity. The de novo synthesized cGMP decreases the concentration of intracellular calcium ions thereby inducing relaxation of smooth muscle cells (1). Again, NO increased the cellular levels of cGMP by mechanisms which were enhanced by lowering oxygen tension (Fig. 8) (37).

III. CONCLUSIONS AND PERSPECTIVES

The present work shows that the lifetime and biological functions of NO, such as the modulation of mitochondrial metabolism, cGMP-dependent and Ca^{2+}-

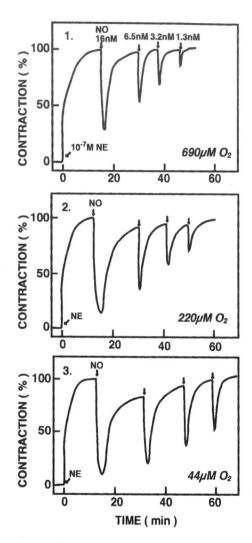

Figure 7 NO-induced relaxation of endothelium-denuded aorta at different oxygen tensions. Prior to each experiment, the oxygen tension of HEPES-buffered physiological saline solution was kept to >700 µM at 37°C and endothelium-denuded aorta (1 cm) was preincubated for at least 2 h. Contraction and relaxation of the specimen was continuously monitored by an isotonic transducer connected to a direct current amplifier. The specimen was contracted by 0.1 µM norepinephrine (NE) and then relaxed by various concentrations of NO. The oxygen concentration in the medium was controlled at 44 to 690 µM by infusing either O_2 or argon gas. Oxygen concentrations of 1, 2, and 3 were 690, 220, and 44 µM, respectively. Added NO concentrations at each arrow were 16, 6.5, 3.2, 1.3 nM from left to right, respectively.

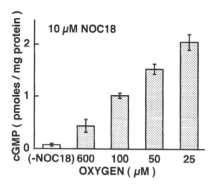

Figure 8 Effect of oxygen concentration on NO-induced formation of cGMP. Endothe-lium-denuded aortic specimens were incubated for 5 min under various oxygen concentra-tions. After incubation for 1 min with 10 μM NOC-18, arterial levels of cGMP were determined.

dependent metabolism including vasorelaxation and apoptosis, were markedly enhanced by physiologically low oxygen tension and by hypoxia. Although NO reacts with various compounds, such as hemeproteins, thiols, and superoxide, interaction with molecular oxygen is critically important. In fact, the present work suggests that molecular oxygen is one of the most important regulators of NO action.

It has been well documented that tumor cells depend on glycolysis for their energy source. However, most cancer cells are also enriched with mitochondria. Hence, they also utilize mitochondrial oxidative phosphorylation as an energy source particularly under aerobic conditions. Thus, unlike normal epithelial cells, tumor cells can survive even under hypoxic conditions. However, they rapidly grow under aerobic conditions depending on the efficient energy metabolism of mitochondria. Most tumors are associated with activated macrophages which ex-press high levels of inducible NO synthase (iNOS). Thus tumor cells might be exposed to relatively high concentrations of NO derived from activated macro-phages. Because NO induces apoptosis of tumor cells, particularly under low oxygen tensions, similar phenomena might also occur in vivo depending on the local concentrations of NO and oxygen in and around tumor cells. In fact, apop-totic cells are often seen in the surface area of macrophage-enriched tumors. It is also well documented that, after neovascularization, tumor cells start to grow rapidly. This might be due to the efficient scavenging of NO by hemoglobin in the circulating erythrocytes. When the inhibitory effect of NO an mitochondrial respiration is eliminated under aerobic conditions, tumor cells might utilize ATP supplied by oxidative phosphorylation, which is 19 times more efficient than glycolysis. Thus oxygen-dependent metabolism and the functions of NO might be of clinical importance for understanding the pathophysiology of cancer cells and for the development of a novel method for treating patients with malignant tumors.

REFERENCES

1. Moncada S, Palmer RM, Higgs EA. Nitric oxide: physiology, pathophysiology, and pharmacology. Pharmacol Rev 1991; 43:109–142.
2. Knowles RG, Moncada S. Nitric oxide as a signal in blood vessels. Trends Biochem Sci 1992; 17:399–402.
3. Lowenstein CJ, Snyder SH. Nitric oxide, a novel biologic messenger. Cell 1992; 70:705–707.
4. Gopalakrishna R, Chen ZH, Gundimeda Y. Nitric oxide and nitric oxide-generating agents induce a reversible inactivation of protein kinase C activity and phorbol ester binding. J Biol Chem 1993; 268:27180–27185.
5. Michetti M, Salamino F, Melloni E, Pontremoli S. Reversible inactivation of calpain isoforms by nitric oxide. Biochem Biophys Res Commun 1995; 207:1009–1014.
6. Kuroshe I, Kato S, Ishii H, et al. Nitric oxide mediates lipopolysaccharide-induced alteration of mitochondrial function in cultured hepatocytes and isolated perfused liver. Hepatology 1993; 18:380–388.
7. Tucker SD, Auzenne EJ, Sivaramakrishnen MR. Inhibition of tumor cell mitochondrial respiration by macrophage cytotoxic mediators distinct from interferon-γ. J Leukocyt Biol 1993; 53:138–143.
8. Stadler J, Curran RD, Ochoa JB, et al. Effect of endogenous nitric oxide on mitochondrial respiration of rat hepatocytes in vitro and in vivo. Arch Surg 1991; 126: 186–191.
9. Granger DL, Taintor RR, Cook JL, Hibbs JB Jr. Injury of neoplastic cells by murine macrophages leads to inhibition of mitochondrial respiration. J Clin Invest 1980; 65:357–370.
10. Drapier J, Hibbs JB Jr. Differentiation of murine macrophages to express nonspecific cytotoxicity for tumor cell results in 1-arginine-dependent inhibition of mitochondrial iron-sulfur enzymes in the macrophage effector cells. J Immunol 1988; 140: 2829–2839.
11. Stadler J, Billiar TR, Curran RD, et al. Effect of exogenous and endogenous nitric oxide on mitochondrial respiration of rat hepatocytes. Am J Physiol (Cell Physiol) 1991; 260:C910–C916.
12. Gorren ACF, Van Gelder BF, Wever R. Photodissociation of cytochrome c oxidase-nitric oxide complexes. Ann NY Acad Sci 1988; 550:139–149.
13. Keilin D. The History of Cell Respiration and Cytochrome. London: Cambridge University Press, 1966: 319–335.
14. Clancy MR, Leszczynska-Piziak J, Abramson SB. Nitric oxide, an endothelial cell relaxation factor, inhibits neutrophil superoxide anion production via a direct action on the NADPH oxidase. J Clin Invest 1992; 90:1116–1121.

15. Rubanyi S, Ho EH, Cantor EH, Lumma WC, Botelho LHP. Cytoprotective function of nitric oxide: inactivation of superoxide radicals produced by human leukocytes. Biochem Biophys Res Commun 1991; 181:1392–1397.

16. Richter C, Gogvadze V, Schlapbach R, Schweizer M, Schlegel J. Nitric oxide kills hepatocytes by mobilizing mitochondrial calcium. Biochem Biophys Res Commun 1994; 205:1143–1150.

17. Stamler JS, Redox signaling: nitrosylation and related target interactions of nitric oxide. Cell 1994; 78:931–936.

18. Stamler JS, Singel DJ, Loscalzo J. Biochemistry of nitric oxide and its redox-activated forms. Science 1992; 258:1898–1900.

19. Schwizer M Richter C. Nitric oxide potently and reversibly deenergizes mitochondria at low oxygen tension. Biochem Biophys Res Commun 1994; 204:169–175.

20. Kong SK, Choy YM, Lee CY. The nitric oxide donor, sodium nitroprusside, increased intranuclear and cytosolic calcium concentration in single pu5-1.8 cells. Biochem Biophys Res Commun 1994; 199:234–240.

21. Sarih M, Souvannavong V, Adam A. Nitric oxide synthase induces macrophage death by apoptosis. Biochem Biophys Res Commun 1993; 191:503–508.

22. Albina JE, Cui S, Mateo RB, Reichner JS. Nitric oxide-mediated apoptosis in murine peritoneal macrophages. J Immunol 1993; 150:5080–5085.

23. Cui S, Reichner JS, Mateo RB, Albina JE. Activated murine macrophages induce apoptosis in tumor cells through nitric oxide-dependent or -independent mechanisms. Cancer Res 1994; 54:2462–2467.

24. Meßmer UK, Ankarcrona M, Nicotera P, Brüne B. p53 expression in nitric oxide-induced apoptosis. FEBS Lett 1994; 355:23–26.

25. Kitajima I, Kawahara K, Nakajima T, Soejima S, Matsuyma T. Nitric oxide-mediated apoptosis in murine mastocytoma. Biochem Biophys Res Commun 1994; 204: 244–251.

26. Kuo M-L, Chau Y-P, Wang J-H, Shiah S-G. Inhibition of poly(ADP-ribose)polymerase block nitric oxide-induced apoptosis but not differentiation in human leukemia HL-60 cells. Biochem Biophys Res Commun 1996; 219:502–508.

27. Robertson BE, Schuber R, Heschelef J, Nelson MT. cGMP-dependent protein kinase activites Ca-activated K channels in cerebral artery smooth muscle cells. Am J Physiol 1993; 265:C299–C303.

28. Taniguchi J, Furukawa K-I, Shigekawa M. Maxi K+ channel are stimulated by cyclic guanosine monophosphate-dependent protein kinase in canine coronary artery smooth muscle cells. Pflugers Arch 1993; 423:167–172.

29. Perrella MA, Heidebrand FL Jr, Margulies KB, Burnett JC Jr. Endothelium-derived relaxing factor in regulation of basal cardiopulmonary and renal function. Am J Physiol 1991; 261:R323–R328.

30. Takehara Y, Kanno T, Yoshioka T, Inoue M, Utsumi K. Oxygen-dependent regulation of mitochondrial energy metabolism by nitric oxide. Arch Biochem Biophys 1995; 323:27–32.

31. Okada S, Takehara Y, Yabuki M, et al. Nitric oxide, a physiological modulator of mitochondrial function. Physiol Chem Phys Med NMR 1996; 28:69–82.

32. Takehara Y, Nakahara H, Inai Y, et al. Oxygen-dependent reversible inhibition of mitochondrial respiration by nitric oxide. Cell Struct Funct 1996; 21:251–258.

33. Inai Y, Takehara Y, Yabuki M, et al. Oxygen-dependent-regulation of Ehrlich ascites tumor cell respiration by nitric oxide. Cell Struct Funct 1996; 21:151–157.

34. Nishikawa M, Sato EF, Utsumi K, Inoue M. Oxygen-dependent regulation of energy metabolism in ascites tumor cells by nitric oxide. Cancer Res 1996; 56:4535–4540.

35. Iha S, Orita K, Kanno T, et al. Oxygen-dependent inhibition of neutrophil respiratory burst by nitric oxide. Free Radic Res 1996; 25:489–498.

36. Yabuki M, Inai Y, Yoshioka T, et al. Oxygen-dependent fragmentation of cellular DNA by nitric oxide. Free Radic Res 1997; 26:245–255.

37. Takehara Y, Nakahara H, Inai Y, et al. Oxygen tension regulates NO-dependent relaxation of rat aorta. Free Rad Res (submitted).

38. Miyahara M, Watanabe S, Okimasu E, Utsumi K. Charge-dependent regulation of NADPH oxidase activity in guinea-pig polymorphonuclear leukocytes. Biochim Biophys Acta 1987; 929:253–262.

39. Berkow RL, Dodson RW, Kraft AS. The effect of a protein kinase C inhibitor, H-7, on human neutrophil oxidative burst and degranulation. J Leuk Biol 1987; 41:441–446.

40. Miyahara M, Okimasu E, Uchida H, et al. Charge-dependent regulation of NADPH oxidase activities in intact and subcellular systems of polymorphonuclear leukocytes. Biochim Biophys Acta 1988; 971:46–54.

41. Rubanyi S, Ho EH, Cantor EH, Lumma WC, Botelho LHP. Cytoprotective function of nitric oxide: inactivation of superoxide radicals produced by human leukocytes. Biochem Biophys Res Commun 1991; 181:1392–1397.

42. Vaux DL, Haecker G, Strasser A. An evolutionary perspective on apoptosis. Cell 1994; 76:777–779.

43. Arends MJ, Wyllie AH. Apoptosis: mechanisms and roles in pathology. Int Rev Exp Pathol 1991; 32:223–254.

44. Chou CC, Laom CY, Yung BYM. Intracellular ATP is required for actinomycin D-induced apoptotic cell death in HeLa cells. Cancer Lett 1995; 96:181–187.

45. Wyllie AH. Glucocorticoid-induced thymocyte apoptosis is associated with endogenous endonuclease activation. Nature 1980; 284:555–556.

46. Hockenbery DM, Oltvai ZN, Yin X-M, Milliman CL, Korsmeyer SJ. Bcl-2 functions in an antioxidant pathway to prevent apoptosis. Cell 1993; 75:241–251.

47. Kane DJ, Sarafian TA, Anton R, et al. Bcl-2 inhibition of neural death: decreased generation of reactive oxygen species. Science 1993; 262:1274–1277.

48. Wolfe JT, Ross D, Cohen GM. A role for metals and free radicals in the induction of apoptosis in thymocytes. FEBS Lett 1994; 352:58–62.

49. Wolvtang EJ, Johnson KL, Krauer K, Ralph SJ, Linnane AW. Mitochondrial respiratory chain inhibitors induce apoptosis. FEBS Lett 1994; 339:40–44.

50. Richter C, Schweizer M, Cossarizza A, Franceschi C. Control of apoptosis by the cellular ATP level. FEBS Lett 1996; 378:107–110.

51. Feigl EO. Coronary physiology. Physiol Rev 1983; 63:1–205.

52. Berne RM. Cardiac nucleotides in hypoxia. Possible role in regulation of coronary blood flow. Am J Physiol 1963; 204:317–322.

53. Pohl U, Busse R, Hypoxia stimulates release of endothelium-derived relaxant factor. Am J Physiol 1989; 255:H1595–H1600.

54. Park KH, Rubin LE, Gross SS, Levi R. Nitric oxide is a mediator of hypoxic coro-

nary vasodilatation. Relation to adenosine and cyclooxygenase derived metabolites. Circ Res 1992; 71:992–1001.

55. Rengasamy P, Hohns RA. Characterization of endothelium-derived relaxing factor nitric oxide synthase from bovine cerebellum and mechanism of modulation by high and low oxygen tensions. J Pharmacol Exp Ther 1991; 259:310–316.

56. Kalsner S. Hypoxic relaxation in functionally intact cattle coronary artery segments involves K^+-ATP channels. J Pharmacol Exp Ther 1995; 275:1219–1226.

57. Roberts AM, Messina EJ, Kaley G. Prostacyclin (PGI_2) mediates hypoxic relaxation of bovine coronary artery strips. Prostaglandins 1981; 21:555–569.

58. Yang BC, Mehta JL. Prior eposide of anoxia attenuates vasorelaxation in response to subsequent episode of anoxia. Am J Physiol 1994; 266:H974–H979.

59. Pohl U. Endothelial cells as part of a vascular oxygen-sensing system: hypoxia-induced release of autacoids. Experientia 1990; 46:1175–1179.

60. Bolothina VM, Najibi S, Palacino JJ, Pagano PJ, Cohen RA. Nitric oxide directly activates calcium-dependent potassium channels in vascular smooth muscle. Nature 1994; 368:850–853.

5

Production of Reactive Oxygen Species in Mitochondria and Age-Associated Pathophysiology: A Reality Check

Henry Jay Forman
University of Southern California, Los Angeles, California

Angelo Azzi
University of Bern, Bern, Switzerland

I. INTRODUCTION

Mitochondria are an important intracellular source of reactive oxygen species, although alternative sources in other organelles and the cytoplasm are present in cells. Based upon the large body of work, it is likely that reactive oxygen species participate in aging and aging-associated pathology. However, whether mitochondrial or other sources play a more important role in aging is an unsolved matter. In both compartments, the mitochondrial and the cytosol enzymatic and nonenzymatic systems are available, which apparently serve to decrease the reactive oxygen species steady-state concentration. Nevertheless, DNA, proteins, and lipids are subject to oxidative damage that increases with age (1–7).

Whether mitochondrial DNA (mtDNA) is subject to more severe oxidative damage than nuclear DNA (nDNA), due to the large mitochondrial production of reactive oxygen species, is still an open question. The fact that mtDNA damage is more extensive than nDNA damage in human cells following H_2O_2 treatment suggests that mtDNA is simply more sensitive to damage relative to nDNA (8).

In any case, due to its sensitivity, mtDNA is a critical cellular target for reactive oxygen species (8).

Accumulation of various mutations in the mitochondrial genome is proposed as an important contributor to aging (9–15). The frequency of nucleotide substitutions in the striatum of Parkinson's disease patients is significantly higher than that in control tissues, but at the same time increased protein modification by 4-hydroxy-2-nonenal is observed (9). These results have been taken to document that a primary mitochondrial mutation induces a mitochondrial respiratory defect, which increases leakage of reactive oxygen species that trigger accumulation of secondary mtDNA mutations, leading to an aggravating vicious cycle.

Such a mechanism may be supported by the data showing that mitochondrial complex I deficiency leads to increased production of superoxide radicals (O_2^-) and induction of superoxide dismutase (SOD) (16). However, an increase in oxygen radical production when complex I activity is diminished would not actually occur, due to the presence of sufficient SOD and of inducibility of this enzyme. The data in fact show that, due to mitochondrial superoxide dismutase (Mn-SOD) induction, the actual level of O_2^- production may decrease, sometimes below that seen in control fibroblast mitochondria (16).

A number of observations support the hypothesis that mtDNA damage does accumulate with age and indicate that respiratory stress greatly elevates mitochondrial damage (10–13). In the brain cortex, the deleted: total mtDNA ratio ranged from 0.00023 to 0.012 in 67 to 77-year-old brains and was up to 0.034 in subjects over 80 years of age. In the putamen, the deletion level ranged from 0.0016 to 0.010 in 67 to 77 year olds up to 0.12 in individuals over the age of 80 years. The cerebellum remained devoid of mtDNA deletions (11). It is puzzling that a general radical injury spares certain tissues, which remain totally devoid of mutations. It is also unclear how such a low level of mutation can be the cause of age-dependent devastating processes. Alternatively, are such mutations just an independent marker of age?

One more intriguing question is the apparent lack of inheritance of the actual mother's age (if age is a function of the number of mitochondrial mutations) by the offspring. As in somatic cells, oocytes from older women contain more deletions of mtDNA (17). It has also been found that oocytes harbor measurable levels (up to 0.1%) of the so-called common deletion, an mtDNA molecule containing a 4977 bp rearrangement that is present in high amounts in many patients with "sporadic" Kearns-Sayre syndrome (KSS) and progressive external ophthalmoplegia (PEO) (18). Why these and other mitochondrial defects are maternally inherited, and yet the mutations considered at the basis of the aging process are not, remains to be clarified. Possibly the answer will come from understanding the relationships between mtDNA damage, nDNA damage, protein and lipid damage, and the aging process. Obviously, environment-independent genetically determined events must be taken into consideration.

Another correlation has been shown between DNA oxidative damage and life expectancy in houseflies. Exposure of live flies to X rays and hyperoxia elevated the level of 8-hydroxy-2'-deoxyguanosine in mitochondrial as well as total DNA and of protein carbonyl content. The 8-hydroxy-2'-deoxyguanosine and protein carbonyl content levels were found to be inversely associated with the life expectancy of houseflies (19). Although mtDNA was reported to be three times more susceptible to age-related oxidative damage than nuclear DNA, the data do not directly indicate that damage of the former rather than the latter is the basis of aging. It is possible that both mitochondrial and nuclear DNA damage play a role in the aging process.

In apparent conflict with the DNA damage hypothesis, in *Drosophila melanogaster*, mtDNA remains intact during aging, staying relatively constant and representing roughly 1% of the total DNA at all ages (20). Nonetheless, an apparently important decrease in the steady-state levels of all mitochondrial transcripts has been found, which correlates with the shape of the life span curve (20). These results suggest that, at least in *Drosophila*, the main effect of aging on the mitochondrial genetic system is downstream from mtDNA itself.

On the other hand, it has been suggested that variations in longevity among different species inversely correlate with the rates of mitochondrial generation of the O_2^- anion radical and H_2O_2 (21). Thus overexpression of antioxidant enzymes should retard the age-related accumulation of oxidative damage and extend the maximum life span of transgenic *D. melanogaster*. This has indeed been achieved by simultaneous overexpression of copper- and zinc-containing superoxide dismutase (Cu,Zn-SOD) and catalase, with a delay of age-related oxidative damage and increased metabolic potential in *D. melanogaster* (22). Nonetheless, it is not clear how increased expression of the cytosolic SOD was beneficial if the primary damage was to mtDNA. In other words, if O_2^- produced by mitochondria is responsible for mtDNA damage, it is not clear how an increase in cytosolic SOD would be able to make any difference. Rather, the results in *D. melanogaster* suggest that O_2^- contributes to age-related damage in the cytosol or other compartments of the cell containing Cu,Zn-SOD.

Furthermore, it has been shown that nuclear but not mitochondrial genome is involved in human age-related mitochondrial dysfunction. Age-related diminution in the activity of cytochrome-*c* oxidase in human skin fibroblasts was neither attributable to a decrease in the copy number of mtDNA molecules nor an increase in the copy number of deletion mutant mtDNA molecules, but rather to a significant decrease in overall polypeptide synthesis in the mitochondria (23). Moreover, intercellular mtDNA transfer experiments showed that fibroblast mtDNA from elderly donors is functionally intact. By contrast, intercellular transfer of HeLa nuclei to fibroblasts from aged donors restored cytochrome-*c* oxidase activity, suggesting that the age-related phenotype was nuclear recessive. These observations support the idea that accumulation of nuclear recessive somatic mutations, but not mtDNA mutations, is responsible for the in vivo age-related mitochondrial dysfunction observed in human skin fibroblasts (23). A progressive decline in muscle performance has also been found to correlate with a decrease in the rate of mitochondrial protein synthesis in vivo (24).

As correctly pointed out by Richter (25), major unsettled issues pertaining to the role of mtDNA modifications in aging have still to be addressed: Is there a cause-and-effect relationship? What is the real extent of DNA damage? What are the functional consequences of mtDNA oxidation? Are reactive oxygen species the cause of the DNA changes observed in vivo? What is the connection between DNA injury and modification of RNAs and proteins? The clarification of the possible causal relationships among reactive oxygen species production, mtDNA modifications, mitochondrial alterations, and their role in aging awaits further thorough investigation.

II. MECHANISM OF MITOCHONDRIAL O_2^- AND H_2O_2 GENERATION

When discussing the mechanism of the onset of aging, other aspects have to be taken into consideration, such as the genetic components. As indicated by recent observations, the *klotho* gene, which encodes a membrane protein similar to β-glucosidase, is involved in the suppression of several aging phenotypes, including a short life span, infertility, arteriosclerosis, skin atrophy, osteoporosis, and emphysema (26).

Regardless of these other factors, production of reactive oxygen species appears to be a factor in aging. Proposing any mechanism through which mitochondrial O_2^- and H_2O_2 generation could play a role in aging requires answers to two major questions. First, what mechanisms underlie the generation of O_2^- and H_2O_2 in mitochondria? Second, what circumstances cause an alteration in the rate of reactive oxygen species production?

A. Cellular H_2O_2 and O_2^- Production

In aerobic eukaryotic cells, the principal consumption of oxygen is its four-electron reduction to water by cytochrome-c oxidase. The cytochrome-c oxidase reaction occurs without release of any intermediates in the reduction of oxygen. The reactions that produce O_2^- and H_2O_2 include both enzymatic and nonenzymatic processes. Numerous oxidases in the cytosol, endoplasmic reticulum, and outer membrane of mitochondria also contribute to oxygen consumption. Probably, as part of normal intermediary metabolism, these enzymes produce O_2^- and H_2O_2 as products (27–29). During metabolism of xenobiotics, some of these enzymes can cause increased production of O_2^- either through the direct reduction of oxygen by the enzyme or through the reduction of the xenobiotic to a form that autooxidizes (29–31). Phagocytic cells are well known for their production of O_2^- by a plasma membrane NADPH oxidase upon stimulation (32), and accumulating evidence suggests that other cell types also produce O_2^- on their surfaces in response to stimuli (33–37). A small percentage of oxygen consumption results from nonenzymatic oxidation of cellular constituents, such as catecholamines, catalyzed by transition metals (38,39). The subject of this chapter, however, is the production of O_2^- and H_2O_2 by leaks from the electron transport chain.

Hydrogen peroxide has long been recognized as a normal component of cellular metabolism. Indeed, the first enzyme to be named was catalase, which dismutes H_2O_2:

$$2H_2O_2 \rightarrow 2H_2O + O_2$$

Nonetheless, the concept of O_2^- as a normal metabolite was not accepted until much later. While Fridovich and Handler (40) showed that xanthine oxidase reduces oxygen to O_2^- and H_2O_2, it was not until after McCord and Fridovich (41) isolated SOD that O_2^- production in a biological system was acknowledged as plausible. Superoxide dismutases, which catalyze the reaction

$$2O_2^- + 2H^+ \rightarrow H_2O_2 + O_2$$

have been found in the cytosol, mitochondria, and extracellular space as distinct gene products (41–43).

B. Mitochondrial Production of H_2O_2

Hydrogen peroxide production by submitochondrial particles was first described by Jensen (44) and Hinkle et al. (45), and in intact mitochondria by Chance and others (31,46,47). Subsequently, work in our laboratories during the early 1970s using submitochondrial particles established that O_2^- was produced as a leak from the electron transport chain (48,49). Shortly thereafter, it became clear that all mitochondrial H_2O_2 production resulted from dismutation of O_2^- (50,51). A debate concerning the actual sites of O_2^- production began, which today is still not settled. One thing that is agreed upon is that O_2^- is produced as a leak of electrons from two complexes of the respiratory chain.

The electron transport chain consists of a series of redox active components (pyridine nucleotides, flavoproteins, iron-sulfur proteins, ubiquinone, and cytochromes) located in the inner membrane of mitochondria. In terms of electron flow, these components are arranged in order of their redox potentials. Physically

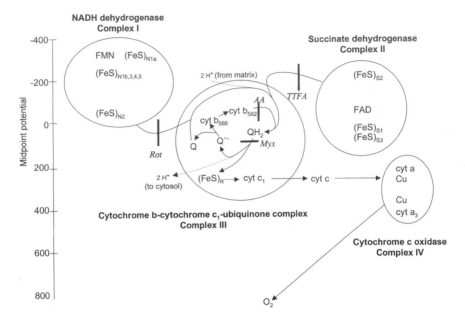

Figure 1 The electron transport chain. The flow of electrons is from high-potential complexes I and II through the ubiquinone–cytochrome–cytochrome-c_1 complex (complex III) to complex IV where four electrons are passed to O_2 to form water. The details of the Q cycle in complex III and the site of inhibition by rotenone (Rot), thenoyltrifluoroacetone (TTFA), myxothiazole (Myx), and antimycin A (AA) are shown.

the components are arranged in complexes as illustrated in Figure 1. This organization allows efficient coupling for the transfer of electrons from NADH or substrates of dehydrogenases in mitochondrial intermediary metabolism to oxygen with production of ATP. Substances that "uncouple" phosphorylation, such as pentachlorophenol, produce a dramatic increase in oxygen consumption along with inhibiting ATP production. Likewise, substances that inhibit electron transport, such as rotenone, antimycin A, and cyanide, also inhibit ATP production. To determine the sites and mechanism of H_2O_2 production in mitochondria, investigators have used such inhibitors and uncouplers. In addition, measurement of O_2^- production requires preparation of inner membranes (submitochondrial particles) from which the endogenous SOD is removed by repeated washing. While use of inhibitors and uncouplers and preparation of submitochondrial particles has allowed useful qualitative interpretation, most of these manipulations affect the rate of O_2^- production (50,52). Obviously, one should be wary of quantitative interpretation of data in which these agents have been used. This caution, how-

Table 1 Respiratory States and Hydrogen Peroxide Production of Mitochondria

| | Metabolite status | | | Rate (relative within column) | | |
State	Substrate	Oxygen	ADP	ATP production	Oxygen consumption	H_2O_2 production
1	Low	Normal	Low	Low	Low	Very low
2	None	Normal	High	None	None	None
3	High	Normal	High	High	High	Low
3u	High	Normal	Irrelevant	None	High	Low
3a	High	Normal	Irrelevant	None	None	High
3a+u	High	Normal	Irrelevant	None	None	Very high
4	High	Normal	Low	Low	Low	Medium
5	High	None	High	None	None	None

The relative rates of ATP production, oxygen consumption, and H_2O_2 production are shown for varoius respiratory states of mitochondria. State 3u is the presence of an uncoupler. State 3a is the presence of antimycin A. State 3a+u is the presence of antimycin A and an uncoupler. The rates are relative to each other only within the particular column as H_2O_2 production under state 3a is only a small fraction of the oxygen consumption in state 3. The results are a compilation of data obtained with mitochondria from several tissues and species (31,52,61,62,64,111,112).
Source: Modified from Refs. 56 and 110.

ever, has not been given sufficient weight in studies of mitochondria and submitochondrial particles from aging tissues until quite recently (53–55).

Regardless of the state of the evidence concerning aging, let us examine what these agents have revealed about the sites and mechanism of mitochondrial O_2^- production. In this discussion, it is useful to categorize the state of mitochondria in terms of their ATP production and oxygen consumption during experimental manipulation (Table 1).

Intact mitochondria produce H_2O_2 at rates that depend upon the mitochondrial metabolic state (56) (Table 1). From initial studies of the effects of respiratory chain inhibitors and uncouplers, it was concluded that at least one electron transport chain carrier, between the rotenone and antimycin A sensitive sites, most likely ubiquinone, reacts in its reduced form with molecular oxygen (48–50). While O_2^- is the initial species produced from the reduction of oxygen, it is rapidly dismuted to H_2O_2 and oxygen by mitochondrial SOD. Dismutation is in competition with other reactions of O_2^-, although only the reaction with nitric oxide (NO; more properly called nitrogen monoxide) is truly competitive (57). Although there have been some reports of O_2^- escaping from intact mitochondria, O_2^- can actually only be measured with submitochondrial particles and then only after removing mitochondrial SOD. The reasons for this are clarified in detail below. The H_2O_2 generated can then diffuse from the mitochondria, although some percentage will be reduced in the glutathione peroxidase reaction (see below) and perhaps, in some tissues, by catalase (58). Thus measurement with intact mitochondria of H_2O_2 generation reflects the net result of O_2^- production, O_2^- dismutation and reaction with NO, and competition between destruction and diffusion of H_2O_2. Nonetheless, in the absence of exogenous inhibitors or uncouplers, the rate of H_2O_2 generation is highest in state 4, where the absence of ADP gives a slower respiratory rate and the components of the respiratory chain are mostly reduced (see Table 1 and associated references). In state 3 or state 3u, where there is fast oxygen uptake, the respiratory carriers are mostly oxidized. Experiments usually begin with mitochondria that are in state 1 prior to addition of substrate (state 4). This results in a marked increase in measurable H_2O_2. Further addition of ADP (state 3) or an uncoupler (state 3u) decreases H_2O_2 production as a result of the increased oxidation of the respiratory carriers. To some extent, one might look at O_2^- production as being in competition with cytochrome-c oxidase for electrons. Addition of antimycin A, which stops the flow of electrons within complex III, markedly increases production of H_2O_2 when added to state 4 mitochondria; however, even state 1 mitochondria produce significantly more H_2O_2 with the addition of antimycin A.

Probably all H_2O_2 production by mitochondria is the result of the dismutation of O_2^- rather than direct two-electron transfer to oxygen (50,59). Several theoretical issues suggest this. First, although two-electron reduction of oxygen is thermodynamically favored, one-electron reduction of oxygen is kinetically favored because two-electron requires spin inversion (60). Second, production of H_2O_2 occurs in the same section of the electron transport chain as does the transition from two- to one-electron transfer.

C.   Sites of O_2^- Production

Early studies with pigeon heart mitochondria using either succinate or NADH-linked substrates showed inhibition of H_2O_2 production by rotenone but increased production with addition of antimycin (61,62). Rotenone blocks flow from complex I into complex III. Antimycin A blocks reduction of ubiquinone by cytochrome-b_{562} to ubiquinol. It also causes a change in the midpoint potentials of the b cytochromes resulting in creation of an isopotential pool composed of cytochrome-b_{566}, cytochrome-b_{562}, and ubisemiquinone (63,64). With antimycin A, an increase occurs in the reduced content of all the members of this pool but no electron flows beyond (Fig. 1). Thus it was concluded initially that the sole source of H_2O_2 production occurred somewhere within complex III. Subsequently, it was demonstrated that oxidation of the FMN of NADH dehydrogenase can also generate O_2^- prior to the rotenone block. Myxothiazole inhibits ubiquinol oxidation to ubisemiquinone in complex III (Fig. 1) and prevents O_2^- production with succinate (65). Nonetheless, myxothiazole increases O_2^- production with NADH-linked substrates, also suggesting production from complex I (65). Thus, although most O_2^- production from NADH-linked substrates results from passage of electrons to complex III, a small amount apparently comes directly from complex I (66). Recently structural evidence has shown that separate binding sites for antimycin A and myxothiazole exist in complex III (67). Further consideration of the effects of uncouplers and inhibitors are described below.

Both ubisemiquinone and cytochrome-b have been proposed as the component within the complex III that reacts with oxygen to produce O_2^- (48,68,69). Although neither b cytochrome forms complexes with CO, oxygen, or NO, one may transfer an electron to oxygen through an outer sphere interaction. Arguments favoring ubisemiquinone are the well-established chemistry of quinone redox cycling (70), the observation of parallel increases in O_2^- production in ubiquinone-depleted and then reconstituted membranes (68), and the finding that complex I also supports O_2^- production through its interaction with ubiquinone (71,72). In addition, several studies involving separation of components or inhibition with thenoyltrifluoroacetone (TTFA) have supported the ubiquinone hypothesis (73–75). Nonetheless, it is rather difficult to identify the O_2^- generator in intact mitochondrial membranes because the rate of O_2^- production is approximately 10^3 times slower than the rate of equilibration between components of complex III.

Other sources of H_2O_2 and O_2^- generation in mitochondria are monoamine oxidase and cytochrome P-450. The activities of these enzymes are not directly associated with the electron transport chain and their distribution among mitochondria from different tissues varies greatly. Although this chapter is not concerned with these sources, one could imagine that variation of these activities with age should be examined.

D. Kinetics and Thermodynamics of O_2^- and H_2O_2 Production

Generation of O_2^- depends upon a reaction that is slightly unfavorable thermodynamically. This was recognized even before the argument began concerning whether ubisemiquinone or cytochrome-b donated the electron to oxygen (49). Most arguments favoring the ubisemiquinone hypothesis suggest that the reaction that produces O_2^- is

$$O_2 + Q^- \leftrightarrow O_2^- + Q$$

a thermodynamically favorable reaction [$E(Q/Q^-) = -220 \pm 20$ mV and $E(O_2/O_2^-) = -156$ mV]. If, however, the ubisemiquinone that reacts with oxygen is bound to protein, the midpoint potential and/or its protonation could be affected such that the reaction becomes less favorable. Indeed, the following discussion would be equally applicable to any other reaction suggested as the cause of O_2^-

production in mitochondria. Forman and Kennedy (49) demonstrated that the oxidation of dihydroorotate was completely inhibited in submitochondrial particles treated with cyanide in the absence of an electron acceptor that could react with O_2^-. When Mn-SOD, which is not inhibited by cyanide, was added to this system, dihydroorotate oxidation recommenced. For the purposes of argument, we will assume that protein-bound ubiquinone is involved so that the rate of O_2^- generation is governed by the equation:

$$d[O_2^-]/dt = k_1[Q^-][O_2] - k_2[Q][O_2^-]$$

where $k_2 > k_1$. In this case, the reaction can be pulled forward by the removal of O_2^-, as was observed with the addition of Mn-SOD to submitochondrial particles. Of course, any other reaction that could remove O_2^-, such as the formation of peroxynitrite (see below), would also pull this reaction forward. In intact mitochondria, where ferricytochrome-c, Mn-SOD, and NO are present, the reaction would proceed. Steady-state O_2^- concentration is negligible, while the relative rates of production of ferrocytochrome-c, H_2O_2 peroxynitrite, and other competing reactions is determined by their relative reaction rates. The absolute rate-limiting step for all of these is the oxidation of ubisemiquinone. Nonetheless, if some component other than ubisemiquinone were the actual molecule that reacts with oxygen, then the same kinetic and thermodynamic considerations would apply to that reaction. From this, it can be expected that an increase in either oxygen or ubisemiquinone would cause a faster rate of production of H_2O_2 and the other products resulting from O_2^- production. Indeed, as described below, increased H_2O_2 production has been demonstrated to occur under various circumstances. An important question to keep in mind is whether these circumstances resemble any process contributing to aging. To this end, several specific questions are raised as a challenge for future investigation.

E. O_2 Dependence of H_2O_2 Production

As stated earlier, one factor that can elevate the rate of O_2^- generation is increased oxygen concentration. As oxygen increases from near anaerobiosis to well below the mixed venous oxygen concentration in normoxia (40 mmHg or 60 μM), an increase occurs in the saturation of cytochrome-c oxidase with no measurable production of H_2O_2 (62). Within the physiological range of tissue oxygen concentration, from near anaerobiosis in the center of exercising fast-twitch muscle to approximately 100 mmHg (150 μM) in lung epithelium, H_2O_2 production increases gradually to a small percentage of total oxygen consumption (62).

This suggests that more poorly perfused tissue could have less oxygen but remain in the same mitochondrial energy state, such that there would be no increase in ubisemiquinone and thus an actual decrease in the production of H_2O_2. During ischemia, transition to a fully reduced complex III can occur. Upon reperfusion, there may be a short-term burst of H_2O_2 production as the restored oxygen oxidizes the reduced components of the electron transport chain, particularly with NADH-dependent substrates (76). However, some argue that the first and major source of oxidants produced within cells during ischemia/reperfusion is that from conversion of xanthine dehydrogenase to xanthine oxidase and accumulated substrates, while mitochondrial components generate O_2^- only after damage and release of the components (77,78). Higher than physiologic oxygen concentration, a state that occurs during oxygen therapy or diving, produces an increase in mitochondrial H_2O_2 production that, above 60 mmHg, is nearly linear with oxygen concentration (62,79,80). In fact, the usual reasons for oxygen therapy—cardiac or pulmonary disease—tend to lessen oxygen delivery. Therefore, aside from the transient production during ischemia/reperfusion occurring in age-related pathologies, there seems to be little to suggest a contribution of increased mitochondrial H_2O_2 as a result of oxygen delivery to tissue. On the other hand, it is conceivable that decreases in oxygen consumption within a tissue could elevate the oxygen concentration available for production of H_2O_2 in a subtler manner (55).

F. Inhibitors, Uncouplers, and Slips

In the previous sections, reference has been made to the use of the electron transport inhibitors rotenone, antimycin A, TTFA, and myxothiazole. These agents have been useful in examining the potential sites of O_2^- production. Nonetheless, like many probes, the inhibitors perturb the system in a manner that can cause, or at least substantially increase, O_2^- production. For example, antimycin A traps reducing equivalents in complex III; alters the midpoint potentials of the cytochrome-b's in complex III, enhancing ubisemiquinone formation; and stops oxygen consumption, thereby increasing oxygen concentration at the mitochondrial inner membrane (55,64).

Is there anything that physiologically mimics antimycin A, and does it increase with age? During state 4 there is more H_2O_2 produced than during state 3, and there is some resemblance between addition of antimycin A and the transition from state 3 to state 4 in regard to increased reduced complex III components and decreased oxygen consumption (31,56). Whether there is any change in state 4 versus state 3 respiration with age remains an open question.

Uncouplers decrease H_2O_2 production. It has been suggested by Papa and Skulachev (55) that "mild uncoupling" is a physiological reality that prevents mitochondrial H_2O_2 production in what has been called "slips." It is proposed that these slips, or "intrinsic proton-pump decoupling," prevent formation of ubisemiquinone and are turned on when $\Delta\mu H^+$, the proton gradient across the inner mitochondrial membrane created by the Q cycle (Fig. 1), exceeds the state 3 level. Furthermore, the proposal suggests that thyroid hormone causes mild uncoupling (81), while steroid hormones cause "recoupling," opposing thyroid hormone and increasing efficiency of respiration (55). Changes in hormone production occur with aging, suggesting a possible physiological basis for alteration of mitochondrial H_2O_2 production with age. The direction of this change needs to be examined.

Recent studies have demonstrated the existence of two mitochondrial uncoupling proteins (82,83). One of these uncoupling proteins is associated with heat generation in brown adipose tissue, while the other is more widely distributed. The uncoupling proteins decrease both mitochondrial membrane potential and H_2O_2 generation (84). Di- and triphosphate purine nucleotides inhibit these proteins and thereby maintain mitochondrial potential and H_2O_2 production. Thus there is the potential for regulation of H_2O_2 generation related to the energy status of the cell. This presents another area for potential investigation.

G.   Combined Inhibition and Uncoupling

One of the most fascinating but largely overlooked findings in this area is the observation that addition of uncouplers to antimycin A–treated mitochondria results in a very dramatic increase in H_2O_2 generation (52). The mechanism for this phenomenon is likely to be a direct consequence of the dissolution of the proton gradient across the inner mitochondrial membrane combined with an increase in ubisemiquinone. Transfer of electrons from ubisemiquinone to oxygen is favored when ubisemiquinone is in the anionic form (Q^-) or perhaps when Q^- is dissociated from protein. When the proton gradient is disrupted, changes favoring ubisemiquinone oxidation could occur. With an uncoupler alone, this affect on ubisemiquinone protonation or dissociation is overshadowed by the decrease in total ubisemiquinone $(Q^- + QH)$. In the presence of antimycin A, the ubisemiquinone pool is increased. In addition, inhibition of respiration by antimycin A results in an increase in oxygen concentration at the inner mitochondrial membrane. Thus the addition of an uncoupler now dramatically increases O_2^- production rather than depressing it. Are there physiologically relevant agents that alone or in combination produce this phenomenon? Do these agents increase with age? Alternatively, does mitochondrial damage itself, regardless of its cause, promote more or less H_2O_2 production? For example, is the permeability transition related to this (see below)?

H.   Nitric Oxide

Nitric oxide (NO) is an enzymatically generated free radical. It is produced in cells to act as a stimulant in important physiological processes such as smooth muscle relaxation and neurotransmission, and as part of the antibacterial armamentarium during inflammation. Nitric oxide reacts with O_2^- at near diffusion rate to produce peroxynitrite $(ONOO^-)$ (57):

$$NO^{\cdot} + O_2^{-} \rightarrow ONOO^{-}$$

This reaction is the only biologically relevant reaction that is known to be faster than the enzymatic dismutation of O_2^{-}, and it could compete with SOD. Nonetheless, low levels of NO bind tightly to cytochrome-c oxidase and therefore tend to increase O_2^{-} production through reduction of complex III and the increase of oxygen at the mitochondrial inner membrane (85). If more than sufficient NO to inhibit cytochrome-c oxidase is present, then peroxynitrite is produced. Both NO and peroxynitrite have been suggested to inhibit complex III in a similar manner to antimycin A and thereby promote O_2^{-} production (85,86). If NO production is significant, and recent evidence shows that it can even be produced in mitochondria, then much of the electron leak through O_2^{-} will result in peroxynitrite production (see below).

III. MECHANISMS OF MITOCHONDRIAL O_2^{-} AND H_2O_2 REDUCTION

As was described earlier, most O_2^{-} production is a result of the displacement of a thermodynamically unfavorable reaction by endogenous NO, ferricytochrome-c, aconitase, or mitochondrial SOD with O_2^{-}. Indeed, any O_2^{-} produced on the intermembrane side of the mitochondrion would be intercepted by the high concentration of ferricytochrome-c in the intermembrane space. The amount of cytochrome-c in rat liver mitochondria is 0.34 nmol/mg protein (87). The intermembrane volume is variable, but always considerably smaller than the matrix volume. The total water content of mitochondria (referred to 1 mg protein) is 1 μl, of which the intermembrane space may be between 1% and 5%. It therefore follows that the cytochrome-c concentration would be 7 to 34 mM. As most of this is normally in the oxidized state, essentially no O_2^{-} can escape from mitochondria into the cytosol. On the matrix side, O_2^{-} could react with peroxynitrite, aconitase, or mitochondrial SOD. The function of mitochondrial SOD therefore appears to be prevention of the production of peroxynitrite or inactivation of aconitase rather than preventing leakage of O_2^{-} into the cytosol.

It is therefore interesting that transfecting flies so that they overexpressed cytosolic SOD (along with catalase) prolonged life span and increased the abilities of the aged flies (22). This suggests that O_2^{-} generation outside of mitochondria is a contributing factor to aging of these insects. Catalase, which was also increased in these flies, was also required. However, sources of H_2O_2 other than from the dismutation of the cytosolic O_2^{-} could not be established by such studies.

The action of the mitochondrial SOD does, however, result in the production of H_2O_2. Mitochondria contain enzymatic mechanisms for removal of H_2O_2.

Mitochondria actively take up glutathione (GSH) synthesized in the cytosol (88,89) and use it to reduce H_2O_2 through the action of glutathione peroxidase:

$$2GSH + H_2O_2 \xrightarrow{\text{Glutathione peroxidase}} GSSG + 2H_2O$$

The glutathione disulfide (GSSG) produced in this reaction is reduced to GSH by glutathione reductase:

$$GSSG + NADPH_2 \xrightarrow{\text{Glutathione reductase}} 2GSH + NADP$$

NADPH is maintained by transhydrogenase, which requires ATP:

$$NADH_2 + NADP + ATP \xrightarrow{\text{Transhydrogenase}} NADH + NADPH_2 + ADP + P_i$$

Thus another consequence of mitochondrial H_2O_2 production is the additional requirement of NADH and ATP to support its elimination.

There have been reports of the presence of catalase in heart mitochondria (58); however, it does not appear to be present in mitochondria from most tissues.

Thus H_2O_2 production by mitochondria is a balance between its production and reduction. Increases in H_2O_2 could be caused by decreases in the enzymatic system that normally maintains it at a very low steady state. Such may be the case in aging; however, to study this requires examination of much more than the H_2O_2 production of isolated mitochondria.

IV. mtDNA MUTATION AND IMPAIRMENT OF MITOCHONDRIAL FUNCTIONS

Rapid progress has been made in unraveling the molecular changes of mtDNA that apparently increase with aging. Many types of mutations of the mitochondrial genome impair the function of the respiratory and oxidative phosphorylation systems. Are these mutations responsible for the age-dependent decline in mitochondrial respiratory functions? To what extent and through what mechanisms is age-dependent decline in cellular functions related to mtDNA mutations? Reactive oxygen species produced by mutated mitochondria may cause oxidative damage to mtDNA in previously normal mitochondria, further impairing cellular functions and thereby increasing the rate of aging. Still the total mutated mtDNA amount remains low ($< 5\%$) (90) and each mitochondrion contain multiple copies of mtDNA.

Hydroperoxides induce loss of pyridine nucleotides and release of calcium from rat liver mitochondria (91). Prooxidant-induced calcium release would result in excessive Ca^{2+} "cycling" and may be the basis for prooxidant-induced cell toxicity (92). Cyclosporin A inhibits these phenomena (93). Cyclosporin A or inhibition of Ca^{2+} cycling completely prevents the mitochondrial damage induced by oxygen free radicals. These findings suggest that the deleterious effect of free radicals on mitochondria in the described experimental system was triggered by the cyclosporin A–sensitive and Ca^{2+}-dependent membrane transition and not by direct impairment of the mitochondrial inner-membrane enzymes (94).

The mitochondrial permeability transition pore (a cyclosporin A–sensitive channel) is tuned by the oxidation-reduction state of vicinal thiols (95), an oxidation bringing about an increased "open" probability of the permeability transition pore. This structure is thus the molecular target of externally added oxidants. Thus the respiratory chain originated H_2O_2 is a good candidate for the mitochondrial endogenous oxidant and is possibly responsible for the channel regulation.

Fluorescence labeling and electron microscopy have produced evidence of immunocytochemical localization of NO synthase (endothelial form) in brain and liver mitochondria (96,97). Nitric oxide inhibits mitochondrial respiration and induces generation of O_2^- (85,86,98). Several proteins have been indicated as possible targets of NO such as aconitase, NADH–ubiquinone oxidoreductase, and succinate-ubiquinone oxidoreductase of the mitochondrial electron transport chain (98). As described in a previous section, NO and O_2^- can combine to produce peroxynitrite. There is evidence that inactivation of aconitase in cell systems may not be simple, due to direct reactions of NO and O_2^- with the FeS cluster, but may rely on the formation of peroxynitrite (99). Nitric oxide is also capable of rapid but reversible inhibition of the mitochondrial respiratory chain at the level of cytochrome-c oxidase, which may be implicated in the cytotoxic effects of NO in the CNS and other tissues (100,101). Nitric oxide at low concentrations can potently deenergize mitochondria at oxygen concentrations that prevail in cells and tissues resulting in a release of Ca^{2+} (102,103).

Nitric oxide and peroxynitrite induce calcium efflux from mammalian mitochondria by modulating the mitochondrial permeability transition. This efflux is blocked by cyclosporin A (102,104,105). Nitric oxide is highly efficient in inducing mitochondrial permeability transition, thereby causing the liberation of apoptogenic factors from mitochondria, which can induce nuclear apoptosis (105).

Mitochondria may contain an apoptosis-inducing factor whose release occurs through the mitochondrial permeability transition. Bcl-2 belongs to a family of apoptosis-regulatory proteins which incorporate into the outer mitochondrial as well as nuclear membranes. The mechanism by which Bcl-2 inhibits apoptosis is thus far elusive, although it may be by favoring the retention of an apoptogenic protease in mitochondria (106). Such an event may occur by antagonizing the proapoptotic effects of Bax, possibly elicited through an intrinsic pore-forming activity (107).

Thus it appears that O_2^-, NO, and the product of their combination, peroxynitrite, in concert with Ca^{2+}, proteases, and nucleases, may be considered mediators of apoptosis. It has also been proposed that the cellular ATP level is also an important determinant for cell death (108). Before assuming that mtDNA mutations have a causative role in aging, the cascade of events following mitochondrial protein damage should be more closely analyzed.

In summary, mitochondria produce several agents that could easily account for increased damage to these organelles and others during aging. The questions raised here need to be addressed in order to determine whether this intriguing hypothesis can be supported by stronger evidence that has a sound mechanistic basis. Others have similarly called for a more critical evaluation of these aspects of the free radical theory of aging (109).

REFERENCES

1. Szweda LI, Uchida K, Tsai L, Stadtman ER. Inactivation of glucose-6-phosphate dehydrogenase by 4-hydroxy-2-nonenal. Selective modification of an active-site lysine. J Biol Chem 1993; 268:3342–3347.
2. Stadtman ER. Protein oxidation and aging. Science 1992; 257:1220–1224.
3. Yim HS, Kang JH, Chock PB, Stadtman ER, Yim MB. A familial amyotrophic lateral sclerosis-associated A4V Cu,Zn-superoxide dismutase mutant has a lower K_m for hydrogen peroxide. J Biol Chem 1997; 272:8861–8863.
4. Markesbery WR. Oxidative stress hypothesis in Alzheimer's disease. Free Radic Biol Med 1997; 23:134–147.
5. Wolff SP, Jiang ZY, Hunt JV. Protein glycation and oxidative stress in diabetes mellitus and ageing. Free Radic Biol Med 1991; 10:339–352.
6. Li D, Devaraj S, Fuller C, Bucala R, Jialal I. Effect of alpha-tocopherol on LDL oxidation and glycation: in vitro and in vivo studies. J Lipid Res 1996; 37:1978–1986.
7. Chappey O, Dosquet C, Wautier MP, Wautier JL. Advanced glycation end products, oxidant stress and vascular lesions. Eur J Clin Invest 1997; 27:97–108.
8. Yakes FM, Van Houten B. Mitochondrial DNA damage is more extensive and persists longer than nuclear DNA damage in human cells following oxidative stress. Proc Natl Acad Sci USA 1997; 94:514–519.
9. Tanaka M, Kovalenko SA, Gong JS, et al. Accumulation of deletions and point mutations in mitochondrial genome in degenerative diseases. Ann NY Acad Sci 1996; 786:102–111.
10. Corral-Debrinski M, Stepien G, Shoffner JM, et al. Hypoxemia is associated with mitochondrial DNA damage and gene induction. Implications for cardiac disease. J Am Med Assoc 1991; 266:1812–1816.
11. Corral-Debrinski M, Horton T, Lott MT. Mitochondrial DNA deletions in human brain: regional variability and increase with advanced age. Nat Genet 1992; 2:324–329.
12. Corral-Debrinski M, Shoffner JM, Lott MT, Wallace DC. Association of mitochondrial DNA damage with aging and coronary atherosclerotic heart disease. Mutat Res 1992; 275:169–180.

13. Corral-Debrinski M, Horton T, Lott MT, et al. Marked changes in mitochondrial DNA deletion levels in Alzheimer brains. Genomics 1994; 23:471–476.
14. Nagley P, Mackay IR, Baumer A, et al. Mitochondrial DNA mutation associated with aging and degenerative disease. Ann NY Acad Sci 1992; 673:92–102.
15. Liu VW, Zhang C, Linnane AW, Nagley P. Quantitative allele-specific PCR: demonstration of age-associated accumulation in human tissues of the A→G mutation at nucleotide 3243 in mitochondrial DNA. Hum Mutat 1997; 9:265–271.
16. Pitkänen S, Robinson BH. Mitochondrial complex I deficiency leads to increased production of superoxide radicals and induction of superoxide dismutase. J Clin Invest 1996; 98:345–351.
17. Keefe DL, Niven-Fairchild T, Powell S, Buradagunta S. Mitochondrial deoxyribonucleic acid deletions in oocytes and reproductive aging in women. Fertil Steril 1995; 64:577–583.
18. Chen X, Prosser R, Simonetti S, et al. Rearranged mitochondrial genomes are present in human oocytes. Am J Hum Genet 1995; 57:239–247.
19. Agarwal S, Sohal RS. DNA oxidative damage and life expectancy in houseflies. Proc Natl Acad Sci USA 1994; 91:12332–12335.
20. Calleja M, Pena P, Ugalde C, et al. Mitochondrial DNA remains intact during *Drosophila* aging, but the levels of mitochondrial transcripts are significantly reduced. J Biol Chem 1993; 268:18891–18897.
21. Sohal RS, Weindruch R. Oxidative stress, caloric restriction, and aging. Science 1996; 273:59–63.
22. Sohal RS, Agarwal A, Agarwal S, Orr WC Simultaneous overexpression of copper- and zinc-containing superoxide dismutase and catalase retards age-related oxidative damage and increases metabolic potential in *Drosophila melanogaster*. J Biol Chem 1995; 270:15671–15674.
23. Hayashi J, Ohta S, Kagawa Y, et al. Nuclear but not mitochondrial genome involvement in human age-related mitochondrial dysfunction. Functional integrity of mitochondrial DNA from aged subjects. J Biol Chem 1994; 269:6878–6883.
24. Rooyackers OE, Adey DB, Ades PA, Nair KS. Effect of age on in vivo rates of mitochondrial protein synthesis in human skeletal muscle. Proc Natl Acad Sci USA 1996; 93:15364–15369.
25. Richter C. Oxidative damage to mitochondrial DNA and its relationship to ageing. Int J Biochem Cell Biol 1995; 27:647–653.
26. Kuro-o M, Matsumaura Y, Aizawa H, Mutation of the mouse *klotho* gene leads to a syndrome resembling ageing. Nature 1997; 390:45–51.
27. Fridovich I, Handler P. Xanthine oxidase. IV. Participation of iron in internal electron transport. J Biol Chem 1958; 233:1581–1585.
28. Massey V, Strickland S, Mayhew SG, et al. The production of superoxide anion radicals in the reaction of reduced flavins and flavoproteins with molecular oxygen. Biochem Biophys Res Commun 1969; 36:891.
29. Strobel HW, Coon MJ. Effect of superoxide generation and dismutation on hydroxylation reactions catalyzed by liver microsomal cytochrome P-450. J Biol Chem 1971; 246:7826.
30. Thurman RG, Ley H, Scholz R. Hepatic microsomal ethanol oxidation: hydrogen peroxide formation and the role of catalase. Eur J Biochem 1972; 25:420–430.

31. Boveris A, Oshino N, Chance B. The cellular production of hydrogen peroxide. Biochem J 1972; 128:617–630.

32. Babior BM, Kipnes RS, Curnutte JT. The production by leukocytes of superoxide, a potential bactericidal agent. J Clin Invest 1973; 52:741.

33. Cross AR, Jones OTG. Enzymic mechanisms of superoxide production. Biochim Biophys Acta 1991; 1057:281–298.

34. Mohazzab HKM, Kaminski PM, Wolin MS. NADH oxidoreductase is a major source of superoxide anion in bovine coronary artery endothelium. Am J Physiol 1994; 266:H2568–H2572.

35. Kliubin IV, Gamalei IA. NADPH oxidase—a specialized enzyme complex for the formation of active oxygen metabolites. Tsitologiia 1997; 39:320–340.

36. Jones SA, Wood JD, Coffey MJ, Jones OT. The functional expression of p47-phox and p67-phox may contribute to the generation of superoxide by an NADPH oxidase-like system in human fibroblasts. FEBS Lett 1994; 355:178–182.

37. Irani K, Xia Y, Zweier JL. et al. Mitogenic signaling mediated by oxidants in Ras-transformed fibroblasts [see comments]. Science 1997; 275:1649–1652.

38. Misra HP, Fridovich I. The role of superoxide anion in the autoxidation of epinephrine and a simple assay for superoxide dismutase. J Biol Chem 1972; 247:3170.

39. Cohen G, Heikkila RE. The generation of hydrogen peroxide, superoxide radical, and hydroxyl radical by 6-hydroxydopamine, dialuric acid, and related cytotoxic agents. J Biol Chem 1974; 249:2447–2452.

40. Fridovich I, Handler P. Xanthine oxidase. V. Differential inhibition of the reduction of various electron acceptors. J Biol Chem 1962; 237:916–921.

41. McCord JM, Fridovich I. Superoxide dismutase: an enzymic function for erythrocuprein (hemocuprein). J Biol Chem 1969; 244:6049–6055.

42. Weisiger RA, Fridovich I. Mitochondrial superoxide dismutase. Site of synthesis and intramitochondrial localization. J Biol Chem 1973; 248:4793–4796.

43. Marklund SL. Properties of extracellular superoxide dismutase from human lung. Biochem J 1984; 220:269–272.

44. Jensen PK. Antimycin-insensitive oxidation of succinate and reduced nicotinamide-adenine dinucleotide in electron-transport particles. I. pH dependency and hydrogen peroxide formation. Biochim Biophys Acta 1966; 122:157–166.

45. Hinkle PC, Butow RA, Racker E, Chance B. Partial resolution of the enzymes catalyzing oxidative phosphorylation. XV. Reverse electron transfer in the flavin-cytochrome beta region of the respiratory chain of beef heart submitochondrial particles. J Biol Chem 1967; 242:5169–5173.

46. Loschen G, Azzi A. On the formation of hydrogen peroxide and oxygen radicals in heart mitochondria. Recent Adv Stud Cardiac Struct Metab 1975; 7:3–12.

47. Chance B, Oshino N. Kinetics and mechanisms of catalase in peroxisomes of the mitochondrial fraction. Biochem J 1971; 122:225–233.

48. Loschen G, Azzi A, Richter C, Flohe L. Superoxide radicals as precursors of mitochondrial hydrogen peroxide. FEBS Lett 1974; 42:68.

49. Forman HJ, Kennedy JA. Role of superoxide radical in mitochondrial dehydrogenase reactions. Biochem Biophys Res Commun 1974; 60:1044–1050.

50. Boveris A, Cadenas E. Mitochondrial production of superoxide anions and its relationship to the antimycin insensitive respiration. FEBS Lett 1975; 54:311.

51. Boveris A. Mitochondrial production of superoxide radical and hydrogen peroxide. In: Reivich M, Coburn R, Lahiri S, eds. Tissue Hypoxia and Ischemia.

52. Cadenas E, Boveris A. Enhancement of hydrogen peroxide formation by proto-phores and ionophores in antimycin-supplemented mitochondria. Biochem J 1980; 188:31–37.

53. Sohal RS, Dubey A. Mitochondrial oxidative damage, hydrogen peroxide release, and aging. Free Radic Biol Med 1994; 16:621–626.

54. Hansford RG, Hogue BA, Mildaziene V. Dependence of H_2O_2 formation by rat heart mitochondria on substrate availability and donor age. J Bioenerg Biomembr 1997; 29:89–95.

55. Papa S, Skulachev VP. Reactive oxygen species, mitochondria, apoptosis and aging. Mol Cell Biochem 1997; 174:305–319.

56. Forman HJ, Boveris A. Superoxide radical and hydrogen peroxide in mitochondria. In: Pryor WA, ed. Free Radicals in Biology, vol. V. New York: Academic Press, 1982:65–90.

57. Koppenol WH, Moreno JJ, Pryor WA, Ischiropoulos H, Beckman JS. Peroxynitrite, a cloaked oxidant formed by nitric oxide and superoxide. Chem Res Toxicol 1992; 5:834–842.

58. Radi R, Turrens JF, Chang LY, et al. Detection of catalase in rat heart mitochondria. J Biol Chem 1991; 266:22028–22034.

59. Boveris A, Cadenas E, Stoppani AOM. Role of ubiquinone in the mitochondrial generation of hydrogen peroxide. Biochem J 1976; 156:435.

60. Taube H. Mechanisms of oxidation with oxygen. J Gen Physiol 1965; 2:29.

61. Loschen G, Flohe L, Chance B. Respiratory chain linked H_2O_2 production in pigeon heart mitochondria. FEBS Lett 1971; 18:261.

62. Boveris A, Chance B. The mitochondrial generation of hydrogen peroxide. General properties and effect of hyperbaric oxygen. Biochem J 1973; 134:707–716.

63. Ohnishi T, Trumpower BL. Differential effects of antimycin on ubisemiquinone bound in different environments in isolated succinate cytochrome c reductase com-plex. J Biol Chem 1980; 255:3278–3284.

64. Erecinska M, Wilson DF. The effect of antimycin A on cytochromes b_{561}, b_{566}, and their relationship to ubiquinone and the iron-sulfur centers S-1 (+N-2) and S-3. Arch Biochem Biophys 1976; 174:143.

65. Giulivi C, Boveris A, Cadenas E. Hydroxyl radical generation during mitochondrial electron transfer and the formation of 8-hydroxydesoxyguanosine in mitochondrial DNA. Arch Biochem Biophys 316:909–916.

66. Turrens JF, Boveris A. Generation of superoxide anion by the NADH dehydroge-nase of bovine heart mitochondria. Biochem J 1980; 191:421–427.

67. Xia D, Yu CA, Kim H, et al. Crystal structure of the cytochrome bc1 complex from bovine heart mitochondria. Science 1997; 277:60–66.

68. Cadenas E, Boveris A, Ragan CI, Stoppani AO. Production of superoxide radicals and hydrogen peroxide by NADH-ubiquinone reductase and ubiquinol-cytochrome c reductase from beef-heart mitochondria. Arch Biochem Biophys 1977; 180:248–257.

69. Loschen G, Azzi A, Flohe L. Mitochondrial H_2O_2 formation at site II. Hoppe Seyl-ers Z Physiol Chem 1973; 354:791–794.

70. Cadenas E, Hochstein P. Pro- and antioxidant functions of quinones and quinone reductases in mammalian cells. Adv Enzymol Relat Areas Mol Biol 1992; 65:97–146.

71. Kotlyar AB, Sled VD, Burbaev DS, Moroz IA, Vinogradov AD. Coupling site I and the rotenone-sensitive ubisemiquinone in tightly coupled submitochondrial particles. FEBS Lett 1990; 264:17–20.

72. Vinogradov AD, Sled VD, Burbaev DS, et al. Energy-dependent complex I-associated ubisemiquinones in submitochondrial particles. FEBS Lett 1995; 370:83–87.

73. Forman HJ, Kennedy J. Dihydroorotate dependent superoxide production in rat brain and liver; a function of the primary dehydrogenase. Arch Biochem Biophys 1976; 173:219–224.

74. Forman HJ, Kennedy J. Effects of chaotropic agents versus detergents on dihydroorotate dehydrogenase. J Biol Chem 1977; 252:3379–3387.

75. Trumpower BL, Simmons Z. Diminished inhibition of mitochondrial electron transfer from succinate to cytochrome c by thenoyltrifluoroacetone induced by antimycin. J Biol Chem 1979; 254:4608–4616.

76. Guarnieri C, Muscari C, Ventura C, Mavelli I. Effect of ischemia on heart submitochondrial superoxide production. Free Radic Res Commun 1985; 1:123–128.

77. Hearse DJ, Manning AS, Downey JM, Yellon DM. Xanthine oxidase: a critical mediator of myocardial injury during ischemia and reperfusion? Acta Physiol Scand Suppl 1986; 548:65–78.

78. Hess ML, Manson NH. Molecular oxygen: friend and foe. The role of the oxygen free radical system in the calcium paradox, the oxygen paradox and ischemia/reperfusion injury. J Mol Cell Cardiol 1984; 16:969–985.

79. Turrens JF, Freeman BA, Crapo JD. Hyperoxia increases H_2O_2 release by lung mitochondria and microsomes. Arch Biochem Biophys 1982; 217:411–421.

80. Freeman BA, Crapo JD. Hyperoxia increases oxygen radical production in rat lungs and lung mitochondria. J Biol Chem 1981; 256:10986–10992.

81. Horst C, Rokos H, Seitz HJ. Rapid stimulation of hepatic oxygen consumption by 3,5-di-iodo-L-thyronine. Biochem J 1989; 261:945–950.

82. Ricquier D, Casteilla L, Bouillaud F. Molecular studies of the uncoupling protein. FASEB J 1991; 5:2237–2242.

83. Fleury C, Neverova M, Collins S, et al. Uncoupling protein-2: a novel gene linked to obesity and hyperinsulinemia. Nat Genet 1997; 15:269–272.

84. Nègre-Salvayre A, Hirtz C, Carrera G, et al. A role for uncoupling protein-2 as a regulator of mitochondrial hydrogen peroxide generation. FASEB J 1997; 11:809–815.

85. Poderoso JJ, Carreras MC, Lisdero C, et al. Nitric oxide inhibits electron transfer and increases superoxide radical production in rat heart mitochondria and submitochondrial particles. Arch Biochem Biophys 1996; 328:85–92.

86. Radi R, Rodriguez M, Castro L, Telleri R. Inhibition of mitochondrial electron transport by peroxynitrite. Arch Biochem Biophys 1994; 308: 89–95.

87. Chance B, Hess B. Metabolic control mechanisms. Electron transfer in the mammalian cell. J Biol Chem 1962; 234:2404–2412.

88. Meredith MJ, Reed DJ. Status of the mitochondrial pool of glutathione in the isolated hepatocyte. J Biol Chem 1982; 257:3747–3753.

89. Griffith OW, Meister A. Origin and turnover of mitochondrial glutathione. Proc Natl Acad Sci USA 1985; 82:4668–4672.

90. Lee HC, Wei YH. Mutation and oxidative damage of mitochondrial DNA and defective turnover of mitochondria in human aging. J Formos Med Assoc 1997; 96: 770–778.

91. Lotscher HR, Winterhalter KH, Carafoli E, Richter C. Hydroperoxide-induced loss of pyridine nucleotides and release of calcium from rat liver mitochondria. J Biol Chem 1980; 255:9325–9330.

92. Richter C, Schlegel J. Mitochondrial calcium release induced by prooxidants. Toxicol Lett 1993; 67:119–127.

93. Richter C, Theus M, Schlegel J. Cyclosporine A inhibits mitochondrial pyridine nucleotide hydrolysis and calcium release. Biochem Pharmacol 1990; 40:779–782.

94. Takeyama N, Matsuo N, Tanaka T. Oxidative damage to mitochondria is mediated by the Ca^{2+}-dependent inner-membrane permeability transition. Biochem J 1993; 294:719–725.

95. Petronilli V, Costantini P, Scorrano L, et al. The voltage sensor of the mitochondrial permeability transition pore is tuned by the oxidation-reduction state of vicinal thiols. Increase of the gating potential by oxidants and its reversal by reducing agents. J Biol Chem 1994; 269:16638–16642.

96. Tang FR, Tan CK, Ling EA. Light and electron microscopic studies of the distribution of NADPH-diaphorase in the rat upper thoracic spinal cord with special reference to the spinal autonomic region. Arch Histol Cytol 1995; 58:493–505.

97. Bates TE, Loesch A, Burnstock G, Clark JB. Immunocytochemical evidence for a mitochondrially located nitric oxide synthase in brain and liver. Biochem Biophys Res Commun 1995; 213:896–900.

98. Stadler J, Billiar TR, Curran RD, et al. Effect of exogenous and endogenous nitric oxide on mitochondrial respiration of rat hepatocytes. Am J Physiol 1991; 260: C910–C916.

99. Ferraris JD, Williams CK, Jung KY, JJ et al. ORE, a eukaryotic minimal essential osmotic response element. The aldose reductase gene in hyperosmotic stress. J Biol Chem 1996; 271:18318–18321.

100. Brown GC. Nitric oxide regulates mitochondrial respiration and cell functions by inhibiting cytochrome oxidase. FEBS Lett 1995; 369:136–139.

101. Cleeter MW, Cooper JM, Darley-Usmar VM, Moncada S, Schapira AH. Reversible inhibition of cytochrome c oxidase, the terminal enzyme of the mitochondrial respiratory chain, by nitric oxide. Implications for neurodegenerative diseases. FEBS Lett 1994; 345:50–54.

102. Schweizer M, Richter C. Peroxynitrite stimulates the pyridine nucleotide-linked Ca^{2+} release from intact rat liver mitochondria. Biochemistry 1996; 35:4524–4528.

103. Schweizer M, Richter C. Nitric oxide potently and reversibly deenergizes mitochondria at low oxygen tension. Biochem Biophys Res Commun 1994; 204:169–175.

104. Packer MA, Murphy MP. Peroxynitrite formed by simultaneous nitric oxide and

superoxide generation causes cyclosporin-A-sensitive mitochondrial calcium efflux and depolarisation. Eur J Biochem 1995; 234:231–239.

105. Hortelano S, Dallaporta B, Zamzami N, et al. Modulation of the mitochondrial permeability transition by nitric oxide. Eur J Biochem 1997; 246:710–718.

106. Susin SA, Zamzami N, Castedo M, et al. Bcl-2 inhibits the mitochondrial release of an apoptogenic protease. J Exp Med 1996; 184:1331–1341.

107. Antonsson B, Conti F, Ciavatta A, et al. Inhibition of Bax channel-forming activity by Bcl-2. Science 1997; 277:370–372.

108. Richter C, Schweizer M, Cossarizza A, Franceschi C. Control of apoptosis by the cellular ATP level. FEBS Lett 1996; 378:107–110.

109. Yu BP, Yang R. Critical evaluation of the free radical theory of aging. A proposal for the oxidative stress hypothesis. Ann N Y Acad Sci 1996; 786:1–11.

110. Chance B, Williams GR. The respiratory chain and oxidative phosphorylation. Adv Enzymol Relat Subj 1956; 17:65–134.

111. Konstantinov AA, Peskin AV, Popova EY, Khomutov GB, Ruuge EK. Superoxide generation by the respiratory chain of tumor mitochondria. Biochim Biophys Acta 1987; 894:1–10.

112. Boveris A, Docampo R, Turrens JF, Stoppani AO. Effect of beta-lapachone on superoxide anion and hydrogen peroxide production in *Trypanosoma cruzi*. Biochem J 1978; 175:431–439.

6
Functional Roles of Ubiquinone

Jeffery R. Schultz
Pacific Lutheran University, Tacoma, Washington

Catherine Freitag Clarke
University of California, Los Angeles, California

I. UBIQUINONE STRUCTURE AND AN OVERVIEW OF UBIQUINONE FUNCTION

Ubiquinone (coenzyme Q or Q_{10}) is an important redox component of cellular membranes. Its structure consists of two parts: a redox-active benzoquinone nucleus, and a hydrophobic polyprenyl tail comprised of isoprene units linked with a *trans* configuration (Fig. 1). Q is detected in all organisms except gram-positive bacteria and cyanobacteria, and the length of the polyprenyl tail is species specific (1,2). The readily reversible redox chemistry of the benzoquinone head group allows Q to undergo sequential, one-electron reduction to form the ubisemiquinone (QH·) (or the semiquinone anion, $Q^{·-}$) and the hydroquinone (QH_2). As a membrane-soluble molecule, Q is an essential component of the eukaryotic respiratory chain located in the inner mitochondrial membrane of eukaryotes and in the plasma membrane of prokaryotes (3). Linear dichroism, NMR, fluorescence, and EPR studies indicate that ubiquinone is oriented in the membrane bilayer midplane with the tail perpendicular to the lipid chains, allowing the quinone head group to fluidly move through the lipids and react with the respiratory complexes and other molecules at the membrane surface (4–6). Membrane-anchoring of the tail is important for maximizing Q reactivity since Q isoforms with shorter prenylated tails are less able to reduce cytochrome-*c* or ferricyanide

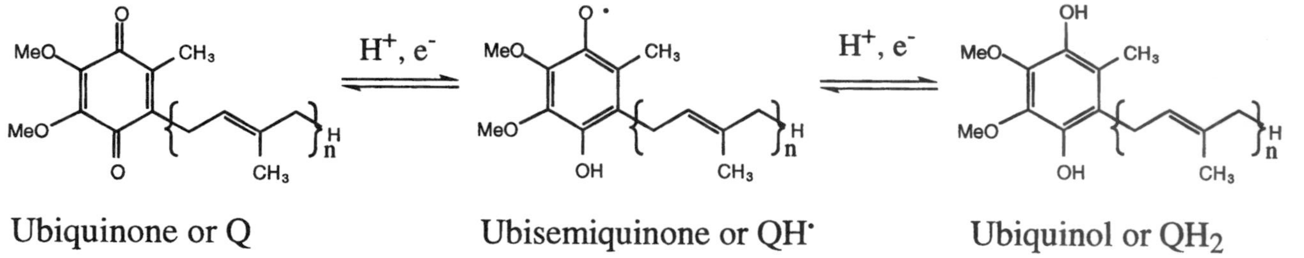

Figure 1   Reversible reduction and oxidation of Q. n designates the number of isoprene units in the polyprenyl tail which can vary from 6 (Q_6, *Saccharomyces cerevisiae*), to 8 (Q_8, *Escherichia coli*), 9 (Q_9, rat), or 10 (Q_{10}, human) (1,2).

in liposomes and also show a lowered capacity to pump protons and generate membrane potential than Q homologues with longer isoprenoid tails (7–9).

The redox chemistry that enables reversible cycling between QH_2, $QH^{.}/Q^{.-}$, and Q in mitochondrial respiration also bestows two other functional roles to Q, its often cited action as either a prooxidant or antioxidant. A prooxidant function is attributed to $QH^{.}$ (or $Q^{.-}$), an intermediate generated during electron transport. Thus $QH^{.}$ or $Q^{.-}$ is proposed to function as a site of reactive oxygen species formation via its reaction with dioxygen to give superoxide (O_2^-) or with H_2O_2 to give H_2O and $OH^{.}$ (10,11). In contrast to its function as a prooxidant, many studies have used in vitro assays to describe the antioxidant properties of QH_2 in mitochondrial and nonmitochondrial membrane fractions, as well as liposome vesicles and lipoprotein particles (as reviewed in 6,12–15). In this capacity QH_2 may act directly as a chain terminator to reduce lipid peroxyl radicals, or indirectly via the reduction of α-tocopheroxyl radicals. There is also evidence that QH_2 functions as an effective antioxidant in vivo (16,17). The majority of total Q (QH_2 + Q) is in the reduced form in rat and human tissues, lipoproteins, and yeast (18–20), indicating that QH_2 is poised to protect against oxidative damage.

While the majority of cellular Q participates in mitochondrial respiration, subcellular fractionation studies find QH_2 present in other intracellular locations such as microsomes, Golgi, lysosomes, peroxisomes, and plasma membranes (21–24). In addition to its proposed function as a lipid-soluble antioxidant, Q participates in several alternative electron transport chains located in the Golgi and plasma membranes (25,26). In the plasma membrane the redox chemistry of Q allows it to ferry electrons from an internal NADH-dehydrogenase to an external-side final acceptor (27). Recent studies suggest that this trans-plasma membrane electron transport system acts to reduce extracellular ascorbate free radicals (23,28). Hence QH_2 is involved in the reduction and regeneration of two crucial antioxidants, ascorbate and vitamin E. This antioxidant function of QH_2 coupled with its key role in energy generation (and the possible role of $QH^{.}$ or $Q^{.-}$ in free radical generation) implicate the $QH_2:Q$ ratio as an indicator of cellular redox status. The focus of this chapter is to review the current knowledge about the cellular functions of Q, with emphasis placed on yeast as a model system.

II. Q BIOSYNTHESIS AND REGULATION

A. The Q Biosynthetic Pathway

Q/QH_2 is the only lipid-soluble antioxidant that can be synthesized by mammalian cells; the other lipid-soluble antioxidants (vitamin E, beta-carotene) must be

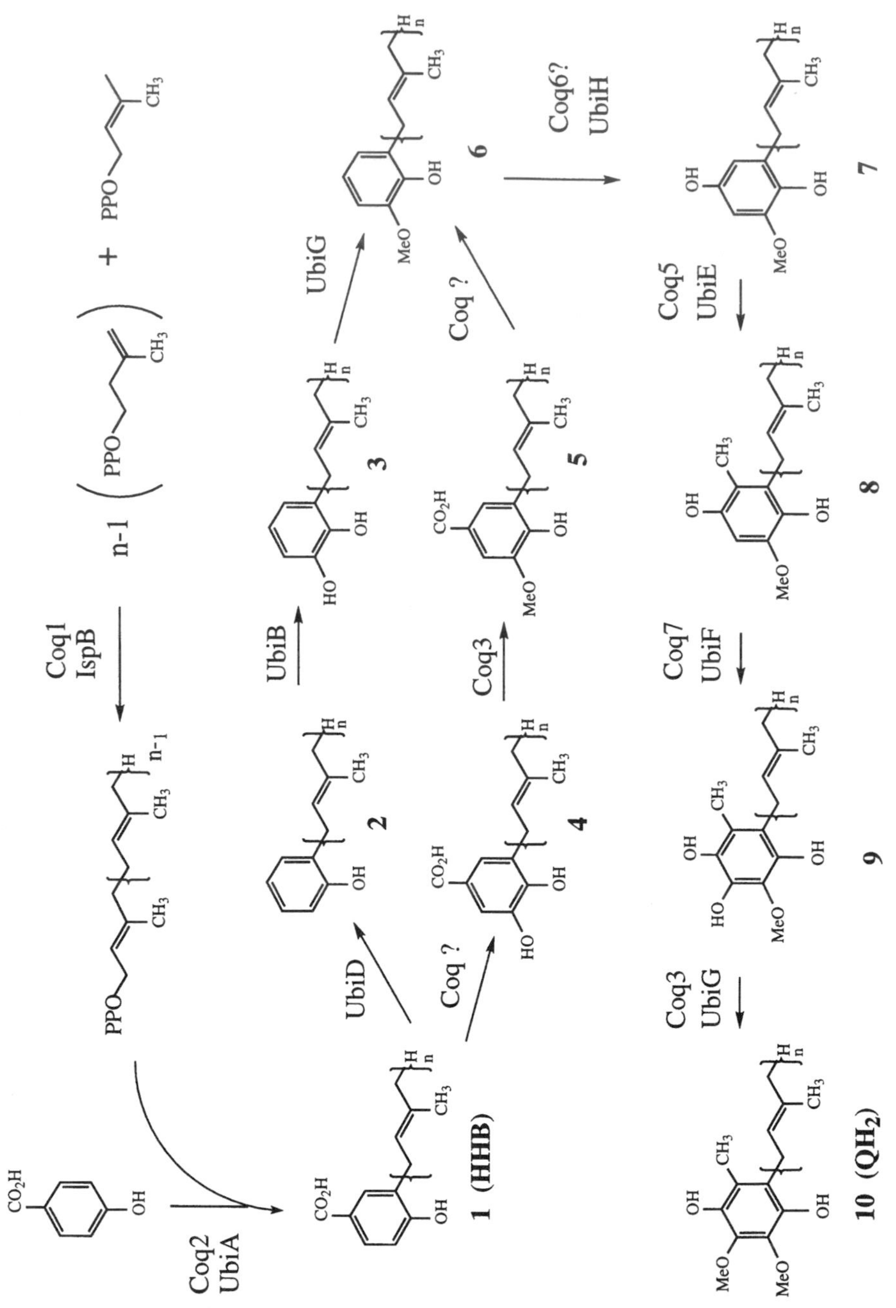
Coq2
UbiA
Coq1
IspB
n-1
UbiD
Coq ?
UbiB
Coq3
UbiG
Coq ?
Coq6?
UbiH
Coq5
UbiE
Coq7
UbiF
Coq3
UbiG
1 (HHB)
2
3
4
5
6
7
8
9
10 (QH2)

derived from the diet. It is necessary to understand how the cell synthesizes and regulates Q in order to characterize how QH_2 functions as an effective antioxidant. Figure 2 shows the putative Q biosynthetic pathway for eukaryotes and prokaryotes. This pathway was proposed from the identification of biosynthetic intermediates of Q that accumulated in Q-deficient mutant strains of *S. cerevisiae* and *E. coli* (1,29,30). Eight complementation groups of yeast mutants completely deficient in Q have been identified and contain mutations in one of eight nuclear genes *(COQ1–COQ8)* (31). The *coq* mutants are classified as nonrespiring, petite strains because the cells cannot grow on nonfermentable carbon sources like ethanol and glycerol, and when grown on glucose the mutants form small colonies compared to wild-type cells (32,33). Strains harboring mutations in the *COQ* genes were shown to be deficient in Q biosynthesis since NADH-reductase activity could be restored to near wild-type levels when either Q_2 or Q_6 was added to mitochondrial extracts (32). These yeast Q-deficient mutants, and similar *E. coli* mutants, have been used to investigate the genes and proteins involved in Q biosynthesis. The yeast genes *COQ1, COQ2, COQ3, COQ5,* and *COQ7* have been characterized (Fig. 2) (34–39). The *COQ4, COQ6,* and *COQ8* genes have also been identified and are currently being examined for function (40). Subcellular fractionation and in vitro mitochondrial import studies indicate that these proteins are localized to the mitochondria (37,38,40,41).

Upon formation of HHB, the eukaryotic and prokaryotic Q biosynthetic pathways diverge, with the prokaryotic pathway (Ubi) first undergoing decarboxylation to form compound 2, while HHB is hydroxylated at the 3-position in the eukaryotic pathway to generate compound 4. Evidence for this divergence comes

---

Figure 2 The proposed pathway of Q biosynthesis. A polyisoprene diphosphate is assembled from dimethylallyl diphosphate and isopentenyl diphosphate by Coq1 (*S. cerevisiae*) or IspB (*E. coli*). After formation of compound (**1**) 3-polyprenyl-4-hydroxybenzoic acid, by the *p*-hydroxybenzoic acid:polyprenyltransferase (Coq2 or UbiA), the proposed biosynthetic pathway for Q in eukaryotes and in prokaryotes is thought to diverge. The other intermediates in the pathway are (**2**) 2-polyprenylphenol; (**3**) 2-polyprenyl-6-hydroxy-phenol; (**4**) 3,4-dihydroxy-5-polyprenylbenzoic acid; (**5**) 3-methoxy-4-hydroxy-5-polyprenylbenzoic acid; (**6**) 2-polyprenyl-6-methoxy-phenol; (**7**) 2-polyprenyl-6-methoxy-1,4-benzoquinol; (**8**) 2-polyprenyl-3-methyl-6-methoxy-1,4-benzoquinol or 5-demethoxyubiquinol; (**9**) 2-polyprenyl-3-methyl-5-hydroxy-6-methoxy-1,4-benzoquinol or demethyl-QH_2; and (**10**) coenzyme Q_nH_2. Compounds 6, 7, and 9 are hypothetical intermediates in *S. cerevisiae* Q biosynthetic pathway, as is compound 3 in *E. coli*. In *S. cerevisiae*, $n = 6$ and compound 1 is referred to as 3-hexaprenyl-4-hydroxybenzoate (HHB). Gene products of different complementation groups are identified as Ubi in *E. coli* and Coq in *S. cerevisiae*.

from observations that *ubiB*-deficient *E. coli* mutants accumulate compound 2 (42), while compound 4 was detected in *coq3* mutants (43), and another yeast mutant (no longer available) accumulates compound 5 (44). However, the predominant intermediate in the yeast strains harboring mutations in the coq3–coq8 genes is HHB (17,39). HHB is also the common predominant intermediate in wild-type yeast strains (45). The accumulation of HHB as a single predominant Q intermediate is very different from the analysis of the Q-deficient *E. coli ubi* mutants. The Q-deficient *E. coli* mutants *ubiH, ubiE, ubiF,* and *ubiG* accumulate compounds 6, 7, 8, and 9 as the respective predominant intermediates (46). Hence the analysis of *E. coli ubi* genes has provided important clues about the function of the eukaryotic *COQ* homologues (37,38,41). Mammalian homologues of the *COQ3* and *COQ7* genes have been isolated by functional complementation of the corresponding yeast mutants (47,48), indicating a shared function and localization of Q biosynthetic enzymes between yeast and higher eukaryotes.

B. Regulation of Q Levels

Yeast Q biosynthesis is regulated by oxygen and glucose availability to the cells (1,49). Biosynthesis is inhibited by low oxygen tension or high glucose (5–10%). When yeast are grown anaerobically, the Q levels are 30- to 300-fold lower compared to cells grown aerobically (50,51). HHB is the predominant intermediate that accumulates in wild-type yeast (45) and in the *coq3*–*coq8* mutants (17). This finding is consistent with either coordinate regulation of the *COQ* gene products, or could also indicate that the Coq proteins form a multisubunit complex required to convert HHB to Q. The hydroxylation of HHB may be the rate-limiting step in Q biosynthesis, or could reflect a high K_m of the enzyme for either O_2 or HHB. Compound 1 ($n = 9$) is also observed as the predominant Q biosynthetic intermediate in rat heart, and accumulates under both normal and hypoxic conditions (52,53).

The synthesis of Q in animal cells also depends on the activity of 3-hydroxy-3-methylglutaryl coenzyme A reductase and polyisoprene availability (52,54–56). For example, when squalene synthase is inhibited by squalestatin-1, Q production is enhanced three- to fourfold (57). Long-term treatment of rat liver with squalestatin also results in a shift in chain length from Q_9 to Q_{10} (58). There is no evidence in tissue culture or animal model systems that Q is able to feedback regulate its own synthesis; however, there appears to be tissue and cell-type control since Q levels are highest in the heart (18,59–62). Additional studies are needed to fully characterize Q biosynthesis and its regulation in eukaryotic cells.

III. FUNCTIONS OF Q

A. Mitochondrial Electron Transport and Energy Transduction

The majority of Q in the cell is localized to the mitochondria where it participates in electron transport and oxidative phosphorylation. Q shuttles electrons from NADH dehydrogenase (complex I) and succinate dehydrogenase (complex II) to ubiquinol:cytochrome-c oxidoreductase (complex III) (63–65). Glycerol-3-phosphate is also oxidized via Q reduction at glycerol-3-phosphate dehydrogenase located on the cytoplasmic face of the mitochondria (66,67). Q is present in the mitochondria at a concentration of approximately 15 to 30 nmol/mg phospholipids (68). This concentration of Q in the mitochondrial inner membrane far exceeds the concentration of other respiratory chain components to ensure that Q is not the limiting step in the electron flux of the pathway (69–71). However, Lenaz and colleagues calculated that while there is enough Q to fully saturate succinate and glycerol-3-phosphate oxidation, NADH oxidation is only partially saturated (72,73). These findings indicate that it is still not completely understood how much Q in the mitochondria is required for physiological function.

In complex III (the cytochrome-bc_1 complex) Q also ferries protons during electron transport from the mitochondrial matrix to the intermitochondrial space via the Q cycle to establish the protonmotive gradient used during ATP production (74,75). The Q cycle is an energy-conserving pathway which involves two distinct Q reaction centers on different sides of the inner mitochondrial membrane (Fig. 3). The first half of the Q cycle is represented by reactions 1–4a in Figure 3 and summarized by equation (1),

$$QH_2 + c_{ox} \rightarrow Q^{\cdot-}{}_n + c_{red} + 2H^+{}_p, \tag{1}$$

where n and p refer to the negative and positive sides of the membrane, and ox and red refer to the oxidized and reduced species (75). The Q cycle is completed when a second molecule of QH_2 is again oxidized at center P and the electrons again bifurcate, one to the Rieske FeS cluster (ISP) and then to cytochrome-c_1 (reactions 1 and 2), and the other electron is sent to cytochromes-b_L and -b_H and then reduces $Q^{\cdot-}{}_n$ in reactions 3 and 4b of Figure 3 [equation (2)]:

$$QH_2 + c_{ox} + Q^{\cdot-}{}_n + 2H^+{}_n \rightarrow c_{red} + QH_2 + Q + 2H^+{}_p \tag{2}$$

The sum of these two reactions represents one complete Q cycle [equation (3)], and results in vectorial proton translocation from the matrix (n side) to the intermembrane space (p side):

$$QH_2 + 2\,c_{ox} + 2H^+{}_n \rightarrow Q + 2c_{red} + 4H^+{}_p \tag{3}$$

·

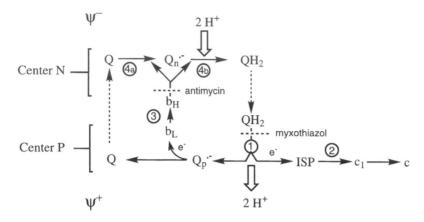

Figure 3 The protonmotive Q cycle. The path of electron transfer from ubiquinol to cytochrome-c through the redox prosthetic groups of the cytochrome-bc_1 complex is depicted as a series of numbered reactions shown by solid arrows. Dashed arrows represent movement of QH_2 and Q between the site where QH_2 is oxidized at the positive side of the membrane (center P) and the site where Q and $Q^{\cdot-}$ are reduced at the negative side of the membrane (center N). Open arrows show the reactions in which protons are released during oxidation of QH_2 and taken up during reduction of Q. Dashed lines show the reactions that are blocked by myxathiazol and antimycin. (From Ref. 75.)

The presence of two Q reaction centers is supported by considerable biochemical evidence (75) and has been corroborated by recent crystal structure analysis of the bc_1 complex from bovine heart mitochondria (76). The oxidation of QH_2 at center P results in a bifurcation of electrons—one to ISP (and hence to cytochrome-c_1 and -c), and the second electron is transferred to heme b_L and then is used to rereduce Q at center N. The chemistry at center N is well understood and involves a stabilized $Q^{\cdot-}$ species (75). However, the chemistry at center P is more speculative and the activation barrier is thought to involve the transient formation of a very unstable $Q^{\cdot-}$ intermediate. This intermediate can be detected in the presence of antimycin (77), where it may leave center P and possibly reduce molecular oxygen to O_2^{-} (75). It is not clear whether the center P unstable $Q^{\cdot-}$ intermediate is produced under normal steady-state conditions (78). Recent evidence supports a proton-gated charge-transfer mechanism (79) in which the reaction at center P is controlled by deprotonation of QH_2 to form the Q^{2-} anion (78).

B. Prooxidant Properties

The prooxidant function of $QH^{\cdot}/Q^{\cdot-}$ has been extensively reviewed (6,80). Autoxidation of $QH^{\cdot}/Q^{\cdot-}$ can generate $O_2^{\cdot-}$ and Q in the presence of O_2 and protons (6). This oxidation is not likely to proceed within the lipid bilayer since protons do not have direct access to the membrane-buried $QH^{\cdot}$ (81,82). Furthermore, any $QH^{\cdot}/Q^{\cdot-}$ formed during the Q cycle is stabilized by ubiquinone-binding proteins to avoid spontaneous dismutation (75,83). However, many in vitro studies suggest that $O_2^{\cdot-}$ and hydroxyl radical ($OH^{\cdot}$) arise from a leak in the electron transport chain at the $QH^{\cdot}$ site (84–87). In fact, more than 80% of the $O_2^{\cdot-}$ generated in yeast, representing approximately 2% of the overall O_2 consumption, is attributed to this reaction (10,86). The $O_2^{\cdot-}$ produced can then undergo spontaneous dismutation to form hydrogen peroxide (H_2O_2) (88) or is converted to H_2O_2 by superoxide dismutase (SOD). $QH^{\cdot}$ may also directly reduce H_2O_2 to form H_2O and $OH^{\cdot}$ (89). Stress imposed by ischemia results in increased submitochondrial production of $O_2^{\cdot-}$ and H_2O_2 that is attributed to the flavin-semiquinone and $QH^{\cdot}$ portion of the NADH dehydrogenase, while reperfusion facilitates ROS production at both the NADH and succinate dehydrogenase complexes (90). During ischemia, mitochondrial NADH accumulation promotes $QH^{\cdot}$ autoxidation due to its proximal orientation at the aqueous surface of the membrane (11).

Thus Q is implicated in the generation of $O_2^{\cdot-}$, H_2O_2, and hydroxyl radicals—reactive oxygen species that have been proposed to contribute to cellular aging. In this oxidative stress theory of aging, mitochondria are considered to be both the main source and the target of oxygen-derived free radicals (91, and this volume). It has been noted however, that the in vitro studies employ conditions that enhance the propensity of radical production. For example, drugs such as antimycin which promote $O_2^{\cdot-}$ production, modify the interaction of $QH^{\cdot}/Q^{\cdot-}$ with proteins, alter the stability of $QH^{\cdot}/Q^{\cdot-}$, and elicit $O_2^{\cdot-}$ production at sites which in vivo may play a very minor role (92). In this regard it is interesting that two recent studies using *sod*Δ yeast mutants demonstrate that the electron leak does occur in vivo at the Q site or at cytochrome-*b* in the respiratory chain (93,94).

C. Antioxidant Effects

Many studies show that ROS, such as $O_2^{\cdot-}$, H_2O_2, $OH^{\cdot}$, and organic peroxyl radicals such as lipid peroxyl radicals ($LOO^{\cdot}$), arise through free radical–mediated reactions, exposure to oxidizing agents, and normal cellular metabolism (95–97). Such oxidative species lead to cellular oxidative damage of nuclear and mitochondrial DNA, proteins, and lipids. Unchecked oxidative damage ultimately results in cell death. Membrane polyunsaturated fatty acids (PUFAs) are particularly prone to oxidative modifications that generate toxic breakdown products which promote additional cellular damage (98,99). Cells are able to respond to oxidative stress through enzymatic (SOD, catalase, and metallothionines) and small molecule antioxidants [glutathione, ascorbate, α-tocopherol (α-TOH), and QH_2] that quench or scavenge free radical propagation. QH_2 was shown to be an effective antioxidant and protect against oxidative damage mediated through lipid peroxidation in both mitochondrial and nonmitochondrial membrane fractions, liposomes, and lipoproteins (6,12,13,15). QH_2 performs this function through the redox chemistry of its head group, such that QH_2 can reduce lipid peroxyl radicals in a similar manner as α-TOH [equations (4) and (5)] (100,101).

$$QH_2 \quad + \; LOO^{\cdot} \; \rightarrow QH^{\cdot} \quad + \; LOOH \tag{4}$$
$$\alpha\text{-TOH} + LOO^{\cdot} \; \rightarrow \alpha\text{-TO}^{\cdot} \; + \; LOOH \tag{5}$$
$$\alpha\text{-TO}^{\cdot} \; + \; QH_2 \quad \rightarrow \alpha\text{-TOH} + \; QH^{\cdot} \tag{6}$$

QH_2 can function independently of α-TOH, or may reduce α-tocopheroxyl radicals to regenerate α-TOH [equation (6)] (100,102–104). α-TOH reacts with peroxyl radicals at a 10-fold faster rate and displays better protection against lipid peroxidation compared to Q_9H_2 or $Q_{10}H_2$ in in vitro studies using rat liver mitochondria or microsomes (105). Since rat liver microsomal membranes contain similar concentrations of both α-TOH and Q_9H_2 (6,106), these studies indicate that α-TOH is a more potent membrane antioxidant than QH_2. The results also suggest that when both α-TOH and QH_2 are present in membranes, QH_2 is a "secondary" antioxidant that acts to preferentially regenerate α-TOH rather than quench peroxyl radicals. QH_2 will also protect against membrane protein carboxylation and mtDNA damage by scavenging peroxyl-mediated ROS (107,108).

The oxidized form of Q also has been found to possess antioxidant properties in liposomes when present in high concentrations (20,102,109). However, it is unlikely this would occur in vivo since QH_2 is the primary form found in cellular membranes and lipoproteins, and the concentrations of Q tested were not physiological (18,19). Interestingly, we observed that high concentrations of Q_6H_2 in liposomes will actually promote lipid peroxidation in the presence of Cu^{2+} through Fenton or Haber-Weiss chemistry (20,95,110). The size of the polyprenyl tail has little effect on the antioxidant properties of QH_2 since Q_3H_2 and Q_7H_2 both inhibit lipid peroxidation in liposomes to the same degree (6,111). Kagan et al. (105) observed that ubiquinol tails with four or less isoprenoid units can protect against lipid peroxidation reactions better than Q homologues containing longer polyprenyl tails (5 to 10 units). Nevertheless, any tail-length effects would be inconsequential for antioxidant efficacy in vivo since all organisms synthesize Q homologues with at least six or more isoprene units.

D. Cellular QH$_2$ Reduction Systems

In order for QH$_2$ to function as an effective membrane antioxidant, it needs to be maintained in the reduced form. Q is reduced in the mitochondria by NADH dehydrogenase, succinate dehydrogenase, and glycerol-3-phosphate dehydrogenase in the respiratory chain (63,64,66), as well as the fatty acid β-oxidation enzyme electron-transferring flavoprotein:ubiquinone reductase (112), the NADPH (quinone) dehydrogenase (113), and the NADH oxidase complex (114,115). Q is also reduced by pyrimidine dihydroorotate dehydrogenase (116,117), the fourth enzyme of pyrimidine de novo synthesis, and the only one located in the inner mitochondrial membrane. Thus dihydroorotate dehydrogenase links mitochondria respiration with the biosynthesis of pyrimidine nucleotides (118). QH$_2$ is also generated through de novo biosynthesis since the last O-methyltransferase step requires demethyl-QH$_2$ (Fig. 2) (119,120). Alternatively, QH$_2$ levels can be maintained outside the mitochondria through systems that either slow QH$_2$ autoxidation (12) or serve as quinone reductases (121–124). Microsomes and Golgi contain NADH reductase/oxidase activity that can generate QH$_2$ (21,105,125). Several cytosolic enzymes have been shown to reduce Q, including DT-diaphorase, the NAD(P)H:(quinone-acceptor) oxidoreductase (126–130), and an NADPH-Q reductase that functions independently of DT-diaphorase (124,131). A human hepatoma cell line and human blood cells also exhibited an ability to reduce extracellular Q, most likely through a transmembrane redox system involving NADH-ascorbate free radical (AFR) reductase activity (26,28,119).

IV. Q-DEFICIENT YEAST AS A MODEL SYSTEM

Yeast mutants lacking Q offer a particularly effective system for studying ubiquinone biosynthesis and function. Yeast metabolize both fermentable (glucose and galactose) and nonfermentable (glycerol, ethanol) carbon sources. Since a functional electron transport chain is required for respiration, respiratory-deficient yeast mutants can be identified by the lack of growth on glycerol or ethanol. The *COQ* genes were isolated by functional complementation that selected for those colonies with restored growth on nonfermentable carbon sources (Sec. II.A). Yeast are simple eukaryotes and contain a complete set of intracellular organelles. Yeast genetics are well understood and molecular biology techniques can be easily employed to generate mutant strains. As already described above, eight complementation groups of yeast mutants have been isolated (31) and can be used to study Q function, regulation, and synthesis.

A.   In Vivo Studies on the Antioxidant Functions of Q

While much in vitro evidence suggests that QH_2 acts as an antioxidant, the most convincing demonstrations of the functional importance of antioxidant defense mechanisms derive from in vivo studies of mutants with defects in such systems. Hence, the possible in vivo function of QH_2 as an antioxidant was tested by subjecting Q-deficient mutant yeast (*coq*) to oxidative stress. The *coq* mutant strains with defects in Q synthesis display very specific phenotypes when challenged with oxidative stress (16). For example, the Q-deficient yeast are hypersensitive to the autoxidation products of PUFAs. Treatment with mono-unsaturated fatty acids, which are much less prone to autoxidation, had no effect. As a result of PUFA exposure, there is a marked accumulation of hydroperoxide and aldehyde products in the *coq* mutant strain, and addition of antioxidants such as trolox or butylated hydroxytoluene rescue the hypersensitivity. These results strongly suggest that under these conditions PUFA autoxidation mediates the cell killing and that QH_2 plays an important role in vivo in protecting eukaryotic cells from these breakdown products (16). Null mutants in all the Q complementation groups (*coq1Δ–coq8Δ*) exhibit sensitivity to PUFA treatment (17).

Other studies with yeast mutants show similar results. A decaprenyl diphosphate synthase (*dps*) *Schizosaccharomyces pombe* yeast mutant was recently identified that is unable to make Q_{10} (132). These mutants do not grow on glycerol or minimal glucose media, but growth is restored by cysteine or glutathione supplementation. Dps is 33% identical to IspB (octaprenyl diphosphate synthase) in *E. coli* and 45% identical to the Coq1 (hexaprenyl diphosphate synthase) in *S. cerevisiae*. Interestingly, this mutant was sensitive to Cu^{2+} and H_2O_2 in the absence of Q, supporting the studies in *S. cerevisiae* that showed QH_2 is an effective antioxidant in vivo.

QH_2 is also important in vivo in scavenging ascorbate free radicals (AFRs), and is required for maintaining the integrity of the plasma membrane electron transport chain in yeast (23,133). *coq3Δ* and *coq7Δ* mutants show a 30% decrease in extracellular ascorbate stabilization when grown in 2% glucose compared to wild-type and the respiratory-defective mutant *atp2Δ*. However, all four strains scavenged AFRs to the same extent when grown in 10% glucose, most likely as a result of inhibited Q biosynthesis at high glucose concentrations (50). The NADH-AFR activity in the *coq3Δ* plasma membrane is 80% lower than the wild-type level, and the activity can be partially restored by exogenous Q_6. Addition of *COQ3* and *COQ7* on single-copy plasmids also restores ascorbate stabilization to near wild-type and *atp2Δ* levels, indicating that QH_2 contains important antioxidant properties in the plasma membrane (133). These experiments with Q-deficient yeast mutants demonstrate that QH_2 is an important cellular defense mechanism that prevents membrane peroxidation by scavenging lipid peroxyl, α-TO, and ascorbyl radicals.

B. Analysis of *COQ7/CAT5/CLK-1*: Functional Implications for Q in Aging and Development

Recent research in the areas of Q biosynthesis, glucose derepression, and aging has converged on the same gene identified as yeast *COQ7* (39), yeast *CAT5* (134), and *Caenorhabditis elegans CLK-1* (135), respectively. This section will summarize the approaches used in the independent isolation of *COQ7/CAT5/CLK-1*, review recent evidence that its primary function is in Q production, and discuss the implications that the different phenotypes associated with mutations in *C. elegans CLK-1* are related to changes in the production of Q.

Yeast *COQ7* was isolated as a gene required for the synthesis of Q (39; Fig. 2). A yeast mutant harboring the *coq7-1* allele (encoding the substitution of Asp for Gly_{104}) was found to accumulate both HHB and a small amount of 2-hexaprenyl-3-methyl-6-methoxy-1,4-benzoquinone (demethoxy-Q), two intermediates in Q biosynthesis (39). However, mutants with deletions in the *COQ7* gene produce only HHB (see Sec. II.A). Transformation of either the *coq7-1* point mutant or the *coq7* null mutant with the yeast *COQ7* gene restored both growth on nonfermentable carbon sources and the synthesis of Q. These results support a proposal that Coq7p provides a component of a multisubunit enzyme complex that is required for the conversion of HHB to Q (17, 39).

The yeast *COQ7* gene was independently isolated as *CAT5*, a gene required for the release of gluconeogenic genes from glucose repression (134). Glucose repression is a global regulatory system in *S. cerevisiae* that affects the transcription of genes involved in gluconeogenesis, alternative sugar metabolism, and respiration (136–138). Upon deletion of *CAT5*, binding of gene activators to the UAS (upstream activating sequence) elements within gluconeogenic promoters was abolished, resulting in a complete loss of gluconeogenic gene activation (134). These data provided support for a role of Cat5p in the cascade regulating gluconeogenic gene activation.

Recently, a homologue of *COQ7/CAT5* was identified as the *C. elegans CLK-1* gene (135). The nematode *C. elegans* has been used as a model system for research into the components responsible for controlling longevity, and several life-extension mutants have been identified (139). Mutations in the *C. elegans CLK-1* gene produce mutants that exhibit a pleiotropic phenotype, characterized by delayed embryonic and postembryonic development, a slowing of adult behaviors such as swimming, pharyngeal pumping, defecation, and an extended life span (140). The *clk-1* mutants also have an increased resistance to stress induced by UV treatment (141). The yeast, rat, and *C. elegans* homologues of *COQ7/CAT5/CLK-1* have a high degree of amino acid sequence identity (from 41–57% identical) (48,135). The homologues also share function—the *COQ7/CAT5/CLK-1* homologue from either rat or *C. elegans* rescued the yeast *coq7/cat5* mutant for growth on nonfermentable carbon sources, suggesting a conservation of function from yeast to animals (48,135).

The apparent dual function of Coq7p/Cat5p in yeast Q biosynthesis and glucose derepression raised the question of whether the observed defect in glucose derepression resulted from a defect in Q biosynthesis or vice versa. Recent work provides evidence for the former (142). Glucose derepression is absent in a broad collection of respiratory-deficient yeast mutants, including those defective in Q production, indicating that such regulation requires an intact mitochondrial respiratory metabolism. Moreover, the growth defect of a *coq7/cat5* mutant under nonfermentable conditions can be restored by external feeding with Q_6. Such Q_6 supplementation also restores the ability to activate gluconeogenic and respiratory enzymes during the transition from glucose to ethanol metabolism (142). The Q_6 rescue data clearly restrict the Coq7p/Cat5p function to Q biosynthesis. Subcellular fractionation studies localize the Coq7 polypeptide to the mitochondrial inner membrane, a site consistent with its role in Q biosynthesis.

The results obtained in the yeast model suggest that the phenotype of delayed development and increased life span in *C. elegans clk-1* mutants may relate to changes in the amount of Q. Three different mutant alleles of *C. elegans clk-1* were isolated: (1) a point mutation resulting in the substitution of a conserved amino acid (E_{148} to K); (2) a 35 aa C-terminal deletion (the wild-type *CLK-1* gene encodes a polypeptide of 187 amino acids); and (3) a large truncation resulting in the loss of 93 aa residues (the C-terminal half of the polypeptide) (135). Despite the drastic nature of the latter mutation, it seems unlikely that this mutation results in a complete loss of function and a lack of Q. This speculation is based on the intolerance of wild-type *C. elegans* to anaerobic incubation conditions (143) and the report that yeast mutants lacking respiration have shorter life spans than wild-type yeast (144). Instead it seems more likely that the delayed development and increased life span might result from changes in either the relative amount or location of Q. The systems potentially impacted by such changes include the diverse cellular functions addressed in section III. However, the influence of Q levels on respiration in *C. elegans*, or its action as a potential pro- or antioxidant, remain to be determined. In addition, changes in Q levels or location would be predicted to affect central metabolic pathways in which it is required, such as the citric acid cycle, fatty acid beta-oxidation or de novo pyrimidine synthesis. Finally, the involvement of Q in the reduction of α-TO· and ascorbyl radical may play a crucial role in slowing lipid peroxidation and cell death through apoptosis (145). It seems likely that the yeast model system will provide an important avenue of investigation in exploring these questions.

REFERENCES

1. Olson RE, Rudney H. Biosynthesis of ubiquinone. Vitam Horm 1983; 40:1–43.
2. Crane FL. Distribution of ubiquinones. In: Morton RA, ed. Biochemistry of Quinones. London: Academic Press, 1965: 183–204.
3. Trumpower BL. New concepts on the role of ubiquinone in the mitochondrial respiratory chain. J Bioenerg Biomembr 1981; 13:1–24.
4. Samorì B, Lenaz G, Battino M, Marconi G, Domini I. On coenzyme Q orientation in membranes: a linear dichroism study of ubiquinones in a model bilayer. J Membr Biol 1992; 128:193–203.
5. Metz G, Howard KP, van Liemt WBS, et al. NMR studies of ubiquinone location in oriented model membranes—evidence for a single motionally-averaged population. J Am Chem Soc 1995; 117:564–565.
6. Kagan VE, Nohl H, Quinn PJ. Coenzyme Q: its role in scavenging and generation of radicals in membranes. In: Cadenas E, Packer L, eds. Handbook of Antioxidants. New York: Marcel Dekker, 1996: 157–201.
7. Lenaz G, Esponsti MD, Bertoli E, et al. Studies on the interactions and mobility of ubiquinone in mitochondrial and model membranes. In: Trumpower BL, ed. Function of Quinones in Energy Conserving Systems. New York: Academic Press, 1982: 111–124.
8. Degli Esposti M, Ngo A, McMullen GL, et al. The specificity of mitochondrial complex I for ubiquinones. Biochem J 1996; 313:327–334.
9. Helfenbaum L, Ngo A, Ghelli A, Linnane AW, Degli Esposti M. Proton pumping of mitochondrial complex I: differential activation by analogs of ubiquinone. J Bioenerg Biomembr 1997; 29:71–80.
10. Chance B, Sies H, Boveris A. Hydroperoxide metabolism in mammalian organs. Physiol Rev 1979; 59:527–605.
11. Nohl H, Gille L, Schonheit K, Liu Y. Conditions allowing redox-cycling ubisemiquinone in mitochondria to establish a direct redox couple with molecular oxygen. Free Radic Biol Med 1996; 20:207–213.
12. Beyer RE. An analysis of the role of coenzyme Q in free radical generation and as an antioxidant. Biochem Cell Biol 1992; 70:390–403.
13. Ernster L, Dallner G. Biochemical, physiological and medical aspects of ubiquinone function. Biochim Biophys Acta 1995; 1271:195–204.
14. Littarru GP, Battino M, Folkers K. Clinical aspects of coenzyme Q: improvement of cellular bioenergetics or antioxidant protection? In: Cadenas E, Packer L, eds. Handbook of Antioxidants. New York: Marcel Dekker, 1996: 203–239.
15. Ernster L, Forsmark-Andrée P. Ubiquinol: an endogenous antioxidant in aerobic organisms. Clin Invest 1993; 71:S60–S65.
16. Do TQ, Schultz JR, Clarke CF. Enhanced sensitivity of ubiquinone deficient mutants of *Saccharomyces cerevisiae* to products of autooxidized polyunsaturated fatty acids. Proc Natl Acad Sci USA 1996; 93:7534–7539.
17. Poon WW, Do TQ, Marbois BN, Clarke CF. Sensitivity to treatment with polyunsaturated fatty acids is a general characteristic of the ubiquinone-deficient yeast *coq* mutants. Mol Aspects Med 1997; 18:s121–s127.

18. Aberg F, Appelkvist E-L, Dallner G, Ernster L. Distribution and redox state of ubiquinones in rat and human tissues. Arch Biochem Biophys 1992; 295:230–234.

19. Stocker R, Bowry VW, Frei B. Ubiquinol-10 protects human low density lipoprotein more efficiently against lipid peroxidation than does α-tocopherol. Proc Natl Acad Sci USA 1991; 88:1646–1650.

20. Schultz JR, Ellerby LM, Gralla EB, Valentine JS, Clarke CF. Autoxidation of ubiquinol-6 is independent of superoxide dismutase. Biochemistry 1996; 35:6595–6603.

21. Takada M, Ikenoya S, Yuzuriha T, Katayama K. Studies on reduced and oxidized coenzyme Q (ubiquinones). Biochim Biophys Acta 1982; 679:308–314.

22. Kalen A, Norling B, Appelkvist EL, Dallner G. Ubiquinone biosynthesis by the microsomal fraction from rat liver. Biochim Biophys Acta 1987; 926:70–78.

23. Santos-Ocaña C, Navas P, Crane FL, Cordoba F. Extracellular ascorbate stabilization as a result of transplasma electron transfer in *Saccharomyces cerevisiae*. J Bioenerg Biomembr 1995; 27:597–603.

24. Takahashi T, Okamoto T, Mori K, Sayo H, Kishi T. Distribution of ubiquinone and ubiquinol homologues in rat tissues and subcellular fractions. Lipids 1993; 28:803–809.

25. Crane FL, Morre DJ. Evidence for coenzyme Q function in Golgi membranes. In: Folkers K, Yamamura Y, eds. Biomedical and Clinical Aspects of Coenzyme Q. Amsterdam: Elsevier, 1977: 3–14.

26. Villalba JM, Navarro F, Cordoba F, et al. Coenzyme Q reductase from liver plasma membrane: purification and role in trans-plasma electron transport. Proc Natl Acad Sci USA 1995; 92:4887–4891.

27. Sun IL, Sun EE, Crane FL, et al. Requirement for coenzyme Q in plasma membrane electron transport. Proc Natl Acad Sci USA 1992; 89:11126–11130.

28. Gomez-Diaz C, Rodriguez-Aguilera JC, Barroso MP, et al. Antioxidant ascorbate is stabilized by NADH-coenzyme Q_{10} reductase in the plasma membrane. J Bioenerg Biomembr 1997; 29:251–257.

29. Gibson F. Chemical and genetic studies on the biosynthesis of ubiquinone by *Escherichia coli*. Biochem Soc Trans 1973; 1:317–326.

30. Meganathan R. Biosynthesis of the isoprenoid quinones menaquinone (vitamin K_2) and ubiquinone (coenzyme Q) In: Neidhardt FC, ed. *Escherichia coli* and *Salmonella typhimurium*—Cellular and Molecular Biology. Washington, DC: ASM Press, 1996:642–656.

31. Tzagoloff A, Dieckmann CL. *PET* genes of S. *cerevisiae*. Microbiol Rev 1990; 54:211–225.

32. Tzagoloff A, Akai A, Needleman RB. Assembly of the mitochondrial membrane system: characterization of nuclear mutants of *Saccharomyces cerevisiae* with defects in mitochondrial ATPase and respiratory enzymes. J Biol Chem 1975; 250:8228–8235.

33. Tzagoloff A, Akai A, Needleman RB. Assembly of the mitochondrial membrane system: isolation of nuclear and cytoplasmic mutants of *Saccharomyces cerevisiae* with specific defects in mitochondrial functions. J Bacteriol 1975; 122:826–831.

34. Ashby MN, Edwards PA. Elucidation of the deficiency in two yeast coenzyme Q

mutants: characterization of the structural gene encoding hexaprenyl pyrophosphate synthetase. J Biol Chem 1990; 265:13157–13164.

35. Ashby MN, Kutsunai SY, Ackerman E, Tzagoloff A, Edwards PA. *COQ2* is a candidate for the structural gene encoding *para*-hydroxybenzoate polyprenyltransferase. J Biol Chem 1992; 267:4128–4136.

36. Clarke CF, Williams W, Teruya JH. Ubiquinone biosynthesis in *Saccharomyces cerevisiae*: isolation and sequence of *COQ3*, the 3,4-dihydroxy-5-hexaprenylbenzoate methyltransferase gene. J Biol Chem 1991; 266:16636–16644.

37. Barkovich RJ, Shtanke A, Shepherd JA, et al. Characterization of the *COQ5* gene from *Saccharomyces cerevisiae*: evidence for a C-methyltransferase in ubiquinone biosynthesis. J Biol Chem 1997; 272:9182–9188.

38. Dibrov E, Robinson KM, and Lemire BD. The *COQ5* gene encodes a yeast mitochondrial protein necessary for ubiquinone biosynthesis and the assembly of the respiratory chain. J Biol Chem 1997; 272:9175–9181.

39. Marbois BN, Clarke CF. The *COQ7* gene encodes a protein in *Saccharomyces cerevisiae* necessary for ubiquinone biosynthesis. J Biol Chem 1996; 271:2995–3004.

40. Lee PT, Hsu AY, Clarke CF. Unpublished work.

41. Hsu AY, Poon WW, Shepherd JA, Myles DC, Clarke CF. Complementation of *coq3* mutant yeast by mitochodrial targeting of the *E. coli* UbiG polypeptide: evidence that UbiG catalyzes both *O*-methylation steps in ubiquinone biosynthesis. Biochemistry 1996; 35:9797–9806.

42. Cox GB, Young IG, McCann LM, Gibson F. Biosynthesis of ubiquinone in *Escherichia coli* K-12: location of genes affecting the metabolism of 3-octaprenyl-4-hydroxybenzoic acid and 2-octaprenylphenol. J Bacteriol 1969; 99:450–458.

43. Goewert RR, Sippel CJ, Olson RE. Identification of 3,4-dihydroxy-5-hexaprenyl-benzoic acid as an intermediate in the biosynthesis of ubiquinone-6 by *Saccharomyces cerevisiae*. Biochemistry 1981; 20:4217–4223.

44. Goewert RR, Sippel CJ, Grimm MF, Olson RE. Identification of 3 methoxy-4-hydroxy-5-hexaprenylbenzoic acid as a new intermediate in ubiquinone biosynthesis by *Saccharomyces cerevisiae*. Biochemistry 1981; 20:5611–5616.

45. Poon WW, Marbois BN, Faull KF, Clarke CF. 3-hexaprenyl-4-hydroxybenzoic acid forms a predominant intermediate pool in ubiquinone biosynthesis in *Saccharomyces cerevisiae*. Arch Biochem Biophys 1995; 320:305–314.

46. Gibson F, Young IG. Isolation and characterization of intermediates in ubiquinone biosynthesis. Methods Enzymol 1978; 53:600–609.

47. Marbois BN, Hsu A, Pillai R, Colicelli J, Clarke CF. Cloning of a rat cDNA encoding dihydroxypolyprenylbenzoate methyltransferase by functional complementation of a *Saccharomyces cerevisiae* mutant deficient in ubiquinone biosynthesis. Gene 1994; 138:213–217.

48. Jonassen T, Marbois BN, Kim L, et al. Isolation and sequencing of the rat *COQ7* gene and mapping of mouse *COQ7* to chromosome 7. Arch Biochem Biophys 1996; 330:285–289.

49. Sippel CJ, Goewert RR, Slackman FN, Olson RE. The regulation of ubiquinone-6 biosynthesis by *Saccharomyces cerevisiae*. J Biol Chem 1983; 258:1057–1061.

50. Gordon PA, Stewart PR. Ubiquinone formation in wild-type and petite yeast: the effect of catabolite repression. Biochim Biophys Acta 1969; 177:358–360.

51. Lester RI, Crane FL. The natural occurrence of coenzyme Q and related compounds. J Biol Chem 1959; 234:2169–2175.

52. Yamamoto T, Shimizu S-I, Sugawara H, Momose K, Rudney H. Identification of regulatory sites in the biosynthesis of ubiquinone in the perfused rat heart. Arch Biochem Biophys 1989; 269:86–92.

53. Sugawara H, Yamamoto T, Shimizu S-I, Momose K. Inhibition of ubiquinone synthesis in isolated rat heart under an ischemic condition. Int J Biochem 1990; 22: 477–480.

54. Nambudiri AM, Ranganathan S, Rudney H. The role of HMG CoA reductase activity in the regulation of ubiquinone synthesis in human fibroblasts. J Biol Chem 1980; 2155:5894–5899.

55. Appelkvist EL, Aberg F, Guan Z, Parmryd I, Dallner G. Regulation of coenzyme Q biosynthesis. Mol Aspect Med 1994; 15:s37–s46.

56. Parmryd I, Dallner G. Organization of isoprenoid biosynthesis. Biochem Soc Trans 1996; 24:677–682.

57. Thelin A, Peterson E, Houston JL, et al. Effect of squalestatin 1 on the biosynthesis of the mevalonate pathway lipids. Biochim Biophys Acta 1994; 1215:245–249.

58. Keller RK. Squalene synthase inhibition alters metabolism of nonsterols in rat liver. Biochim Biophys Acta 1996; 1303:169–179.

59. Maltese WA, Aprille JR, Green RA. Activity of 3-hydroxy-3-methylglutaryl-coenzyme A reductase does not respond to ubiquinone uptake in cultured cells. Biochem J 1987; 246:441–447.

60. Ranganathan S, Ramasarma T. The regulation of the biosynthesis of ubiquinone in the rat. Biochem J 1975; 148:35–39.

61. Elmberger PG, Kalen A, Appelkvist E-L, Dallner G. *In vitro* and *in vivo* synthesis of dolichol and other main mevalonate products in various organs of the rat. Eur J Biochem 1987; 168:1–11.

62. Lin L, Sotonyi P, Somogyi E, et al. Co Q_{10} contents in different parts of the normal human heart. Clin Physiol 1988; 8:391–398.

63. Chance B, Hollunger G. The interaction of energy and electron transfer reactions in mitochondria. I. General properties and nature of the products of succinate-linked reduction of pyridine nucleotide. J Biol Chem 1961; 236:1534–1543.

64. Ernster L. Reaction pathways of succinate-linked acetoacetate reduction in tissue homogenates and isolated mitochondria. Nature 1962; 193:1050–1052.

65. Löw H, Vallin I. Succinate-linked diphosphopyridine nucleotide reduction in submitochondrial particles. Biochim Biophys Acta 1963;69:361–374.

66. Rauchová H, Battino M, Fato R, Lenaz G, Drahota Z. Coenzyme Q-pool function in glycerol-3-phosphate oxidation in hamster brown adipose tissue mitochondria. J Bioenerg Biomembr 1992; 24:235–241.

67. Ernster L. Facts and ideas about the functions of coenzyme Q in mitochondria. In: Folkers K, Yamamura Y, eds. Biomedical and Clinical Aspects of Coenzyme Q. Amsterdam: Elsevier, 1977:15–28.

68. Lenaz G, Fato R, Castelluccio C, et al. An updating of the biochemical function of coenzyme Q in mitochondria. Mol Aspects Med 1994; 15:s29–s36.

69. Ernster L, Lell IY, Norling B, Persson B. Studies with ubiquinone depleted submitochondrial particles. FEBS Lett 1969; 3:21–26.
70. Kröger A, Klingenberg M. Quinones and nicotinamide nucleotides associated with electron transfer. Vitam Horm 1970; 28:533–574.
71. Kröger A, Klingenberg M. The kinetics of the redox reactions of ubiquinone related to the electron-transport activity in the respiratory chain. Eur J Biochem 1973; 34: 358–368.
72. Lenaz G, Fato R, Castelluccio C, et al. The function of coenzyme Q in mitochondria. Clin Investig 1993; 71:S66–S70.
73. Estornell E, Fato R, Castelluccio C, et al. Saturation kinetics of coenzyme Q in NADH and succinate oxidation in beef heart mitochondria. FEBS Lett 1992; 311: 107–109.
74. Mitchell P, Moyle J. Protonmotive mechanisms of quinone function. In: Trumpower BL, ed. Functions of Quinones in Energy Conserving Systems. New York: Academic Press, 1982:553–575.
75. Brandt U, Trumpower B. The protonmotive Q cycle in mitochondria and bacteria. Crit Rev Biochem Mol Biol 1994; 29:165–197.
76. Xia D, Yu C-A, Kim H, et al. Crystal structure of the cytochrome bc1 complex from bovine heart mitochondria. Science 1997; 277:60–66.
77. De Vries S, Albracht SPJ, Berden JA, Slater EC. A new species of bound ubisemiquinone anion in QH2:cytochrome c oxidoreductase. J Biol Chem 1981; 256: 11996–11998.
78. Brandt U. Energy conservation by bifurcated electron-transfer in the cytochrome bc_1 complex. Biochim Biophys Acta 1996; 1275:41–46.
79. Brandt U, Okun JG. Role of deprotonation events in ubihydroquinone: cytochrome c oxidoreductase from bovine heart and yeast mitochondria. Biochemistry 1997; 36:11234–11240.
80. Beyer RE, Nordenbrand K, Ernster L. The function of coenzyme Q in free radical production and as an antioxidant: a review. Chem Scripta 1987; 27:145–153.
81. Nohl H. Is redox-cycling ubiquinone involved in mitochondrial oxygen activation? Free Radic Res Commun 1990; 8:307–315.
82. Nohl H, Stolze K. Ubisemiquinones of the mitochondrial respiratory chain do not interact with molecular oxygen. Free Radic Res Commun 1992; 16:409–419.
83. King TE. Ubiquinone proteins. In: Lenaz G, ed. Coenzyme Q. New York: John Wiley, 1985:391–408.
84. Loschen G, Azzi A, Richter C, Flohe L. Superoxide radicals as precursors of mitochondrial hydrogen peroxide. FEBS Lett 1974; 42:68–72.
85. Boveris A, Cadenas E. Mitochondrial production of superoxide anions and its relationship to the antimycin insensitive respiration. FEBS Lett 1975; 54:311–314.
86. Boveris A, Cadenas E. Production of superoxide radicals and hydrogen peroxide in mitochondria. In: Oberley LW, ed. Superoxide Dismutases, vol. 2. Boca Raton, FL: CRC Press, 1982: 15–30.
87. Forman HJ, Boveris A. Superoxide radical and hydrogen peroxide in mitochondria. In: Pryor WA, ed. Free Radicals in Biology. New York: Academic Press, 1982: 65–90.
88. Turrens JF, Alexandre A, Lehninger AL. Ubisemiquinone is the electron donor for

superoxide formation by complex III of heart mitochondria. Arch Biochem Biophys 1985; 237:408–414.

89. Nohl H, Jordan W. The involvement of biological quinones in the formation of hydroxyl radicals via the Haber-Weiss reaction. Bioorg Chem 1987; 15:374–382.

90. Gonzalez-Flecha B, Boveris A. Mitochondrial sites of hydrogen peroxide production in reperfused rat kidney cortex. Biochim Biophys Acta 1995; 1243:361–366.

91. Ames BN, Shigenaga MK, Hagan TM. Mitochondrial decay in aging. Biochim Biophys Acta 1995; 1271:165–170.

92. Forman HJ, Azzi A. On the virtual existence of superoxide anions in mitochondria: thought regarding its role in pathophysiology. FASEB J 1997; 11:374–375.

93. Longo VD, Gralla EB, Valentine JS. Superoxide dismutase activity is essential for stationary phase survival in *Saccharomyces cerevisiae*. Mitochondrial production of toxic oxygen species *in vivo*. J Biol Chem 1996; 271:12275–12280.

94. Guidot DM, McCord JM, Wright RM, Repine JE. Absence of electron transport (Rho 0 state) restores growth of a manganese-superoxide dismutase-deficient *Saccharomyces cerevisiae* in hyperoxia. Evidence for electron transport as a major source of superoxide generation *in vivo*. J Biol Chem 1993; 268:26699–26703.

95. Halliwell B, Gutteridge JMC. Free Radicals in Biology and Medicine. New York: Oxford University Press.

96. Kappus H. A survey of chemicals inducing lipid peroxidation in biological systems. Chem Phys Lipids 1987; 45:105–115.

97. Sies H, de Groot H. Role of reactive oxygen species in cell toxicity. Toxicol Lett 1992; 64–65:547–551.

98. Porter NA. Mechanisms for the autoxidation of polyunsaturated lipids. Accounts Chem Res 1986; 19:262–268.

99. Esterbauer H, Schaur RJ, Zollner H. Chemistry and biochemistry of 4-hydroxynonenal, malonaldehyde and related aldehydes. Free Radic Biol Med 1991; 11:81–128.

100. Forsmark P, Aberg F, Norling B, et al. Inhibition of lipid peroxidation by ubiquinol in submitochondrial particles in the absence of vitamin E. FEBS Lett 1991; 285:39–43.

101. Matsura T, Yamada K, Kawasaki T. Antioxidant role of cellular reduced coenzyme Q homologs and alpha-tocopherol in free radical-induced injury of hepatocytes isolated from rats fed diets with different vitamin E contents. Biochim Biophys Acta 1992; 1127:277–283.

102. Kagan VE, Serbinova EA, Koynova EA, Antioxidant action of ubiquinol homologues with different isoprenoid chain length in biomembranes. Free Radic Biol Med 1990; 9:117–126.

103. Mukai K, Kikuchi S, Urano S. Stopped-flow kinetic study of the regeneration reaction of tocopheroxyl radical by reduced ubiquinone-10 in solution. Biochim Biophys Acta 1990; 1035:77–82.

104. Stoyanovsky DA, Osipov AN, Quinn PJ, Kagan VE. Ubiquinone-dependent recycling of vitamin E radicals by superoxide. Arch Biochem Biophys 1995; 323:343–351.

105. Kagan VE, Serbinova E, Packer L. Antioxidant effects of ubiquinones in micro-

somes and mitochondria are mediated by tocopherol recycling. Biochem Biophys Res Commun 1990; 169:851–857.

106. Lang J, Gohil K, Packer L. Simultaneous determination of tocopherols, ubiquinols, and ubiquinones in blood, plasma, tissue homogenates, and subcellular fractions. Anal Biochem 1986; 157:106–116.

107. Forsmark-Andrée P, Dallner G, Ernster L. Endogenous ubiquinol prevents protein modification accompanying lipid peroxidation in beef heart submitochondrial particles. Free Radic Biol Med 1995; 19:749–757.

108. Forsmark-Andrée P, Lee C-P, Dallner G, Ernster L. Lipid peroxidation and changes in the ubiquinone content and the respiratory chain enzymes of submitochondrial particles. Free Radic Biol Med 1997; 22:391–400.

109. Landi L, Fiorentini D, Stefanelli C, Pasquali P, Pedulli GF. Inhibition of autoxidation of egg yolk phosphatidylcholine in homogeneous solution and in liposomes by oxidized ubiquinone. Biochim Biophys Acta 1990; 1028:223–228.

110. Beyer RE. The role of ascorbate in antioxidant protection of biomembranes: interaction with vitamin E and coenzyme Q. J Bioenerg Biomembr 1994; 26:349–358.

111. Fiorentini D, Cabrini L, Landi L. Ubiquinol-3 and ubiquinol-7 exhibit similar antioxidant activity in model membranes. Free Radic Res Commun 1993; 18:201–209.

112. Frerman FE. Acyl-CoA dehydrogenases, electron transfer flavoprotein and electron transfer flavoprotein dehydrogenase. Biochem Soc Trans 1988; 16:416–418.

113. Frei B, Winterhalter KH, Richter C. Menadione- (2-methyl-1,4-naphthoquinone-) dependent enzymatic redox cycling and calcium release by mitochondria. Biochemistry 1986; 25:4438–4443.

114. Rasmussen UF, Rasmussen HN. The NADH oxidase system (external) of muscle mitochondria and its role in the oxidation of cytoplasmic NADH. Biochem J 1985; 229:631–641.

115. Nohl H. Demonstration of the existence of an organo-specific NADH dehydrogenase in heart mitochondria. Eur J Biochem 1987; 169:585–591.

116. Nagy M, Lacroute F, Thomas D. Divergent evolution of pyrimidine biosynthesis between anaerobic and aerobic yeasts. Proc Natl Acad Sci USA 1992; 89:8966–8970.

117. Jones ME. Pyrimidine nucleotide biosynthesis in animals: genes, enzymes, and regulation of UMP biosynthesis. Annu Rev Biochem 1980; 49:253–279.

118. Loffler M, Jockel J, Schuster G, Becker C. Dihydroorotate-ubiquinone oxidoreductase links mitochondria in the biosynthesis of pyrimidine nucleotides. Mol Cell Biochem 1997; 174:125–129.

119. Stocker R, Suarna C. Extracellular reduction of ubiquinone-1 and -10 by human Hep G2 and blood cells. Biochim Biophys Acta 1993; 1158:15–22.

120. Houser RM, Olson RE. 5-demethylubiquinone-9-methyltransferase from rat liver mitochondria: characterization, localization, and solubilization. J Biol Chem 1977; 252:4017–4021.

121. Tappel AL. Vitamin E and free radical peroxidation of lipids. Ann NY Acad Sci 1972; 203:12–21.

122. Villalba JM, Navarro F, Cordoba F, et al. Coenzyme Q reductase from liver plasma membrane:purification and role in trans-plasma electron transport. Proc Natl Acad Sci USA 1995; 92:4887–4891.

123.   Navarro F, Villalba JM, Crane FL, Mackellar WC, Navas P. A phospholipid-depen-
        dent NADH-coenzyme Q reductase from liver plasma membrane. Biochem Bio-
        phys Res Commun 1995; 212:138–143.
124.   Takahashi T, Yamaguchi T, Shitashige M, Okamoto T, Kishi T. Reduction of ubi-
        quinone in membrane lipids by rat liver cytosol and its involvement in the cellular
        defence system against lipid peroxidation. Biochem J 1995; 309:883–890.
125.   Crane FL, Sun IL, Barr R, Morre DJ. Coenzyme Q in Golgi apparatus membrane
        redox activity and protein uptake. In: Folkers K, Yamamura Y, ed. Biomedical and
        Clinical Aspects of Coenzyme Q. Amsterdam: Elsevier Science, 1984:77–86.
126.   Landi L, Fiorentini D, Galli MC, Segura-Aguilar J, Beyer RE. DT-diaphorase main-
        tains the reduced state of ubiquinones in lipid vesicles thereby promoting their
        antioxidant function. Free Radic Biol Med 1997; 22:329–335.
127.   Beyer RE, Segura-Aguilar J, Di Bernardo S, et al. The role of DT-diaphorase in
        the maintenance of the reduced antioxidant form of coenzyme Q in membrane sys-
        tems. Proc Natl Acad Sci USA 1996; 93:2528–2532.
128.   Cadenas E. Antioxidant and prooxidant functions of DT-diaphorase in quinone me-
        tabolism. Biochem Pharmacol 1995; 49:127–140.
129.   Cadenas E, Hochstein P, Ernster L. Pro- and antioxidant functions of quinones and
        quinone reductases in mammalian cells. Adv Enzymol Relat Areas Mol Biol 1992;
        65:97–146.
130.   Lind C, Hochstein P, Ernster L. DT-diaphorase as a quinone reductase: a cellular
        control device against semiquinone and superoxide radical formation. Arch Bio-
        chem Biophys 1982; 216:178–185.
131.   Takahasi T, Okamoto T, Kishi T. Characterization of NADPH-dependent ubiqui-
        none reductase activity in rat liver cytosol: effect of various factors on ubiquinone-
        reducing activity and discrimination from other quinone reductases. J Biochem
        1996; 119:256–263.
132.   Suzuki K, Okada K, Kamiya Y, et al. Analysis of the decaprenyl diphosphate syn-
        thase (*dps*) gene in fission yeast suggests a role of ubiquinone as an antioxidant.
        J Biochem 1997; 121:496–505.
133.   Santos-Ocana C, Cordoba F, Crane FL, Clarke CF, Navas P. Coenzyme Q_6 and
        iron reduction are responsible for the extracellular ascorbate stabilization at the
        plasma membrane of *Saccharomyces cerevisiae*. J Biol Chem 1998; 273:8099–
        8105.
134.   Proft M, Kotter P, Hedges D, Bojunga N, Entian K-D. *CAT5*, a new gene necessary
        for derepression of gluconeogenic enzymes in *Saccharomyces cerevisiae*. EMBO
        J 1995; 14:6116–6126.
135.   Ewbank JJ, Barnes TM, Lakowski B, et al. Structural and functional conservation
        of the *Caenorhabditis elegans* timing gene *clk-1*. Science 1997; 275:980–983.
136.   Johnston M, Carlson M. Regulation of carbon and phosphate utilization. In: Jones
        EW, Pringle JR, Broach JR, eds. The molecular and cellular biology of the yeast
        Saccharomyces cerevisiae Plainview, NY: Cold Spring Harbor Press, 1992: 193–
        281.
137.   Entian K-D, Barnett JA. Regulation of sugar utilization by *Saccharomyces cerevis-
        iae*. Trends Biochem Sci 1992; 17:506–510.
138.   Ronne H. Glucose repression in fungi. Trends Genet 1995; 11:12–17.

139. Lithgow GJ. Invertebrate gerontology: the age mutations of *Caenorhabditis elegans*. Bioessays 1996; 18:809–815.
140. Wong A, Boutis P, Hekimi S. Mutations in the *clk-1* gene of *Caenorhabditis elegans* affect developmental and behavioral timing. Genetics 1995; 139:1247–1259.
141. Murakami S, Johnson TE. A genetic pathway conferring life extension and resistance to UV stress in *Caenorhabditis elegans*. Genetics 1996; 143:1207–1218.
142. Jonassen T, Proft M, Randez-Gil F, et al. Yeast Clk-1 homologue (Coq7/Cat5): a mitochondrial protein in coenzyme Q synthesis. J Biol Chem 1998; 273:3351–3357.
143. Cooper AF, Van Gundy SD. Metabolism of glycogen and neutral lipids by *Aphelenchus avenae* and *Caenorhabditis* sp. in aerobic, microaerobic, and anaerobic environments. J Nematol 2:305–315.
144. Austriaco NR. To bud until death: the genetics of ageing in the yeast *Saccharomyces*. Yeast 1996; 12:623–630.
145. Barroso MP, Gomez-Diaz C, Villalba JM, et al. Plasma membrane ubiquinone controls ceramide production and prevents cell death induced by serum withdrawal. J Bioenerg Biomembr 1997; 29:259–267.

7

Mitochondrial Generation of Reactive Oxygen Species and Oxidative Damage During Aging: Roles of Coenzyme Q and Tocopherol

Rajindar S. Sohal, Achim Lass, Liang-Jun Yan, and Linda K. Kwong
Southern Methodist University, Dallas, Texas

The life cycle of virtually all multicellular animals follows the common pattern where the initial stages, dominated by growth and maturation, culminate in the reproductive phase, which is followed by an era of gradual, progressive, and irreversible decline in the efficiency of homeostatic mechanisms, eventually resulting in death. The purpose of current gerontological research is not only to understand the nature of the deleterious senescent changes but, more importantly, to determine what causes or initiates the alterations. An increasing body of evidence suggests that mitochondria play a dual role in the aging process stemming from being the generators of both reactive oxygen species (ROS) and ATP. It is hypothesized that mitochondria, the main intracellular sites of oxygen consumption and generation of ROS, are also the main targets of ROS-inflicted damage, resulting in a progressively decreasing ability to synthesize ATP, which consequentially lowers the homeostatic ability to adapt to the destabilizing effects of external and/or internal stresses. Several lines of evidence collectively provide credence to this hypothesis. This chapter focuses on the effect of aging on mitochondrial ROS generation and protein oxidative damage, and the roles of mitochondrial coenzyme Q and vitamin E in protection against ROS.

I. MITOCHONDRIAL GENERATION OF ROS DURING AGING

Oxidative phosphorylation in mitochondria accounts for approximately 90% of the intracellular oxygen consumed (1), of which approximately 2–3% is estimated to be chemically converted to $O_2{}^{\cdot-}$ as a result of autoxidation of certain components of the respiratory chain. The stoichiometric product of $O_2{}^{\cdot-}$ dismutation, H_2O_2, if not eliminated, is freely permeable through the mitochondrial and other cellular membranes. In vitro rate of mitochondrial $O_2{}^{\cdot-}/H_2O_2$ generation is highest during state 4 respiration (high substrate concentration and no ADP), when the respiratory chain is fully reduced, and lowest during state 3 (with substrate and ADP) (2). Two main sites of $O_2{}^{\cdot-}/H_2O_2$ generation have been identified—ubiquinone at complex III (3) and NADH dehydrogenase at complex I (4). There is increasing evidence suggesting the presence of a third site, α-glycerophosphate dehydrogenase (5–8). The relative contribution from each of the sites to the total amount of ROS generation varies among different tissues and species. It should, however, be borne in mind that in vitro estimation of the mitochondrial rate of ROS generation depends on the substrate and sometimes the respiratory inhibitor (Fig. 1) (8). In housefly flight muscle mitochondria, little H_2O_2 was generated with the NADH-linked substrates pyruvate, malate, and proline with or without the respiratory inhibitor rotenone (7,9). Maximum rate was observed with α-glycerophosphate as a substrate in the presence of rotenone and antimycin.

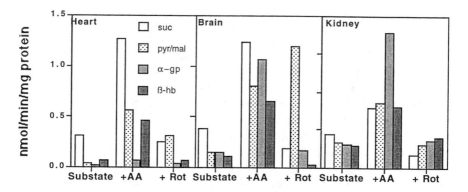

Figure 1 Substrate and respiratory inhibitor specificity of H_2O_2 generation in mouse mitochondria. Mitochondria were isolated from heart, brain, and kidney of 13–15-month-old C57B1/6NNia mice. The rate of H_2O_2 generation was measured as described in Kwong and Sohal (8). Abbreviations used: AA, antimycin; rot, rotenone; suc, succinate; pyr/mal, pyruvate/malate; α-gp, α-glycerophosphate; β-hb, β-hydroxybutyrate. Representative values are shown. (Adapted from Ref. 8.)

Similar findings were reported in granary weevil mitochondria (5), suggesting that in insects NADH dehydrogenase is not a notable site of ROS generation. Thus the major site of ROS generation in insects is at complex III. In the bovine heart submitochondrial particles, using succinate as the substrate, complex III also appeared to be the major site of ROS generation (3). However, results from studies of rat heart (10) and brain (11) mitochondria suggested that more H_2O_2 is generated at the complex I site than at the complex III site via reverse electron flow. Recent results from our laboratory showed that in mitochondria from mouse heart, brain, and kidney, quantitatively similar rates of H_2O_2 generation occur at both sites with succinate as a substrate (Fig. 2) (8). Although the in vitro sites and rates of mitochondrial ROS generation may be biased by the experimental conditions, evidence strongly supports the existence of in vivo mitochondrial ROS generation (12,13).

It has been amply demonstrated that various ROS can cause specific types of damage to different macromolecules resulting in the loss of enzymic activity, mutations, and membrane damage, among others. The oxidative stress hypothesis of aging postulates that cells have an innate imbalance between prooxidants and antioxidants resulting in a state of chronic oxidative stress, manifested as oxidative macromolecular damage. While no clear age-related changes in the level of

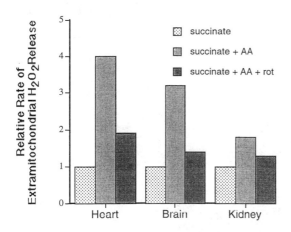

Figure 2   Effect of sequential addition of respiratory inhibitors on the rate of H_2O_2 generation in mouse mitochondria. In the presence of antimycin (AA) using succinate as the substrate, the maximum rate of H_2O_2 generated at both complex I and complex III sites was measured. After the sequential addition of rotenone (rot), which inhibited the reverse flow of electron to the complex I site, the measured rate of H_2O_2 generation was presumed to be mainly from the ubiquinone site. Representative values are shown. (Adapted from Ref. 8.)

enzymatic antioxidative defenses are evident (14,15), an increase in the rate of ROS generation has been observed in a variety of species and organs. In housefly flight muscle mitochondria, the rate of H_2O_2 generation increased approximately twofold with aging (7,16). Similar age-related increases in mitochondrial H_2O_2 generation were noted in *Drosophila melanogaster* (17). In mitochondria from mouse heart, brain, and kidney, the rate of O_2^-/H_2O_2 generation also increased with age (Fig. 3) (15). Of the organs studied, the heart had the highest increase in the rate of O_2^- and H_2O_2 generation. These observations have been confirmed in rat (18,19) and gerbil (20). Caloric restriction in mice (by 40%), which has been shown to increase life span by approximately 35–40%, attenuated the age-related increase in ROS generation (Fig. 4) (15,21). Thus the findings lend support to the concept that the level of oxidative stress/damage increases with age primarily because of an increase in the rate of mitochondrial O_2^-/H_2O_2 generation.

Not only is there an age-related increase in both the rate of ROS generation and the level of oxidative damage, the two phenomenon appear to be interrelated. In various oxidative stress models, the direction of the change in the level of oxidative damage corresponds to that of the rate of ROS generation (12,13). In the calorie restricted mice, both the level of oxidative damage to DNA and protein, as well as the rate of ROS generation (Fig. 4), were less than those in the ad libitum fed group (22). In the housefly, prior exposure to hyperoxia increased both the mitochondrial protein carbonyl content and H_2O_2 generation (16). Overexpres-

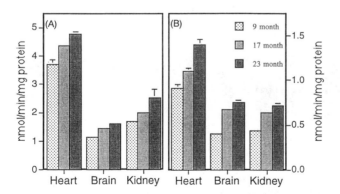

Figure 3 Effect of age on the rate of ROS generation in mitochondria from mouse heart, brain, and kidney. (**A**) Rate of O_2^- generation was measured in SMPs as SOD-inhibitable reduction of acetylated cytochrome-*c*. Values are mean ± SE of four to eight determinations. (**B**) Rate of H_2O_2 generation was measured in isolated mitochondria according to Sohal et al. (15). Values are mean ± SE of four to six determinations. (From Ref. 15.)

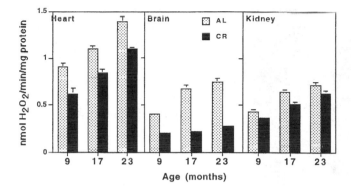

Figure 4   Effect of caloric restriction on the rate of H_2O_2 generation in mitochondria isolated from heart, brain, and kidney from 9-, 17-, and 23-month-old mice. Calorie restricted (CR) mice were fed 60% of the food intake of the ad libitum (AL) fed mice starting at 4 months of age. In all three age groups studied, the rates of H_2O_2 generation were lower in the mitochondria isolated from the CR mice when compared to those isolated from the AL mice. Values are mean ± SEM of four to six measurements. (From Ref. 15.)

sion of superoxide dismutase (SOD) and catalase resulted in an increase in life span as well as a slower rate of increase in 8-OHdG content in DNA and mitochondrial H_2O_2 production in *Drosophila* (17).

Experimental studies indicate that as mitochondria sustain oxidative damage, the rate of ROS generation increases. Exposure of isolated housefly mitochondria to ROS-generating systems resulted in an increase in H_2O_2 generation (7). The rate of ROS generation appeared to reflect the amount of oxidative damage to the mitochondria. The underlying cause of the increase may be related to oxidative damage to the inner mitochondrial membrane, since submitochondrial particles subjected to X-ray irradiation had elevated protein carbonyl content as well as H_2O_2 generation (Fig. 5) (16). Furthermore, treatment of housefly mitochondria with glutaraldehyde also resulted in an elevated rate of H_2O_2 generation (7). Thus ROS causes oxidative damage to mitochondria, which in turn results in an elevation in ROS generation. This positive feedback of oxidative damage and ROS generation may explain the observed exponential increase in oxidative stress during aging.

As discussed earlier, the mechanism underlying the observed increase in ROS generation may be caused by the loss of mitochondrial function in the inner membrane. This loss of function would quite likely impede the flow of electrons through the chain resulting in autoxidation of certain components of the chain (electron leakage). The activity of cytochrome-c oxidase (complex IV), which is

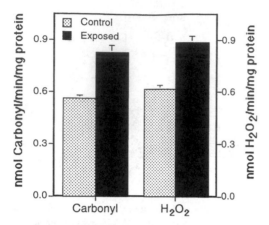

Figure 5 Effect of X rays on the concentration of protein carbonyls and the rate of H_2O_2 generation in submitochondrial particles. Submitochondrial particles isolated from 5-day-old houseflies were exposed to 4 kr of X rays. Values are mean $\pm$ SE of four to six determinations for H_2O_2 and four determinations for carbonyls. (From Ref. 16.)

responsible for the final transfer of four electrons to oxygen, forming water, decreases in insects and mammals with age (9,23). In human skeletal muscle, there was an age-linked decrease in mitochondrial respiration associated with a decrease in cytochrome-c oxidase activity (24). In addition, in rhesus monkey brain mitochondria, an age-related decrease in both complex I and complex IV activities was noted (25). However, in housefly mitochondria, age-related perturbation in the oxidoreductases included an increase in the activity of some of the oxidoreductases and a decrease in cytochrome-c oxidase (7,9). Nevertheless, an imbalance in activity levels of the electron transport components can result not only in decreased efficiency of the electron flow, but also in an increased amount of electron leakage resulting in a greater generation of ROS.

II. MITOCHONDRIAL PROTEIN OXIDATIVE DAMAGE DURING AGING

Oxidative damage to proteins during the aging process has been widely thought to be random. However, on the basis of our previous studies (26) and those by Levine and coworkers (27,28), we postulated that such damage may be selective rather than random. This hypothesis was tested in the flight muscle mitochondria of the housefly because their mitochondria produce a high rate of O_2^-/H_2O_2 generation and flight muscles undergo dysfunctional changes during senescence; flies lose flight ability toward the end of life (7,19,29). Furthermore, it is possible to identify flies of the same chronological age which differ in physiological age or life expectancy. In a cohort population, flies that only walk ("crawlers") and are unable to fly will die within a few days while those that can fly ("flyers") will live considerably longer (30,31).

Mitochondrial matrix proteins were reacted with the carbonyl reagent 2,4-dinitrophenylhydrazine and resolved by SDS-PAGE, and oxidatively modified proteins were detected with anti-2,4-dinitrophenyl antibodies (27,32). While a relatively large number of protein bands were visualized by Coomassie blue staining, carbonyl immunostain intensity was differential; only one major carbonyl-containing band was observed (Fig. 6A). The protein, exhibiting strong immunostain in Figure 6, had a molecular weight of approximately 84 kDa and a pI of 7.3. It exhibited a negative immunostain when the sample was not treated with DNPH and showed an increasing carbonyl content with advancing age (Fig. 6B). It was therefore purified and subjected to automated Edman sequencing which allowed its identification as mitochondrial aconitase (Fig. 6C).

The carbonyl content of mitochondrial aconitase was measured densitometrically (33) in 5-, 10-, and 15-day-old flies, representing young, middle-aged, and old animals (16,34). Aconitase carbonyl content in 15-day-old flies was 0.71 mol carbonyl/mol protein, approximately 50% higher than that of 5- or 10-day-old flies (0.46 and 0.49 mol/mol, respectively). The catalytic activity of mitochondrial aconitase was measured in flies of different ages using two different methods (35,36), which yielded the same result (Fig. 7A). Aconitase activity decreased 62% between 7 and 15 days of age, with a precipitous decline after 10 days of age. Thus the drop in functional activity matched the increase in oxidative modification of the protein. A comparison of carbonyl content and aconitase activity was also made between 15-day-old flies that had lost the ability to fly (crawlers) and their cohorts which retained the ability to fly (flyers). Aconitase carbonyl content was distinctly lower in flyers compared to crawlers (Fig. 7B) while enzyme activity was 70% higher in the flyers (Fig. 7C). Thus oxidative inactivation of aconitase was inversely related to the physiological age of the files.

To determine whether aconitase inactivation had any effect on housefly life span, two strategies, namely exposure to hyperoxia and administration of fluoroacetate, were used to manipulate the activity of aconitase during aging. In response to hyperoxic exposure, the intensity of immunostaining of aconitase carbonyls increased progressively (Fig. 8A), and aconitase activity was remarkably decreased, with less than 10% activity remaining after 72 h exposure (Fig. 8B). This was not a global effect, as citrate synthase activity was only slightly affected by hyperoxia, demonstrating that oxidative damage to mitochondrial matrix proteins is selective. When houseflies were fed with fluoroacetate, a specific aconitase inhibitor (37), in their drinking water, the life span was shortened from

(A)

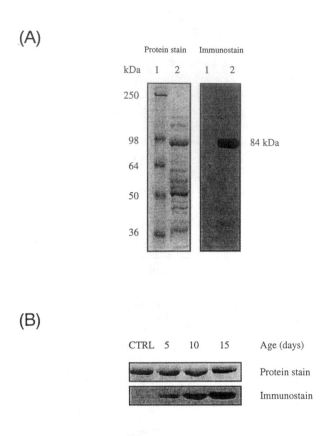

(B)

(C)

A **KVA**LSKFDNDV YLPYDKLAKTLDVVKDRL   The 84 kD protein from housefly mitochondria

SKVAISKFEPKS YLPYE KLSQTVKIVKDRL   *C. elegans* mitochondrial aconitase

AKVAMSHFEPHEYIR YDLLEKNIDIVRKRL   Bovine heart mitochondrial aconitase

Figure 6 Identification of housefly mitochondrial aconitase as a target of oxidative damage. (**A**) Immunochemical detection of protein carbonyls in mitochondria from flight muscles of houseflies (15 days old). DNPH-treated mitochondrial matrix proteins were electrophoresed on SDS-PAGE (8.5% resolving gel) and transferred to Immobilon-P membrane. Oxidized proteins were detected immunochemically. Lane 1 shows protein standard markers, and lane 2 contains mitochondrial matrix proteins. Protein was stained with Coomassie blue. An 84 kD protein in the matrix exhibited a strong immunoreaction for the carbonyl

47 days to 38, 30, and 17 days by 1, 50, and 100 µM fluoroacetate. Therefore, experimental inactivation of aconitase by either hyperoxia or fluoroacetate shortened life span, suggesting a causal relationship between aconitase activity and life span.

A considerable number of studies have been performed on the inactivation of aconitase by oxidation. It is believed that the particular susceptibility of mitochondrial aconitase to oxidative damage may be related to the presence of the iron-sulfur cluster [4Fe-4S] in its active site (38). In vitro studies have established that aconitase is particularly sensitive to reaction with O_2^- (39–43), which causes release of one iron atom from the cluster (44). The released iron may in turn initiate an amplification of protein oxidation through metal-dependent, site-specific oxidation (45).

The identification of a specific protein target of oxidative damage during aging indicates that oxidative damage can inactivate certain enzymes, contributing to the aging process. The identified protein can be clinically useful as a biomarker in assessing interventions designed to retard the aging process.

III. RELATIONSHIP BETWEEN MITOCHONDRIAL COENZYME Q AND VITAMIN E CONTENT, RADICAL GENERATION, AND AGING

As described earlier autoxidation of mitochondrial ubisemiquinone appears to be the predominant source of O_2^-/H_2O_2 generation in cells. Recent studies in this laboratory suggest that variations in ubiquinone or coenzyme Q (CoQ) and vitamin E affect the rate of O_2^-/H_2O_2 generation and oxidative damage to mitochondria (14,46,47).

A. Levels of Coenzyme Q and Vitamin E in Different Mammalian Species

Coenzyme Q plays a dual role in the mitochondrial respiratory chain: it transfers electrons to complex III and it translocates protons across the inner mitochondrial

---

groups. (**B**) Increase in carbonyl content of the 84 kD protein in housefly at different ages. The top panel shows the protein band stained with amido black, the bottom panel shows the immunostain. The control (CTRL) shown is 5-day-old fly mitochondrial matrix protein without DNPH treatment. Twenty-five microgram protein was applied onto SDS-PAGE in both A and B. (**C**) A computer-assisted search from the protein database identified the 84 kD protein as mitochondrial aconitase. The underlined amino acids show the N-terminal amino acid sequence homology in mitochondrial aconitase from the housefly and *C. elegans*, whereas the in-bold amino acids show the homology with bovine heart mitochondrial aconitase. (From Ref. 70.)

(A)

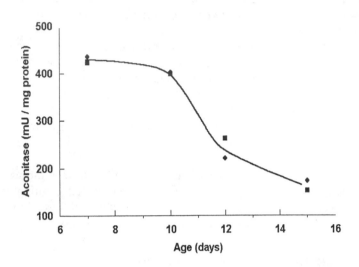

(B)

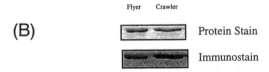

(C)

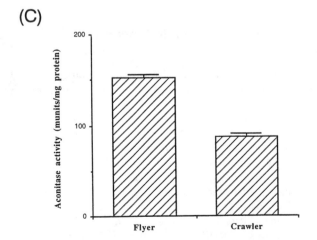

membrane. More recently, it has been suggested that CoQ may also act as a free radical scavenger, thereby preventing oxidative damage to mitochondrial membranes (48–52). It is intriguing that ubiquinones are also present in membranes of other cellular organelles where their role in electron transport has not been demonstrated. Like CoQ, α-tocopherol is also a lipoid membrane component. The latter has been unambiguously demonstrated to be a potent inhibitor of lipid peroxidation in cellular membranes (53–55) by undergoing one-electron oxidation to form the tocopheroxyl radical (54), which prevents the propagation of lipid peroxidation chain reaction.

Measurements of CoQ and α-tocopherol levels in cardiac mitochondria have revealed the existence of remarkable differences among species (Figs. 9 and 10). Although natural CoQ is a quinone derivative with a chain of 1–12 isoprene units (CoQ_n), mammals predominantly exhibit either CoQ_9 or CoQ_{10}. Similarly, α-tocopherol rather than γ-tocopherol is the predominant form. The CoQ_9 content in heart mitochondria of relatively short-lived mammals like the mouse or the rat is almost 10-fold greater than their CoQ_{10} content (Fig. 9A). In contrast, the CoQ_9 content in heart mitochondria of long-lived mammals like the horse or the cow is almost 60-fold lower than that of CoQ_{10} (Fig. 9B). In sharp contrast, α-tocopherol levels in heart mitochondria are relatively high in short-lived mammals like the mouse and up to 40-fold lower in long-lived mammals like the horse, obviously following the same trend as CoQ_9 content (compare Fig. 10 with Fig. 9A).

B. Effect of Coenzyme Q Content on the Rate of $O_2{}^-$ Generation

We have shown previously (46) that the content of CoQ_9 in heart mitochondria of different mammals correlates directly and that of CoQ_{10} correlates inversely with the rate of $O_2{}^-$ generation, best fitted with a sigmoid shaped curve (Fig. 11).

Figure 7 (A) Mitochondrial aconitase activity in the flight muscles of houseflies at different ages. Two different methods were used to determine aconitase activity. Method 1 (■) is a coupled assay in which the formation of NADPH was followed at 340 nm, using citrate as a substrate. Method 2 (♦) is based on the formation of both citrate and isocitrate from *cis*-aconitate, measured as the decrease in absorbance at 240 nm. (B) Immunochemical quantitation of aconitase carbonyls in 15-day-old flyers and crawlers, collected from the same population of flies. The procedure was the same as in Figure 6A. (C) Aconitase enzyme activity in mitochondria from the flight muscles of 15-day-old flyers and crawlers collected from the same population of flies. Enzyme activity was determined by method 1. (From Ref. 70.)

(A)

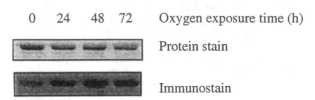

(B)

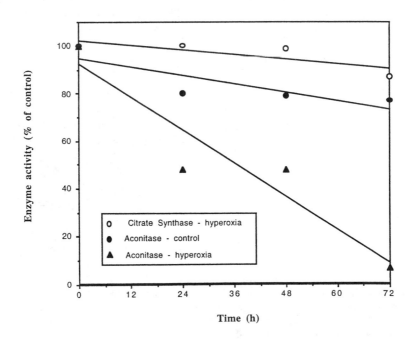

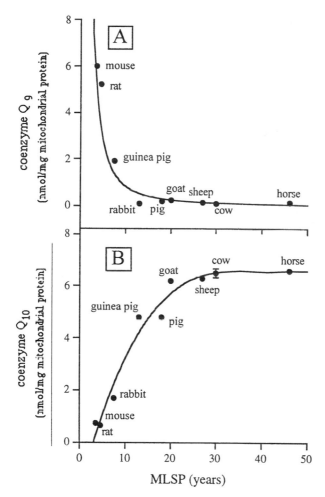

Figure 9 Relationship between the amounts of CoQ_9 (**A**) and CoQ_{10} (**B**) in cardiac mitochondria and the maximum life span potential of different mammalian species.

Figure 8 (**A**) Immunochemical detection of aconitase carbonyls in houseflies exposed to 100% ambient oxygen. The assay was performed as described in Figure 6. (**B**) Effect of 100% ambient oxygen on the activities of mitochondrial aconitase and citrate synthase. Six-day-old flies were exposed to 100% oxygen for indicated times. Aconitase activity was determined by method 1, as described in the legend of Figure 7. (From Ref. 70.)

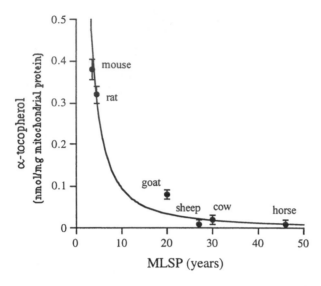

Figure 10 Relationship between the amounts of α-tocopherol in cardiac mitochondria and the maximum life span potential of different mammalian species.

This suggests that there might exist a direct relationship between the ubiquinone homologue present in the electron transport chain and the rate of O_2^- generation. In accordance with this hypothesis, elevated levels of α-tocopherol in heart mitochondrial membranes were found to be correlated with higher rates of O_2^- production. Indeed, comparison of the rate of oxygen consumption and O_2^-/H_2O_2 generation between rat and cow heart mitochondria revealed approximately fivefold lower rates in the latter (14,46). On this basis, it would seem that there is a relatively lesser requirement for tocopherol in mitochondria of animals with lower rates of metabolism. Notwithstanding, other unexplored possibilities that may invalidate this interpretation are the differences in antioxidative defenses and membrane phospholipid composition.

To test the hypothesis that CoQ_9 might be causally related to elevated O_2^- generation, comparisons were made between heart submitochondrial particles (SMPs) of long- and short-lived mammalian species (i.e., rat and cow, respectively). SMPs were first depleted of their natural CoQ content with pentane and thereafter repleted with either CoQ_9 or CoQ_{10} of various concentrations. The results, however, indicated that in the physiological range of CoQ_9 or CoQ_{10} content, the rates of O_2^- generation were not significantly different (46) (Fig. 12). Nonetheless, at concentrations of CoQ (CoQ_9 or CoQ_{10}) that were fivefold greater than the physiological level, SMPs of both rat and cow heart mitochondria (see Fig. 9 for respective CoQ content) exhibited higher rates of O_2^- production when

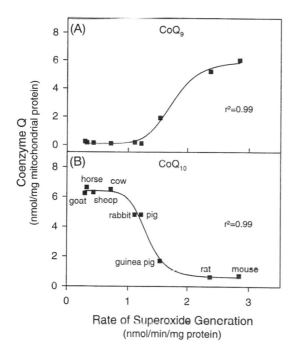

Figure 11   Relationship between the rates of O_2^- generation by cardiac submitochondrial particles from different species and the amounts of CoQ$_9$ and CoQ$_{10}$. The rates of O_2^- generation were measured as SOD-inhibitable reduction of acetylated ferricytochrome-c and are plotted against the content of CoQ$_9$ (**A**) and CoQ$_{10}$ (**B**) of different mammalian species. (From Ref. 46.)

augmented with CoQ$_9$ homologue than with CoQ$_{10}$ (Fig. 12). Although at first glance these experimental data do not seem to suggest a direct link between CoQ homologues and O_2^- production within the physiological range, another interpretation is also tenable. This is so because pentane extractions of SMPs cause a 30% loss in O_2^- generation and a 60% decrease in oxygen consumption. These are irreversible effects on the functional state of SMPs. Therefore, it is quite probable that the efficacy of mitochondrial membranes of such pentane-treated SMPs will be greatly diminished compared to the in vivo situation. This logic is supported by the fact that the rate of oxygen consumption in reconstituted SMPs reaches a plateau at concentrations of CoQ that are higher than the physiological level. This indicates that in pentane-extracted SMPs, higher than the natural amount of CoQ levels are needed to restore the diminished function. Despite this caveat, studies on reconstituted SMPs do provide useful information on the relationship between CoQ homologues and O_2^- generation. It is noteworthy that

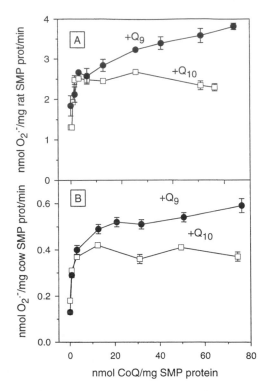

Figure 12   Rates of $O_2{}^-$ generation in CoQ-depleted/reconstituted rat (**A**) and bovine (**B**) heart SMPs. Freeze-dried SMPs were depleted of native CoQ homologues by pentane extractions and reconstituted with specific amounts of CoQ_9 or CoQ_{10} in pentane. The reconstituted SMPs were dried and suspended in phosphate buffer, and rates of $O_2{}^-$ generation were measured as described in the legend of Figure 11. (Adapted from Ref. 46.)

only SMPs reconstituted with CoQ_9, and not with CoQ_{10}, showed a steady increase in the rate of $O_2{}^-$ generation with an increasing amount of CoQ_9. Furthermore, the increase in the rate of $O_2{}^-$ generation in rat heart SMPs, augmented with CoQ_9, was one order of magnitude higher than that observed in cow heart SMPs (Fig. 12). This difference would seem to depend on the species-specific structural differences in the SMPs, as the rates of $O_2{}^-$ generation in unreconstituted SMPs also differed five fold between rat and cow. Such differences in $O_2{}^-$ generation would be unlikely if autoxidation alone was the cause for increased rates of $O_2{}^-$ generation. These findings suggest that the structural organization of the inner mitochondrial membrane, from which the SMPs are derived, plays a major role in determining the rates of CoQ autoxidation, leading to the formation of O_2^-.

C. Maximum Life Span is Inversely Related to Rate of O_2^- Generation

Previous studies in this laboratory have indicated that the rate of mitochondrial O_2^- generation varies greatly, even in the same type of tissue, among different mammalian species, and is inversely related to the maximum life span potential (MLSP) of the species (14,46,56). This inverse relationship between the rate of O_2^- generation and MLSP was found to hold in a sample of mammalian species as well as a group of dipteran insect species (14,56,57). Most recently we have investigated the rate of O_2^- generation in cardiac SMPs of different mammals and found that the rates were highest in the short-lived mammals (mouse and rat) and decreased exponentially with MLSP (Fig. 13). The lowest rates of O_2^- generation were found in heart SMPs of rather long-lived mammals such as cow or horse, which were five times lower than that in mouse or rat. CoQ_{10} content in cardiac SMPs is thus directly related and CoQ_9 inversely related to MSLP in mammals. This supports the hypothesis that during evolution functionality of mitochondria in different mammalian species was optimized by lowering the generation of ROS through shifting from CoQ_9 to CoQ_{10}. As a consequence, high

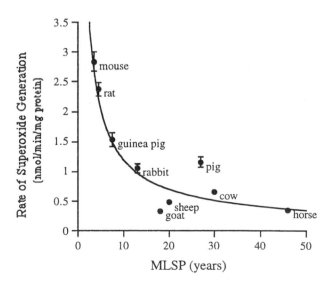

Figure 13   Relationship between the rate of O_2^- generation and the maximum life span potential of different mammalian species.

levels of α-tocopherol to prevent membrane damage would presumably not be needed in the long-lived mammalian species.

D. Interaction of Coenzyme Q and Vitamin E as Antioxidants

The respective roles of CoQ and α-tocopherol as antioxidants in the inner mitochondrial membrane have been a subject of considerable controversy. It is widely accepted that α-tocopherol is the main antioxidant in various membrane systems. It prevents propagation of chain reactions during lipid peroxidation by scavenging radicals. Although the antioxidative activity of CoQ has been demonstrated in vitro, there are conflicting views about its role in vivo. One issue is whether or not CoQ acts as an antioxidant by directly scavenging radicals. Since in homogeneous solutions the reactivity of ubiquinone (the oxidized form of CoQ) toward peroxyl radicals is two orders of magnitude lower [ubiquinone-9: $k_1 = 0.33 \times 10^3/M/s$, (58)] than that of ubiquinol [the reduced form of CoQ or ubiquinol-9: $k_1 = 3.4 \times 10^5/M/s$, (58)], the latter is thought to be the most likely radical scavenger. Indeed, several studies have demonstrated that ubiquinol is an effective inhibitor of lipid peroxidation in mitochondrial membranes in tissues such as liver, heart, and brain (59,60). Ubiquinol was shown to inhibit lipid peroxidation in mitochondria and SMPs depleted of the natural tocopherol content by pentane extraction (61). Another view is that CoQ acts indirectly as an antioxidant by regenerating tocopherol, with the latter being the active antioxidant (59). ESR studies on model systems of homogeneous solutions (62) as well as membranes (63) have demonstrated that ubiquinols can prevent α-tocopherol from oxidation, suggesting recycling of α-tocopherol from its phenoxyl radicals. Maguire et al. (64) reported that NADH and succinate can cause reduction of tocopheroxyl radicals in mitochondrial membranes, which suggests that ubiquinols are involved in recycling of α-tocopherol. On the basis of these studies, it still remained unresolved whether α-tocopherol is the primary radical scavenger and whether CoQ acts directly and/or indirectly as an antioxidant in the mitochondrial membranes.

The interaction between CoQ and α-tocopherol was examined in a recent study on cardiac mitochondria from rat and cow which differ 15-fold in α-tocopherol content. Autoxidation of rat and cow mitochondria was monitored in the absence or presence of succinate, which allowed assessment of antioxidative activity of the reduced (quinol) and oxidized (quinone) form of coenzyme Q under the same conditions. In the cow, which has a relatively low amount of α-tocopherol, there was no detectable loss in ubiquinone or ubiquinol content during autoxidation in the presence or the absence of succinate (see Figs. 9 and 10 for respective levels of CoQ and α-tocopherol). Whereas in the rat heart mitochondria, which contain relatively high amounts of α-tocopherol, there was a depletion of both quinol and α-tocopherol during autoxidation in the absence of succinate.

However, in the presence of succinate, where CoQ was predominantly present in the reduced form, there was no loss of α-tocopherol, while ubiquinol showed a time-dependent loss. Since there was no depletion of quinols in cow heart mito chondria under the same conditions, the depletion of quinols and α-tocopherol in rat heart mitochondria suggested an interaction between the quinols and α-tocoperoxyl radical. This view was bolstered by studies on SMPs, augmented with different amounts of exogenous α-tocopherol (47). Under conditions of low α-tocopherol:quinol molar ratios, there was no loss of α-tocopherol during au toxidation, whereas in SMPs with equimolar ratios, α-tocopherol was depleted, probably due to the presence of insufficient amounts of quinols for recycling of α-tocopherol. Collectively these data support the hypothesis that the antioxidative role of ubiquinols in mitochondria is due to their interaction with tocopherols rather than their ability to directly scavenge radicals.

The above study (47) also indicated that recycling of tocopherol depends on the respiratory activity of mitochondria, which provides the reducing equivalents for the conversion of ubiquinones to ubiquinols. The involvement of succinate-ubiquinone reductase in the reduction of ubiquinones for the consequent recycling of α-tocopherol in mitochondrial membranes was first demonstrated by Maguire et al. (64). Our finding that succinate-dependent reduction of ubiquinone in SMPs was abolished by 4,4,4-trifluoro-1-(2-thienyl)-1,3-butanedione, an inhibitor of succinate dehydrogenase, also indicated the ultimate dependence of ubiquinol-tocopherol interaction on the respiratory activity of mitochondria. The amount of reducible ubiquinone is apparently dependent on the availability of different substrates (65).

Whether or not CoQ acts as a direct scavenger of radicals under physiological conditions remains unclear. However, ubiquinol, a highly hydrophobic molecule present in the middle of the phospholipid bilayer, has been shown to have considerably less intramembrane mobility than α-tocopherol (66), which would almost certainly attenuate its radical scavenging potential. Furthermore, the rate constants of reactions of ubiquinol with peroxyl radicals (58) and α-tocopheroxyl radicals (67) are of the same order of magnitude. The rate constant of the reaction between α-tocopheroxyl radicals and ascorbate (68,69), which recycles tocopherol, is also of the same order of magnitude as that between α-tocopheroxyl radicals and ubiquinol. Since the reactivity of α-tocopherol with peroxyl radicals far exceeds that of peroxyl radicals with ubiquinol, and the latter being a relatively less mobile molecule, it would seem that ubiquinol is unlikely to be a direct radical scavenger except perhaps under pathological conditions where α-tocopherol is scarce. The mechanism by which CoQ regenerates α-tocopherol has been hypothesized to involve an interaction between semiquinone of CoQ and tocopheroxyl radical (62).

Altogether, current evidence in literature suggests that ubiquinol and α-tocopherol act in concert to scavenge radicals during autoxidation of mitochondria. Ubiquinol is most likely involved in the recycling of α-tocopherol rather than in directly scavenging the radicals. Its α-tocopherol recycling efficiency seems to depend on the ratio between ubiquinol and α-tocopherol.

ACKNOWLEDGMENTS

This research was supported by grants R01AG7657 and R01AG13563 from the National Institute on Aging of the National Institutes of Health.

REFERENCES

1. Chance B, Sies H, Boveris A. Hydroperoxide metabolism in mammalian organs. Physiol Rev 1979; 59:527–605.
2. Boveris A, Cadenas E. Production of superoxide radicals and hydrogen peroxide in mitochondria. In: Oberley LW, ed. Superoxide Dismutase. Vol. 2. Boca Raton, FL: CRC Press, 1980:16–30.
3. Turrens JF, Alexandre A, Lehninger AL. Ubisemiquinone is the electron donor for superoxide formation by complex III of heart mitochondria. Arch Biochem Biophys 1985; 237:408–414.
4. Cadenas E, Boveris A, Ragan CE, Stoppani AOM. Production of superoxide radicals and hydrogen peroxide by NADH-ubiquinone reductase and ubiquinol cytochrome c reductase from beef heart mitochondria. Arch Biochem Biophys 1977; 180:248–257.
5. Bolter CJ, Chefurka W. Extramitochondrial release of hydrogen peroxide from insect and mouse liver mitochondria using the respiratory inhibitors phosphine, myxothiazol and antimycin and spectral analysis of inhibited cytochromes. Arch Biochem Biophys 1990; 278:65–72.
6. Zoccarato F, Cavallini L, Deana R, Alexandre A. Pathways of hydrogen peroxide generation in guinea pig cerebral cortex mitochondria. Biochem Biophys Res Commun 1988; 154:727–734.
7. Sohal RS, Sohal BH. Hydrogen peroxide release by mitochondria increases during aging. Mech Ageing Dev 1991; 57:187–202.
8. Kwong LK, Sohal RS. Substrate and site specificity of hydrogen peroxide generation in mouse mitochondria. Arch Biochem Biophys 1998; 350:118–126.
9. Sohal RS. Aging, cytochrome oxidase activity, and hydrogen peroxide release by mitochondria. Free Radic Biol Med 1993; 14:583–588.
10. Hansford RG, Hogue BA, Mildaziene V. Dependence of H_2O_2 formation by rat heart mitochondria on substrate availability and donor age. J Bioenerg Biomembr 1997; 29:89–95.
11. Cino M, Del Maestro RF. Generation of hydrogen peroxide by brain mitochondria: the effect of reoxygenation following postdecapitative ischemia. Arch Biochem Biophys 1989; 269:623–638.

12. Sohal RS. Role of mitochondria and oxidative stress in the aging process. In: Beal MF, Howell N, Bodis-Wollner I, eds. Mitochondria and Free Radicals in Neurodegenerative Diseases. New York: Wiley-Liss, 1997:91–107.

13. Weindruch R, Sohal RS. Caloric intake and aging. N Engl J Med 1997; 337:986–994.

14. Ku HH, Brunk UT, Sohal RS. Relationship between mitochondrial superoxide and hydrogen peroxide production and longevity of mammalian species Free Radic Biol Med 1993; 15:621–627.

15. Sohal RS, Ku HH, Agarwal S, Forster MJ, Lal H. Oxidative damage, mitochondrial oxidant generation and antioxidant defenses during aging and in response to food restriction in the mouse. Mech Ageing Dev 1994; 74:121–133.

16. Sohal RS, Dubey A. Mitochondrial oxidative damage, hydrogen peroxide release, and aging. Free Radic Biol Med 1994; 16:621–626.

17. Sohal RS, Agarwal A, Agarwal S, Orr WC. Simultaneous overexpression of copper-and zinc-containing superoxide dismutase and catalase retards age-related oxidative damage and increases metabolic potential in *Drosophila melanogaster*. J Biol Chem 1995; 270:15671–15674.

18. Sohal RS, Arnold LA, Sohal BH. Age related changes in antioxidant enzymes and prooxidant generation in tissues of the rat with special reference to parameters in two insect species. Free Radic Biol Med 1990; 10:495–500.

19. Nohl H, Hegner D. Do mitochondria produce oxygen radicals *in vivo*? Eur J Biochem 1978; 82:563–567.

20. Sohal RS, Agarwal S, Sohal BH. Oxidative stress and aging in the Mongolian gerbil (*Meriones unguiculatus*). Mech Ageing Dev 1995; 81:15–25.

21. Weindruch R, Walford RL, Fligiel S, Guthrie D. The retardation of aging in mice by dietary restriction: longevity, cancer, immunity and lifetime energy intake. J Nutr 1986; 116:641–654.

22. Sohal RS, Agarwal S, Candas M, Forster MJ, Lal H. Effect of age and caloric restriction on DNA oxidative damage in different tissues of C57BL/6 mice. Mech Ageing Dev 1994; 76:215–224.

23. Benzi G, Pastoris O, Marzatico RF, Villa RF, Curti D. The mitochondrial electron transfer alteration as a factor involved in the brain aging. Neurobiol Aging 1992; 13:361–368.

24. Papa S. Mitochondrial oxidative phosphorylation changes in the life span. Molecular aspects and physiopathological implications. Biochim Biophys Acta 1996; 1276:87–105.

25. Bowling AC, Mutisya EM, Walker LC, et al. Age-dependent impairment of mitochondrial function in primate brain. J Neurochem 1993; 60:1964–1967.

26. Agarwal S, Sohal RS. Differential oxidative damage to mitochondrial proteins during aging. Mech Ageing Dev 1995; 85:55–63.

27. Shacter E, Williams JA, Lim M, Levine RL. Differential susceptibility of plasma proteins to oxidative modification: examination by western blot immunoassay. Free Radic Biol Med 1994;17:429–437.

28. Cabiscol E, Levine RL. Carbonic anhydrase III. Oxidative modification in vivo and loss of phosphatase activity during aging. J Biol Chem 1995; 270:14742–14747.

29. Farmer KJ, Sohal RS. Relationship between superoxide anion radical generation and aging in the housefly, *Musca domestica*. Free Radic Biol Med 1989; 7:23–29.

30. Ragland SS, Sohal RS. Mating behavior, physical activity and aging in the housefly, *Musca domestica*. Exp Gerontol 1973; 8:135–145.

31. Sohal RS, Toy PL, Allen RG. Relationship between life expectancy, endogenous antioxidants and products of oxygen free radical reactions in the housefly, *Musca domestica*. Mech Ageing Dev 1986; 36:71–77.

32. Keller RJ, Halmes NC, Hinson JA, Pumford NR. Immunochemical detection of oxidized proteins. Chem Res Toxicol 1993; 6:430–433.

33. Winn LM, Wells PG. Evidence for embryonic prostaglandin H synthase-catalyzed bioactivation and reactive oxygen species-mediated oxidation of cellular macromolecules in phenytoin and benzo[a]pyrene teratogenesis. Free Radic Biol Med 1997; 22:607–621.

34. Sohal RS, Agarwal S, Dubey A, Orr WC. Protein oxidative damage is associated with life expectancy of houseflies. Proc Natl Acad Sci USA 1993; 90:7255–7259.

35. Kennedy MC, Emptage MH, Dreyer JL, Beinert H. The role of iron in the activation-inactivation of aconitase. J Biol Chem 1983; 258:11098–11105.

36. Fasler B, Lowenstein JM. Aconitase from pig heart. Methods Enzymol 1969; 13: 26–30.

37. Lauble H, Kennedy MC, Emptage MH, Beinert H, Stout CD. The reaction of fluorocitrate with aconitase and the crystal structure of the enzyme-inhibitor complex. Proc Natl Acad Sci USA 1996; 93:13699–13703.

38. Kent TA, Dreyer JL, Kennedy MC, et al. Mossbauer studies of beef heart aconitase: evidence for facile interconversions of iron-sulfur clusters. Proc Natl Acad Sci USA 1982; 79:1096–1100.

39. Gardner PR, Fridovich I. Superoxide sensitivity of the *Escherichia coli* aconitase. J Biol Chem 1991; 266:19328–19333.

40. Gardner PR, Nguyen DD, White CW. Aconitase is a sensitive and critical target of oxygen poisoning in cultured mammalian cells and in rat lungs. Proc Natl Acad Sci USA 1994; 91:12248–12252.

41. Castro L, Rodriguez M, Radi R. Aconitase is readily inactivated by peroxynitrite, but not by its precursor, nitric oxide. J Biol Chem 1994; 269:29409–29415.

42. Verniquet F, Gaillard J, Neuburger M, Douce R. Rapid inactivation of plant aconitase by hydrogen peroxide. Biochem J 1991; 276:643–648.

43. Gardner PR, Raineri I, Epstein LB, White CW. Superoxide radical and iron modulate aconitase activity in mammalian cells. J Biol Chem 1995; 270:13399–13405.

44. Flint DH, Smyk-Randall E, Tuminello JF, Draczynska-Lusiak B, Brown OR. The inactivation of dihydroxy-acid dehydratase in *Escherichia coli* treated with hyperbaric oxygen occurs because of the destruction of its Fe-S cluster, but the enzyme remains in the cell in a form that can be reactivated. J Biol Chem 1993; 268:25547–25552.

45. Stadtman ER. Metal ion-catalyzed oxidation of proteins: biochemical mechanism and biological consequences. Free Radic Biol Med 1990; 9:315–325.

46. Lass A, Agarwal S, Sohal RS. Mitochondrial ubiquinone homologues, superoxide radical generation, and longevity in different mammalian species. J Biol Chem 1997; 272:19199–19204.

47. Lass A, Sohal RS. Electron transport-linked ubiquinone-dependent recycling of alpha-tocopherol inhibits autoxidation of mitochondrial membranes. Arch Biochem Biophys 1998; 352:229–236.
48. Booth RFG, Galanopoulou DG, Quinn PJ. Protection by ubiquinone and ubiquinol against lipid peroxidation in egg yolk. Biochem Int 1982; 5:151–156.
49. Bindoli A, Cavallini L, Siliprandi N. Protective effect of epomediol on calcium release induced by palmitoyl CoA in the sarcoplasmic reticulum. Pharmacol Res Commun 1984; 16:647–652.
50. Bindoli A, Valente M, Cavallini L. Inhibition of lipid peroxidation by alpha-tocopherolquinone and alpha-tocopherolhydroquinone. Biochem Int 1985; 10:753–761.
51. Beyer RE. An analysis of the role of coenzyme Q in free radical generation and as an antioxidant. Biochem Cell Biol 1992; 70:390–403.
52. Ernster L, Dallner G. Biochemical, physiological and medical aspects of ubiquinone function. Biochim Biophys Acta 1995; 1271:195–204.
53. Witting LA. Vitamin E and lipid antioxidants in free-radical-initiated reactions. In: Pryor, W., ed. Free Radicals in Biology. Vol. 4. New York: Academic Press, 1980: 295–319.
54. McCay PB. Vitamin E: interactions with free radicals and ascorbate. Annu Rev Nutr 1985; 5:323–340.
55. Niki E, Yamamoto Y, Takahashi M, Komuro E, Miyama Y. Inhibition of oxidation of biomembranes by tocopherol. Ann NY Acad Sci 1989; 570:23–31.
56. Sohal RS, Svensson I, Sohal BH, Brunk UT. Superoxide anion radical production in different animal species. Mech Ageing Dev 1989; 49:129–135.
57. Sohal RS, Sohal BH, Orr WC. Mitochondrial superoxide and hydrogen peroxide generation, protein oxidative damage, and longevity in different species of flies. Free Radic Biol Med 1995; 19:499–504.
58. Naumov VV, Khrapova NG. Study of the interaction of ubiquinone and ubiquinol with peroxide radicals by the chemiluminescent method. Biophysics 1983; 28:774–780.
59. Mellors A, Tappel AL. The inhibition of mitochondrial peroxidation by ubiquinone and ubiquinol. J Biol Chem 1966; 241:4353–4356.
60. Takayanagi R, Takeshige K, Minakami P. NADH- and NADPH-dependent lipid peroxidation in bovine submitochondrial particles. Biochem J 1980; 192:853–860.
61. Forsmark P, Aberg F, Norling B, et al. Inhibition of lipid peroxidation by ubiquinol in submitochondrial particles in the absence of vitamin E. FEBS Lett 1991; 285: 39–43.
62. Stoyanovsky DA Osipov AN, Quinn PJ, Kagan VE. Ubiquinone-dependent recycling of vitamin E radicals by superoxide. Arch Biochem Biophys 1995; 323:343–351.
63. Kagan V, Serbinova E, Packer L. Antioxidant effects of ubiquinones in microsomes and mitochondria are mediated by tocopherol recycling. Biochem Biophys Res Commun 1990; 169:851–857.
64. Maguire JJ, Kagan V, Ackrell BA, Serbinova E, Packer L. Succinate-ubiquinone reductase linked recycling of alpha-tocopherol in reconstituted systems and mitochondria: requirement for reduced ubiquinone. Arch Biochem Biophys 1992; 292: 47–53.

65. Jorgensen BM, Rasmussen HN, Rasmussen UF. Ubiquinone reduction pattern in pigeon heart mitochondria. Identification of three distinct ubiquinone pools. Biochem J 1985; 229:621–629.

66. Fato R, Battino M, Castelli GP, Lenaz G. Measurement of the lateral diffusion coefficients of ubiquinones in lipid vesicles by fluorescence quenching of 12-(9-anthroyl) stearate. FEBS Lett 1985; 179:238–242.

67. Mukai K, Kikuchi S, Urano S. Stopped-flow kinetic study of the regeneration reaction of tocopheroxyl radical by reduced ubiquinone-10 in solution. Biochim Biophys Acta 1990; 1035:77–82.

68. Packer JE, Slater TF, Willson RL. Direct observation of a free radical interaction between vitamin E and vitamin C. Nature 1979; 278:737–738.

69. Scarpa M, Rigo A, Maiorino M, Ursini F, Gregolin C. Formation of alpha-tocopherol radical and recycling of alpha-tocopherol by ascorbate during peroxidation of phosphatidylcholine liposomes. An electron paramagnetic resonance study. Biochim Biophys Acta 1984; 801:215–219.

70. Yan L-J, Levine RL, Sohal RS. Oxidative damage during aging targets mitochondrial aconitase. Proc Natl Acad Sci USA 1997; 94:11168–11172.

8

Mitochondria and Aging: The Mouse's Tale

Steve Esworthy
City of Hope National Medical Center, Duarte, California

I. INTRODUCTION

The free radical aging hypothesis is one of many hypotheses that attempt to reduce the myriad observations about aging to a few simple causes or processes (1). The hypothesis is of the "wear and repair" variety, where the major deterioration mechanism is conceptualized as stochastic and extrinsic to the genetic programming of the organism. But the process can be modified by genes or by environment (2). The free radical aging hypothesis as adapted to mitochondrial processes can be considered as a version of the "rate of living" hypothesis, since both theories relate metabolic activity rates to aging (2). Free radicals and free radical initiated damaging reaction cascades are the inevitable by-product of aerobic respiration. Common physiology among animals means that a more or less fixed percentage of the electrons and oxygen slated for respiration will be diverted to radicals which will produce damage to the animal. Lethal senescence will eventually result from the cumulative damage. Small animals consume oxygen at higher rates per unit body mass and consequently live shorter lives. While rates of oxygen consumption correlate well with much of the life span variation among species, birds and bats are notable exceptions of animals that live longer than species of similar sizes (2).

Genetic influences could modify the rate of oxygen radical diversion from respiration, the rate of scavenging of radicals prior to interaction with vulnerable targets, and the extent of repair to damaged molecules. That would modify the mean total oxygen consumption value attained by the onset of senescence. The rate of oxygen radical diversion from respiration is less in birds than in mammals.

When the rate of oxygen radical generation by mitochondria is used as the rate of living parameter rather than oxygen consumption, birds do not stand out as exceptions to the relationship of size and life span (3,4).

Not all of the damaging chemical species implicated in the above-mentioned aging hypotheses are actually free radicals. Partially reduced oxygen by-products divert electrons from the respiratory chain or xenobiotic metabolism to the reduction and mobilization of iron (or the oxidation and mobilization of heme). In concert with other partially reduced oxygen species, reduced iron (oxidized hemin) initiates damaging cascades in proteins, lipids, DNA, and carbohydrates (5,6). There are many indications that aging is accompanied by accumulated damage consistent with the action of partially reduced oxygen species cascades (1,7–10). The rate of living/free radical theory provides that the inevitable generation of these species may be a root cause of aging (1). But these species may also be just the visible damaging aspect of other underlying causes of aging. Either way, the apparent pervasiveness of age-related damage produced by the action of partially reduced oxygen by-products suggests that controlling their levels will modify the nature of senescence (7,8,11). Studies with flies indicate that more efficient oxidant scavenging can increase vigor in older flies and increase life span (12).

II. OXIDANTS, ANTIOXIDANTS, AND AGING

The sources of partially reduced oxygen by-products are manifold and the processes for quenching or eliminating the by-products are diverse. A few production sources and quenching processes are indicated by pervasiveness or by suggestive changes with aging to actually be involved with the pathophysiology of the aging process. These are the leak of O_2^- (converted to H_2O_2) from the mitochondrial respiratory chain (1,12), the generation of H_2O_2 from mitochondrial monoamine oxidase B (MAO-B) in brain (11), declines in GSH levels (13), increases in the pool of stored iron in ferritin (14), accumulation of iron in nonferritin protein deposits (15), and glycation and related reactions of sugars (2,16).

III. TRANSGENIC MOUSE LINES

Transgenic mouse lines exist in which mitochondrial O_2^- levels and H_2O_2 levels should be constitutively altered from that in wild-type mice, in effect exacerbating the first two specific processes listed previously. This will allow testing of the premise that mitochondrial oxidative stress is an element in the aging process and permit an examination of the critical features of such an aging process (e.g., mutation of mitochondrial DNA, lipid peroxidation, mobilization of iron from [4Fe-4S] protein centers). This chapter will briefly review what is known about the currently available mouse lines that have alterations relevant to the testing of the free radical aging hypothesis. Tables 1 and 2 summarize the enzymes that have been the target of transgenic technology and are immediately useful in tests of the free radical aging hypothesis. Ec-SOD is listed because the mouse line provides an interesting observation about an acute oxidant stress condition used to test the other lines (17). But the Ec-SOD mouse line is not considered as an aging model in this discussion. Other mouse lines with alterations in genes that are responsive to oxidant stress, such as *Hmox1* (heme oxygenase), have been described (18,19). There are no indications that gene knockout lines for *Hmox1* model any condition found in aging (20–22). And like the Mn-SOD gene knockout line, there is an extreme toxicity associated with the oblation of heme oxygenase 1 activity (18,23).

First, some general points about terminology and transgenic technology need to be considered. The term "transgenic" (Tg) will be applied to lines with extra copies of an expressed cDNA or gene. The phenotype of the mouse line in this nomenclature scheme is "overexpression" of the mRNA and protein. In some cases, the overexpression is deliberately tissue specific. In most cases, universal overexpression was sought. But even where the genetic promoter should drive overexpression in all tissues, significant overexpression of enzyme activity doesn't always occur. As an example, liver glutathione peroxidase (GPX) activity in some *Gpx1* Tg lines is not significantly elevated (24). Mice with a targeted disruption of a gene, leading to effective inactivation of the gene, are referred to as "knockout mice" or "KO" lines. Since the general purpose of the disruption is to silence the gene, the symbol "−" refers to the allele that has the targeted disruption. The symbol "+" refers to the normally active gene. The terminology is standardized in the rules for genetic nomenclature (25), although the opposite use of the symbols "+" and "−" can be found in the literature.

Tg and KO technologies come with their own problems. Some of the neonatal or adult KO or Tg mice may have passed developmental hurdles of unknown causes. Both *Sod1* and *Hmox1* KO fetuses (−/−) are preferentially absorbed in utero (18,26). This may mean that the −/− survivors were hardier than the absorbed sibs by virtue of genetic background (many Tg and KO lines are of mixed mouse strain background) or epigenetically enhanced antioxidant status. The timing of the lethal impact of the homozygous *Sod2* gene KO is strain dependent and new pathologies are observed in the longer-lived strains (23,27). The longer life span of one *Sod2* KO line has been attributed to possible compensation by higher expression of other antioxidant genes (27) which may not occur in all strains. In the case of the *Sod1* and *Hmox1* KO lines, fetal death was spotted as a marked decline in litter size from matings of heterozygous mice relative to

Table 1 List of Relevant Tg, KO, or Mutant Mouse Strains

Enzyme affected	Effect on unstressed mice	Altered stress response	References
Ec-SOD KO	None evident	Sensitive to hyperoxia	17
Ec-SOD Tg[a]	None evident	No alteration to radiation[b]	52
Mn-SOD Tg	None evident	Resistant to ischemia,[c] hyperoxia,[d] doxorubicin, radiation,[a,b] paraquat[b]	100 101 45 52 102
Mn-SOD KO	Neonatal death (−/−); mitochondrial aconitase low; respiration affected (+/−)	Sensitive to paraquat[a,h]	
Cu,Zn-SOD Tg	Neurological damage	Resistance in cerebral ischemia; marginal resistance in hyperoxia; no effect with radiation[a,b]	103 104 52
Cu,Zn-SOD KO	Some fetal death; none evident in adult survivors[e]	No alteration in hyperoxia; hypersensitive to paraquat[a,b]	26 26
GPX1 Tg	None evident	Heat sensitive; resistant to ischemia-reflow; sensitive to cancer	105 106 107

GPX1 KO	None evident	Sensitive to ischemia, paraquat, avirulent coxsackievirus; no alteration in hyperoxia	108 88 89 64
Catalase Tg	None evident in heart-specific Tg	Resistant to doxorubicin[f]	99
Acatalasemic Mice	Lipid metabolism altered; earlier onset of mammary virus activation[g]	Sensitive to H_2O_2; no alteration in hyperoxia, radiation; AT sensitive liver tumor formation	36,96 95,98 98

[a] Cell line model.
[b] Paired comparison with Tg or KO with same footnote letter.
[c] Protection from ischemia-reflow may occur in a narrow range of overexpression based on other work (50,51).
[d] Lung-specific Tg.
[e] Observation leaves some doubt about adult normality.
[f] Resistance observed only in a narrow range of overexpression.
[g] Note that the acatalasemic strain was developed from irradiated mice and may harbor other mutations. The mice have catalase peroxidase activity. In early studies, aminotriazole was used to eliminate residual catalase expression during acute oxidative stress experiments. The mutations were developed on a C3H background so that mammary tumor virus can be passed to offspring.
[h] Paired comparison with Tg or KO with same footnote letter.

Table 2　List of Antioxidant Enzymes and Properties

Enzyme	Abbreviation[a]	EC number	Model Rx[b]	Localization[c]	Substrate[d]	Isozymes[e]
Manganese superoxide dismutase	Mn-SOD (*Sod2*)	1.15.1.1	$2O_2 + 2H^+ \rightarrow H_2O_2 + O_2$	Mitochondria	—	Cu,Zn-SOD
Copper, zinc superoxide dismutase	Cu,Zn-SOD (*Sod1*)	1.15.1.1	See above	Cytosol	$ONOO^-$	Mn-SOD
Glutathione peroxidase	GPX1 (*Gpx1*)	1.11.1.9	$H_2O_2 + 2GSH \rightarrow 2H_2O +$ GSSG	Mitochondria Cytosol	ROOH GSNO	GPX2 GPX3 GPX4 GST others[f]
Catalase	CAS1 (*Cas1*)	1.11.1.6	$2H_2O_2 \rightarrow 2H_2O + O_2$	Peroxisome	ROH[g]	

[a] Abbreviation in parentheses is the gene name.

[b] Model reaction. In some cases alternative reactions are possible involving other substrates (see footnote c) and other different mechanisms (see footnote g).

[c] Localization of enzyme in most tissues and under ordinary circumstances. Catalase has been detected in heart mitochondria (89,90). However, data on specific activity in heart mitochondria have not been reported.

[d] Substrate alternatives from the model reactions. Naturally occurring mutations in Cu,Zn-SOD can result in $ONOO^-$ being used as a substrate. An additional side reaction in which Cu,Zn-SOD produces hydroxyl radical has been reported. For GPX1 "R" in ROOH can be a free fatty acid or nucleic acid base. GSNO has been reported to serve as a reducing substrate for GPX1.

[e] Gene products from the same gene families with overlapping reactions and unrelated gene products that can catalyze some of the same reactions. With GPX this has become a real problem not only because of glutathione transferases with GPX activity, but also because some GPX isoenzymes can use alternative reducing substrates, like thioredoxin. Thus thioredoxin peroxidases can be considered as isoenzymes of GPX or as competitors for H_2O_2 as catalase is regarded. Also, see footnote f.

[f] Recently a new class of selenium-independent GPX, distinct from glutathione transferases, has been reported (67). This may be a gene family. Induction during wound healing in skin has been described.

[g] Catalase can mediate peroxidase reactions with alcohols and formic acid as the donor substrates.

homozygous wild-type mice. But fetal death may be occurring to lesser degrees in other lines.

Tg female mice often fail to foster their newborn. This problem is noted by animal husbandry professionals (28,29). It seems to be a random occurrence. It may be that complex behaviors can be disrupted by insertion or deletion of DNA in many parts of the genome or by the hybridization of the particular inbred strain combinations that have been used in Tg and KO production schemes. Reduced fostering behaviors may be a sensitive end-point indicator of all types of behavior problems that such mice might have as a result of the "random" effects of the genetic and other physical manipulations that were used to create the lines. Therefore, alterations in behavior and perhaps learning and memory in Tg lines must be based on evidence that the change was actually a specific effect of the construct. In the case of KO mice with specific targeted disruption of a gene, this should not be a general problem, but an issue only in some specific contracts. For example, the human *gpx1* gene is flanked on the 5′ side by two transcription units, *rho A* and *mid* (30). The transcripts of these adjacent genes terminate within the *gpx1* gene regulatory region. It is conceivable that a KO construct involving exon 1 of *gpx1* would interfere with the expression of *rho A* and *mid*. To the extent that expression of these two genes is an attribute of normal or abnormal human physiology, this could produce quite unexpected outcomes that might be ascribed to specific effects of *gpx1* gene disruption. The compressed distribution of genetic transcription units around the human *gpx1* gene may be unusual, and it is not reported whether it is conserved in mice. The example is strictly cautionary. Some specifically created "control lines" for KOs could create problems by providing an abnormal version of the "wild-type" physiology for comparative purposes. Unwillingness to foster and fetal death can create serious problems for investigations that require large numbers of animals.

The more common procedures for generating KO mice and for sustaining good breeding and fostering in Tg lines, by deliberately interbreeding two or more inbred lines, defeats the original goals of inbred strain developers, limiting the use of the founder KO and Tg lines in cancer, immunology, and aging studies (31). Many standard inbred lines are inherently difficult to breed and many were selected deliberately to harbor viruses and oncogenes that cause cancer and early death. KO mice often start out as hybrids of 129 and C57Bl strains (32). Other strain combinations are used. C3H and CBA strains are often brought into the breeding of Tg and KO lines because the females are fecund and exhibit good fostering behavior (29). Strains 129, C57Bl, and C3H have very different mean life spans (33). Mammary tumor virus can result in much shorter life spans for C3H strain females than for the males. In large commercial facilities, C3H lines are commonly maintained so that the viral incidence is reduced or eliminated. But because the generation and maintenance of most Tg and KO lines is up to individual investigators, virus and other problems may be present. Workers in the aging field generally should transfer the Tg or KO genes to a uniform inbred line background before real studies begin. C57Bl is a more commonly used inbred strain for aging studies. Currently *Gpx1* and *Sod2* (Mn-SOD) gene KO alleles are being transferred to C57Bl strain backgrounds for studies in aging (A. Richardson, and A. Murphy, VA Hospital, San Antonio, TX), cancer studies (F.-F. Chu and S. Esworthy, City of Hope, Duarte, CA), and atherosclerosis studies (A. Lusis and M.-L. Brennan, UCLA, Los Angeles, CA). There is no reported KO mouse line for the *Cas1* gene (catalase). Radiation-induced mutations were used to produce mice in which the catalase multimer is unstable (35,36). Liver catalase activity remains at 25% of wild-type mice. The residual activity is as low as 3% in blood. The acatalasemic mice also have residual peroxidase activity from catalase subunits. There are reports of the mutations impact on a C57Bl background (37), but the author was unable to determine the current availability of the C57Bl subline.

IV. AGING AND TRANSGENIC MICE

A. Specific Transgenic Lines

Table 1 contains a list of the transgenic lines (and the *Cas1*[b] mutant line) that were considered for this review because the target of the genetic manipulation is an antioxidant enzyme. Along with each mouse line is a short summary of acute oxidative stress experiments that have been performed with the lines, the outcome relative to a control mouse line, and observations of the lines under unstressed conditions. This list of observations is provided as a stopgap measure. A real test of aging demands that actual life span studies be made and that the nature of the senescence mechanism be determined. This has not been done. Part of the delay is in producing the inbred strains of Tg and KO on an inbred strain background suitable for aging studies. Second, in most cases direct measurement of the steady-state levels of the substrates of the enzymes or a suitable marker has not been made. Thus, in the absence of life span studies, the long-term impact of the transgenic alteration can only be indirectly inferred from pathology observed after experimentally imposed short-term stress. Third, where more direct measurements are available, the results of experimental oxidative stress or natural pathology can be used to examine how critical the enzyme pathway is to mice. The enzymes affected by the genetic manipulations in the pertinent mouse strains are listed in Table 2, along with a model enzyme reaction scheme, enzyme properties, and isoenzymes.

B. Oxidative Pathways in Mitochondria and the Extramitochondrial Spaces

Figure 1a shows the localization of the antioxidant enzymes subject to transgenic manipulation in relationship to the flux of oxidants. The diagram is chiefly cen-

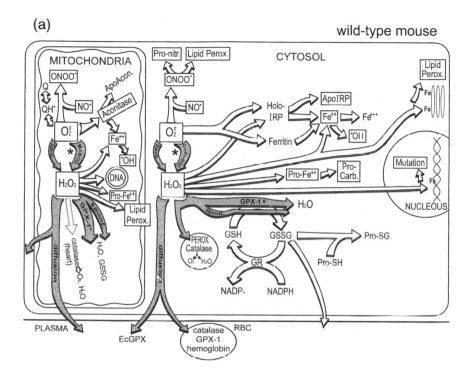

Figure 1 (a) Pathways of O_2^- anion and H_2O_2 metabolism in a wild-type animal. The location of key enzymes in the pathways is indicated by a * in the diagram. The size of a box enclosing a reactive oxidant species indicates a relative pool size that can be compared among Figures a–d to see the conventional view of how the targeted disruption of the *Sod* and *Gpx* genes should affect the pools. The halftone ribbons in (a) only, show antioxidant pathways. The relative thickness of a ribbon among figures provides a rough approximation of the effect of enzyme oblation on the pathways. White ribbons leading away from pools of reactive oxidants indicate potentially damaging reaction pathways. Ribbon width among figures indicates relative flux along the pathway in wild-type and each KO mouse line, based on reactions indicated to occur in the literature. In some cases, pools of reactive intermediates are speculative. $ONOO^-$ in mitochondria may exist. The pool could exist based on in vitro studies. The equilibrium between ubiquinone (Q), the semiquinone (QH), and superoxide (O_2^-) in mitochondria has been disputed. Abbreviations: Acon = aconitase; ApoAcon = aconitase stripped of one iron from the [4Fe-4S] center; GSH = glutathione, reduced; GSSG = glutathione, oxidized. IRP = iron responsive protein. Holo-IRP has an intact [4Fe-4S] center and aconitase activity. Apo-IRP lacks one iron from the [4Fe-4S] center. OH = hydroxyl radical; NO = nitric oxide; $ONOO^-$ = peroxynitrite; Pro-carb = protein carbonyls; Pro-Fe = ferrous iron bound to protein; Pro-SG = thiolated protein; Pro-nitr = nitrated protein.

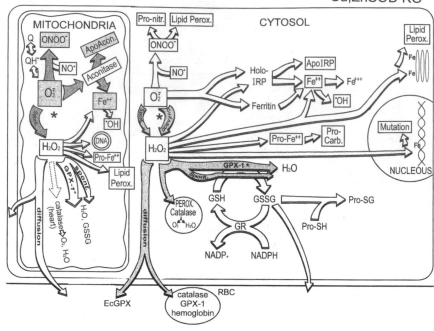

(b) Cu,ZnSOD KO

Figure 1 Continued **(b)** Oxidant pathways in a *Sod2*$^{-/-}$ mouse. Halftone ribbons show the flux from the expanded pool of O_2^-. Primary deleterious reactions are indicated by shading of boxes. Note that the pool of H_2O_2 has shrunk, so that reactions with the expanded pool of ferrous iron mobilized from aconitase and other [4Fe-4S] proteins may not yield greater hydroxyl radical. Likewise, other free iron–initiated events like DNA mutation and protein carbonyl formation may not be accelerated in this scenario. Not shown is the relief of the NADH pool from lowered H_2O_2 generation, which could be beneficial. Increased lipid peroxidation and protein nitration could occur from the larger ONOO pool even though H_2O_2-mediated lipid peroxidation may be less. Reactions like that depicted for aconitase do seem to occur in these mice and affect complex II and III activities as well. The Holo-IRP pool is not affected, justifying the limit of the diagram to the mitochondrion. **(c)** Oxidant flux in a *Sod1*$^{-/-}$ mouse line. See legends to (a) and (b) for details. **(d)** Oxidant flux in a *Gpx¹* gene KO mouse. Most deleterious reactions of H_2O_2 stem from reactions with ferrous iron. As such, many damaging reactions are indicated to be occurring in this mouse based on the premise that ferrous iron pools would remain at the same size as in the wild-type mouse. If this is not the case, then some of the indicated pathways would not be affected. Ferritin induction has been found in some oxidative stress, which could reduce the transit iron pool. Heme oxygenase 1 may be elevated, although we have failed to find mRNA induction. This would effectively transport heme iron out of liver and reduce opportunities for lipid peroxidation in microsomes. The opportunity for H_2O_2 metabolism by a number of old and newly discovered ways including catalase thioredoxin peroxidases, non-GST, selenium-independent GPXs and EcGPX subsequent to diffusion from the cell, means that the hypothetical increase in H_2O_2 pool size or the enlarged flux through deleterious pathways may be exaggerated. NADH would not be as vulnerable to depletion in the mitochondrion of these mice as in wild-type mice, so that the oblation of GPX1 may not be detrimental. On the other hand, oxidation of GSH is not stimulated as much in oxidative stress, so that protein thiolation might not be available as a protective mechanism.

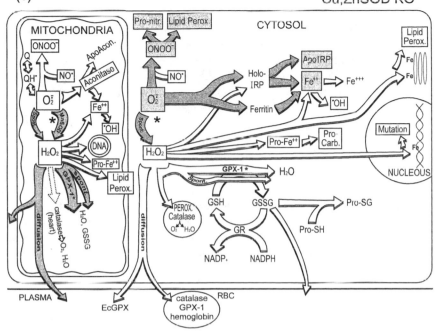

(c) Cu,ZnSOD KO

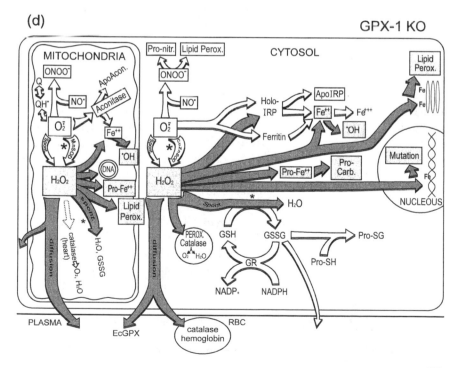

(d) GPX-1 KO

tered on the flux of O_2^-, H_2O_2, and the impact on the targets of iron, protein, DNA, and lipids. Lipid peroxidation and lipid radical pathways, DNA damage, and protein modification have been indicated as end points. Lipid peroxidation is a process equally as complicated as the pathways indicated in Figure 1a. But the discussion of the roles of the antioxidant enzymes currently subject to Tg and KO research can be largely surmised based on simplified Figures 1a–d.

Some of the indicated changes in oxidant intermediate pool sizes are speculative. In fact, the existence of some pools, while based on reactions with high rate constants in vitro, might be considered speculative. The reaction between O_2^- and NO has a huge rate constant. But the concentration of NO in mitochondria is not known (38). Further, if O_2^- or NO concentrations change in mitochondria, there could be negative feedback on the production of the other. The equilibrium between O_2^- and ubiquinone (Q) or ubisemiquinone (QH) has been disputed. An alternate reaction between ubisemiquinone and H_2O_2, producing hydroxyl radical, has been proposed which could be equally detrimental to the mitochondrion (13).

C. Superoxide Dismutases

Figure 1b is a schematic representation of the impact of loss of Mn-SOD. Figure 1c shows the same for Cu,Zn-SOD. In the diagram, the impact of Mn-SOD loss is confined to the mitochondrion and that of Cu,Zn-SOD is confined to the extramitochondrial space. This is simplistic even though the two isoenzymes accumulate in their respective subcellular spaces and the substrate probably cannot traverse the barriers between the two compartments. Superoxide in the unprotonated anionic form (dominant form at neutral pH) does not diffuse through subcellular membranes (unless it can use anionic channels). The mitochondria could communicate the effects of oxidative stress to other cellular compartments by simple processes such as the diffusion of H_2O_2 or alterations in ATP output. More elaborate means of oxidant-initiated mitochondria–extramitochondrial space communication are the subjects of other chapters in this book. But the immediate impact of the loss of Mn-SOD would be damage to mitochondrial components. Aconitase, a vulnerable target for O_2^--initiated damage and possibly H_2O_2-mediated damage, is abundant in mitochondria and cytoplasm (39,40). In the cytosol, aconitase has a primary role as the iron response protein (IRP) (41). The loss of aconitase activity in $Sod2^{-/+}$ mice is confined to the activity in the mitochondrial inner space (34). Aconitase is of particular interest since its inactivation has the potential to deprive the cells of energy production and mobilize ferrous iron for further damaging cascades (42–44).

Sod2 Tg mice and cell lines exhibit enhanced resistance to all types of oxidative stress (Table 1). Overexpression of Mn-SOD in mice does not seem to cause pathology as does the overexpression of Cu,Zn-SOD, although there are indicated limits to the beneficial effects and the possible risk of increased pathology in cardiac ischemia-reflow (45–51). The broad range of protection of Mn-SOD suggests that mitochondria are affected in all types of free radical stress. This is possible since procedures commonly used to induce oxidative stress in animals cannot be precisely selective for the mitochondrial compartment or the extramitochondrial compartments. Comparison of effects in *Sod1* (Cu,Zn-SOD) Tg mouse lines (Table 1) and transgenic cell lines suggest that the protective benefits of Cu,Zn-SOD are not as potent as Mn-SOD in enhancing survival with similar stresses. The study of Wong (52) shows that is primarily due to the localization of the isoenzymes. When enhanced O_2^- generation is largely confined to the microsomes, as in paraquat toxicity, Cu,Zn-SOD is seen to have a larger but not an exclusive protective effect on cell survival over Mn-SOD (53).

$Sod2^{-/-}$ mice die at or before weaning, depending on the strain background (23,27). The shorter-lived strains show a depression in aconitase levels (sum of mitochondria plus cytosolic activity) in the heart (42%), liver (37%), and brain (22%), and these strains have a significant depression of succinate dehydrogenase (complex II of mitochondria) activities in the heart and skeletal muscles (23). This is consistent with the damaging action of O_2^- on the [4Fe-4S] centers of enzymes (42), although the observation of Yan et al. (54) that the aconitase protein backbone is a target of oxidation might also apply. The $Sod2^{-/-}$ animals show lipid accumulation in the livers, they are hypothermic, and they have reduced growth rates. The hearts are dilated and have thinner walls. The cause of death is presumed to be heart failure due to mitochondrial failure. The longer-lived strains exhibit additional pathology. The mitochondria show abnormal morphology; the animals exhibit skeletal muscle weakness, bone marrow abnormalities which may indicate anemia, and neuronal degeneration of the basal ganglia and brain stem (27).

Heterozygous mice have about one-half the mitochondrial aconitase levels of the $Sod2^{+/+}$ mice without reductions in cytosolic aconitase levels (34). The aconitase enzyme level measurements in $Sod2^{+/-}$ mice must be made on the semi-purified mitochondria to observe the difference from $Sod2^{+/+}$ mice. In the original report by Li et al. (23) the total cell aconitase levels between the $Sod2^{+/+}$ and $Sod2^{+/-}$ mice were not statistically different and the pooled data were reported in their tables (T.-T. Huang, personal communication). Since the levels of aconitase are supposed to reflect O_2^- levels, based on the work of Gardner and Fridovich (39), the observations would indicate greater O_2^- steady-state levels relative to oxygen consumption rates in the $+/-$ mice relative to the $+/+$ mice. The observation of depressed mitochondrial aconitase activity is interesting since it implies that the induction of Mn-SOD is not triggered by elevations of O_2^- capable of a measurable damaging impact.

Mitochondrial respiration is reduced in the $Sod2^{+/-}$ mice due, at least in part, to depressed activities of aconitase and of complexes I and III (34). Unlike the study on $Sod2^{-/-}$ mice, no change was found in complex II activity. It is possible that differences in assay methods for complex II are producing different results. Alternatively, the differences in O_2^- concentration between the $Sod2^{-/-}$ and the $Sod2^{-/+}$ strains may produce different manifestations of damage. Reports by two groups agree that complex IV is not affected in mouse lines with reduced levels of Mn-SOD (23,34).

Mitochondrial complex IV activity is noted to decline with age (55–59). Superoxide toxicity in young and adolescent mice does not seem to closely model the pathology of mitochondrial aging, since declines in complex IV activity are not occurring, while declines in other complexes are occurring. Since complex IV does not have a Fe-S center (60) and has been shown to be resistant to O_2^- (61), the decline has been attributed to the encoding of two subunits in mtDNA which is hypothesized to be vulnerable to mutation (62). Such phenomena may be manifest when the mice are old enough. Alternatively, complex IV activity is more sensitive to GSH depletion than complexes I–III and mitochondrial GSH levels decline with age (13,63). Finally, the elevation of O_2^- concentration without an elevation in H_2O_2 concentration is an artificial condition which could be occurring in $Sod2^{-/+}$ mice (Fig. 1b). It is possible that prototypical mitochondrial aging is triggered by a simultaneous rise in O_2^- and H_2O_2 levels. The $Sod2^{-/-}$ and $Sod2^{-/+}$ mice appear to be suffering from mitochondrial O_2^- toxicity, with manifestations more akin to iron deficiency than aging (56–60).

It is unclear whether the mtDNA of the $Sod2^{-/+}$ mice will show a higher mutation rate than that of wild-type mice. Iron mobilization from aconitase and complexes I and II should occur. But H_2O_2 levels may be lower than in wild-type mice, negating the impact of potentially higher free iron levels (Fig. 1b). If the mutation rate is enhanced in the knockout mouse, this might provide a real test of the power of free radical–induced mtDNA mutation to promote declines in mitochondrial function (55,62). Specifically, mutation rates for DNA encoding complex IV subunits could be compared with complex IV activity in $Sod2^{+/+}$ versus $Sod2^{-/+}$ mice. The $Sod2^{-/+}$ mice would also seem suited to studies of lipofusin accumulation and disposition with age (see Chap. 12).

The $Sod2^{-/+}$ line is of interest because it probably does provide an animal model in which mitochondrial energy output will decline at an accelerated rate relative to wild-type mice. Thus the mouse line may be useful to aging studies where declines in mitochondrial performance are relevant even if the immediate pathology does not precisely model aging phenomenology.

D.   Glutathione Peroxidase 1

Glutathione peroxidase 1 (GPX1) is the major cytosolic and mitochondrial selenium-dependent glutathione peroxidase activity in most tissues (64,65). The location of GPX1 in the antioxidant system is indicated in Figure 1a, and the impact of GPX1 deficiency is indicated in Figure 1d. Note that diffusion of H_2O_2 out of the mitochondrion could be a major detoxification route for this organelle. GSH concentrations in mitochondria could be high enough for the nonenzymatic reduction to be significant. The nonenzymatic reduction of H_2O_2 by GSH has been invoked as one possible explanation for high H_2O_2 metabolism in the lens of mice deficient in GPX1 (66). It is also plausible that one of the non-GST, selenium-independent GPX activities recently described could be acting in lens tissue (66,67).

Many studies suggest that GPX1 is the major cytosolic and mitochondrial hydroperoxidase activity in tissues at physiological hydroperoxide concentrations (68,69). But animals evolved in ecosystems in which adequate selenium supplies were not assured. There are many regions on earth where the soil content of selenium is insufficient to provide an adequate supply of the trace element to animals. This is based on the incidence of selenium-dependent pathological conditions (70). In animal models, liver and kidney GPX1 is the most vulnerable of the selenoproteins to selenium deprivation. However, organs rather than individual selenoproteins may be the major focus of selenium regulation. Brain, testes, ovaries, and endocrine organs can preserve selenium stores against deprivation, while liver and kidney do not seem to do this (71,72). Below this level of organ regulation is a hierarchy of individual selenoprotein regulation. GPX1 seems to be at the bottom of this hierarchy (73–76).

Many workers have invoked the idea of compensation to explain the nearly normal physiology of laboratory animals when challenged with selenium deprivation at or after weaning. Catalase metabolizes more H_2O_2 when conditions allow leakage from the cytosol into peroxisomes (77). However, only two of the many reports on catalase activity during selenium depletion report an increase in activity. Most reports note no change or a modest decline in activity (78–81). Catalase activity in $Gpx1^{-/-}$ mice is not higher than in wild-type mice (64), nor are the levels of other GPX isoenzymes elevated (65,82). GSTs with GPX activity can reduce some lipid hydroperoxides (69). The kinetic parameters that describe reactions of GSTs with model and physiological hydroperoxides suggest that they are relatively poor at this process (83). It is the massive quantities of GSTs in the liver and kidney that make the proposal that GSTs metabolize significant quantities of lipid hydroperoxides seem plausible. The supposed compensation by GST for GPX1 decline in selenium deprivation was not supported by critical experiments performed in mice by Reiter and Wendel (84). By two criteria, the induction of GSTs was shown to be triggered by low or declining selenium levels and not by low GPX1 activity levels. In the author's limited experience, GST levels were found to be lower in animals with GPX1 activities of less than 0.2% of the GPX activity supported by 0.1 ppm selenium. The selenium content of the diet that allowed GPX1 to decline to the 0.2% level was 0.006 ppm. Most likely, active compensation for declines in GPX1 activity is not required since there is sufficient capacity elsewhere to remove hydroperoxides, unless stress is imposed on an animal.

The full range of hydroperoxidase proteins is just now being realized. Non-GST, selenium-independent GPX activities and thioredoxin-dependent hydroperoxidase activities have been recently discovered (85–87). Based on the claimed potency of some of these alternative systems and the ability of the spontaneous GSH reaction to reduce H_2O_2 at high rates in some tissues, it is amazing that eliminating GPX1 has any effect on an animal. Only by adding stress to the GPX1 oblation can one uncover obvious defects in the $Gpx1^{-/-}$ mice. Paraquat toxicity is enhanced in $Gpx1^{-/-}$ mice and the mice are permissive for the mutation of avirulent coxsackievirus to the virulent form with subsequent inflammation in the heart (88,89). The immune system of the KO mice has been altered, but the mice seem to be deficient only in the TH2 pathway. The odd observation is, as with Cu,Zn-SOD–deficient mice, that the homozygous $Gpx1$ KO mice are not more susceptible to hyperoxia than wild-type mice (64). The pathology of hyperoxia may have to be rethought to accommodate the observations from Tg and KO mouse lines (Table 1).

$Gpx1$ KO mice are comparatively robust up to 1 year of age when observed in the unstressed condition of the laboratory cage (64). Observations beyond this point are not reported. In our colony, 9-month-old female KO mice were routinely giving birth and fostering litters of five to nine pups. It is clear that the effects of BSO-induced mitochondria GSH depletion on mitochondrial morphology are not mediated by GPX1 (90–92). However, as indicated above for coxsackievirus challenge, this lack of visible pathology does not indicate complete normalcy. Four-month-old $Sod2^{-/+}$ mice appear normal on casual observation, but they show signs of O_2^- toxicity when the mitochondria are analyzed or the animals are stressed. Advanced age may be sufficiently stressful that GPX1-deficient mice will not show normal aging capacity. Whether this could be manifest as accelerated senescence with multiple symptomatology or with generally normal senescence and greater risk for fatal infections or other very specific pathology is unpredictable.

Unfortunately we do not have some critical data for $Gpx1$ KO mice. Tissue H_2O_2 concentrations are not published. There are no assays of mitochondrial function published for $Gpx1$ KO mice. Lipid peroxide decomposition products, such as MDA and 4-HNE, do not seem to be elevated in the major tissues of KO mice (64,89), but such assays may not be sensitive enough.

E.   Catalase

Mouse strains with altered catalase expression could be used to contrast the impact of aging with $Gpx1$ KO lines. Catalase deficiency or overexpression could modify the clearance of H_2O_2 in the extramitochondrial space more selectively than modulation of GPX1 levels, since GPX1 levels are high in both spaces. Hydrogen peroxide can diffuse throughout the cell and catalase activity has been reported in heart mitochondria, so this may not be strictly true (93,94). An available acatalasemic line retains some catalase activity in liver and kidney but has very little catalase activity in some other tissues (36,37). The line also has peroxidase activity from catalase monomers. When the mice are administered aminotriazole (AT) to eliminate the residual activities, the mice are still not more susceptible to radiation or hyperoxia than wild-type mice (95). These findings are not so surprising when reviewed after the results obtained from *Sod1* and *Gpx1* KO mice. The acatalasemic mice are more sensitive to H_2O_2 intoxication whether treated with AT or not. The acatalasemic line is hypolipidemic (96). Mammary virus activation (in unfostered C3H lines) occurred earlier in life (97) and AT-induced liver tumors occurred at a greater incidence in the acatalasemic mice (98). The viral activation can be attenuated by supplementing diets with vitamin E (97). Life-span studies may have been performed with acatalasemic mice, but only life span with mice fed AT appear to have been reported in the general literature (98). Aminotriazole induces liver tumors, so the results can't be used to determine if there is any impact of catalase deficiency on life span.

Heart-specific *Cas1* Tg lines were developed (99). They were used to show that increased catalase activity could mitigate the cardiopathology of doxorubicin. Since heart mitochondria seem to be a vulnerable site for oxidant toxicity, the *Cas1* Tg lines could be of interest if *Gpx1* KO mice demonstrate aging-related mitochondrial deterioration.

V. CONCLUSIONS

The value of Tg and KO mouse lines to the understanding of aging lies in their potential rather than with available results. A foreseeable problem is that critical measurements of reactive oxidants will remain uninvestigated in the lines while phenomenology is vigorously pursued. Thus many of the changes indicated to occur in Figure 1b–d will remain speculation. The situation is marginal for the *Sod2* KO lines, where aconitase has been used as a marker for O_2^- levels. The *Sod1* KO lines and the *Gpx1* KO and Tg lines are in need of hard analysis of reactive oxygen species levels.

Life-span studies will commence soon with some of the lines. Already $Sod2^{+/-}$ mice show signs of mitochondrial senescence that may foreshadow an abbreviated life span. But it is not clear that an accelerated version of a natural aging process has been modeled in this line. In this case, perhaps what is needed to clarify the apparent discrepancies is a description of murine mitochondrial aging. This seems to be missing from the literature. In any event, the mice should be useful as models of the impact of mitochondrial senescence on aging and of use in studies of oxidant-driven mitochondrial DNA mutation and its impact on mitochondrial function. Despite the apparent lack of impact on mice at 1 year of age, GPX1 oblation may yet show an impact on senescence. It seems clear that real life-span studies will have to be performed on *Gpx1* KO mice before any conclusions can be made.

ACKNOWLEDGMENTS

The author is indebted to the following people for information and advice used in the preparation of this article: Ye-Shih Ho, Ting-Ting Huang, Arlan Richardson, Balvin H. L. Chua, Melinda Beck, the genome informatics personnel at Jackson Laboratories and the animal resources staff at the City of Hope. The work was supported by NIH grant R29 DK46921.

REFERENCES

1. Harman D. Free radical theory of aging. Mutat Res 1992; 275:257–266.
2. Ricklefs RE, Finch CF. Aging. New York: Scientific American Library, 1995.
3. Barja G, Cadenas S, Rojas C, Perez-Campos R, Lopez-Torres M. Low mitochondrial free radical production per unit O_2 consumption can explain the simultaneous presence of high longevity and high aerobic metabolic rate in birds. Free Radic Res 1994; 21:317–327.
4. Ku H-H, Brunk UT, Sohal RS. Relationship between mitochondrial superoxide and hydrogen peroxide production and longevity of mammalian species. Free Radic Biol Med 1993; 15:621–627.
5. Rice-Evans CA. Formation of free radicals and mechanisms of action in normal biochemical processes and pathological states. In: Rice-Evans CA, Burdon RH, eds. Free Radical Damage and Its Control. New York: Elsevier Science, 1994:131–153.
6. Chaudiere J. Some chemical and biochemical constraints of oxidative stress in living cells. In: Rice-Evans CA, Burdon RH, eds. Free Radical Damage and Its Control. New York: Elsevier Science, 1994:25–66.
7. Butterfield DA, Howard BJ, Yatin S, Allen KL, Carney JM. Free radical oxidation of brain proteins in accelerated senescence and its modulation by N-tert-α-phenyl-nitrone. Proc Natl Acad Sci USA 1997; 94:674–678.
8. Orr WC, Sohal RS. Extension of life-span by over-expression of superoxide dismutase and catalase in *Drosophila melanogaster*. Science 1994; 263:1128–1130.
9. Beckman KB, Ames BN. Oxidants, antioxidants, and aging. In: Scandalios JG, ed. Oxidative Stress and the Molecular Biology of Antioxidant Defenses. Cold Spring Harbor, NY: Cold Spring Harbor Laboratory Press, 1997:201–246.
10. Ames BN, Shigenaga MK, Hagen TM. Mitochondrial decay in aging. Biochim Biophys Acta 1995; 1271:165–170.

11. Knoll J. The striatal dopamine dependency of life span in male rat. Longevity study with (−) deprenyl. Mech Ageing Dev 1988; 46:237–262.

12. Sohal RS, Dubey A. Mitochondrial oxidative damage, hydrogen peroxide release, and aging. Free Radic Biol Med 1993; 16:621–626.

13. Benzi G, Moretti A. Age- and peroxidative stress-related modifications of the cerebral enzymatic activities linked to mitochondria and the glutathione system. Free Radic Biol Med 1995; 19:77–101.

14. Rikans LE, Ardinska V, Hornbrook KR. Age-associated increase in ferritin content of male rat liver: implications for diquat-mediated oxidative injury. Arch Biochem Biophys 1997; 344:85–93.

15. Smith MA, Harris PL, Sayre LM, Perry G. Iron accumulation is a source of redox-generated free radicals. Proc Natl Acad Sci USA 1997; 94:9866–9868.

16. Hunt JV, Smith CCT, Wolff SP. Autoxidative glycosylation and possible involvement of peroxides and free radicals in LDL modification by glucose. Diabetes 1990; 39:1420–1424.

17. Carlsson LM, Edlund JJT, Marklund SL. Mice lacking extracellular superoxide dismutase are more sensitive to hyperoxia. Proc Natl Acad Sci USA 1995; 92: 6264–6268.

18. Poss KD, Tonegawa S. Heme oxygenase 1 is required for mammalian iron reutilization. Proc Natl Acad Sci USA 1997; 94:10919–10924.

19. Poss KD, Tonegawa S. Reduced stress defense in heme oxygenase1-deficient cells. Proc Natl Acad Sci USA 1997; 94:10925–10930.

20. Pantopoulos K, Hentze MW. Rapid responses to oxidative stress mediated by key iron proteins EMBO J 1995; 14:2917–2924.

21. Cook MN, Mark GS, Vreman HJ, et al. Ontology of heme oxygenase activity in the frontal cerebral cortex and cerebellum of the guinea pig. Brain Res Dev Brain Res 1996; 92:18–23.

22. Ewing T, Maines MD. Immunohistochemical localization of biliverdin reductase in rat brain: age related expression of protein. Brain Res 1995; 672:29–41.

23. Li Y, Huang T-T, Carlson EJ, et al. Dilated cardiomyopathy and neonatal lethality in mutant mice lacking manganese superoxide dismutase. Nat Genet 1995; 11:376–381.

24. Esworthy RS. Unpublished work. 1997.

25. Blake JA, Davisson MT, Eppig JT, et al. A report on the international nomenclature workshop held May 1997 at the Jackson Laboratory, Bar Harbor, Maine, USA. Genomics 1997; 45:464–468.

26. Ho Y-S, Gargano M, Cao J, Bronson RT, Heimler I, Hutz RJ. Reduced fertility in female mice lacking copper-zinc superoxide dismutase. J Biol Chem 1998; 273: 7765–7769.

27. Lebovitz RM, Zhang H, Vogel H, et al. Neurodegeneration, myocardial injury, and perinatal death in mitochondrial superoxide dismutase-deficient mice. Proc Natl Acad Sci USA 1996; 93:9782–9787.

28. Pinkert CA. The history and theory of transgenic animals. Lab Anim 1997; 26:29–35.

29. Monastersky GM, Robl JM. Strategies in Transgenic Animal Science. Washington, DC: American Society for Microbiology Press, 1995.

30. Moscow JA, Morrow CS, He R, Mullenbach GT, Cowan KH. Structure and func-

tion of the 5' flanking sequence of the human cytosolic selenium-dependent gluta-thione peroxidase gene (*hgpx1*). J Biol Chem 1992; 267:5949–5958.

31. Hogan B, Costanti F, Lacy E. Manipulating the Mouse Embryo. Cold Spring Harbor, NY: Cold Spring Harbor Laboratory Press, 1986:81–88.

32. Gossler A, Doetschman T, Korn R, Serfling E, Kemler R. Transgenesis by means of blastocyst-derived embryonic stem cell lines. Proc Natl Acad Sci USA 1986; 83:9065–9069.

33. Smith GS, Walford RL, Mickey MR. Life span and incidence of cancer and other disease in selected long-lived inbred mice and their F_1 hybrids. J Natl Cancer Inst 1973; 50:1195–1213.

34. Williams M, Remmen HV, Richardson A. The potential use of the MnSOD knock-out mouse in aging research [abstract]. 50th Annual Meeting of the Gerontological Society of America. J Gerontol Biol 1997; 37:1.

35. Feinstein RN, Howard JB, Braun JT, Seaholm JE. Acatalasemic and hypocata-lasemic mouse mutants. Genetics 1966; 53:923–933.

36. Feinstein RN, Braun JT, Howard J. Reversal of H_2O_2 toxicity in the acatalasemic mouse by catalase administration: suggested model for possible replacement therapy of inborn errors of metabolism. J Lab Clin Med 1966; 68:952–957.

37. Feinstein RN, Fry RJ, Staffeldt EF. Comparative effects of aminotriazole on normal and acatalasemic mice. J Environ Pathol Toxic 1978; 1:779–790.

38. Richter C, Schweizer M. Oxidative stress in mitochondria. In: Scandalios JG, ed. Oxidative Stress in Mitochondria. Cold Spring Harbor, NY: Cold Spring Harbor Laboratory Press, 1997:169–200.

39. Gardner PR, Fridovich I. Inactivation-reactivation of aconitase in *Escherichia coli*. J Biol Chem 1992; 267:8757–8763.

40. Gardner PR, Nguyen D-D, White CW. Aconitase is a sensitive and critical target of oxygen poisoning in cultured mammalian cells and in rat lungs. Proc Natl Acad Sci USA 1994; 91:12248–12252.

41. Kennedy MC, Mende-Mueller L, Blondin GA, Beinert H. Purification and charac-terization of cytosolic aconitase from beef liver and its relationship to the iron-responsive element binding protein. Proc Natl Acad Sci USA 1992; 89:11730–11734.

42. Flint DH, Tuminello JF, Emptage MH. The inactivation of Fe-S clusters containing hydro-lyases by superoxide. J Biol Chem 1993; 268:22369–22376.

43. Keyer K, Imlay JA. Superoxide accelerated DNA damage by elevation of free-iron levels. Proc Natl Acad Sci USA 1996; 93:13635–13640.

44. Henle ES, Linn S. Formation, prevention, and repair of DNA damage by iron/hydrogen peroxide. J Biol Chem 1997; 272:19095–19098.

45. Yen H-C, Oberly TD, Vicitbandha S, Ho Y-S, St. Clair DK. The protective role of manganese superoxide dismutase against adriamycin-induced cardiac toxicity in transgenic mice. J Clin Invest 1996; 98:1253–1260.

46. Reaume AG, Elliot JL, Hoffman EK, et al. Motor neurons in Cu/Zn superoxide dismutase-deficient mice develop normally but exhibit enhanced cell death after axonal injury. Nat Genet 1996; 13:43–47.

47. Bar-Peled O, Korkotian E, Segal M, Groner Y. Constitutive overexpression of Cu/Zn superoxide dismutase exacerbates kainic acid-induced apoptosis of transgenic-

Cu/Zn superoxide dismutase neurons. Proc Natl Acad Sci USA 1996; 93:8530–8535.

48. Epstein CJ, Avarham KB, Lovett M, et al. Transgenic mice with increased Cu/Zn-superoxide dismutase activity: animal model of dosage effects in Down's syndrome. Proc Natl Acad Sci USA 1987; 84:8044–8048.

49. Peled-Kamar M, Wirguin LI, Hermalin WA, Groner Y. Oxidative stress mediated impairment of muscle function in transgenic mice with elevated levels of wild-type Cu/Zn superoxide dismutase. Proc Natl Acad Sci USA 1997; 94:3883–3887.

50. Omar BA, McCord JM. The cardioprotective effect of Mn-superoxide dismutase is lost at high doses in the postischemic isolated rabbit heart. Free Radic Biol Med 1990; 9:473–478.

51. Omar BA, Gad N, Jordan MC, et al. Cardioprotection by Cu,Zn-superoxide dismutase is lost at high doses in the reoxygenated heart. Free Radic Biol Med 1990; 9: 465–471.

52. Wong GHW. Protective roles of cytokines against radiation: induction of mitochondrial MnSOD. Biochim Biophys Acta 1995; 1271:205–209.

53. Huang T-T, Yasunami M, Carlson EL, et al. Superoxide-mediated cytotoxicity in superoxide dismutase-deficient fetal fibroblasts. Arch Biochem Biophys 1997; 344: 424–432.

54. Yan L-J, Levine RL, Sohl RS. Oxidative damage during aging targets mitochondrial aconitase. Proc Natl Acad Sci USA 1997; 94:11168–11172.

55. Papa S. Mitochondrial oxidative phosphorylation changes in the life span. Molecular aspects and physiopathological implications. Biochim Biophys Acta 1996; 1276: 87–105.

56. Boffoli D, Scacco SC, Vergari R, et al. Declines with age of the respiratory chain activity in human skeletal muscle. Biochim Biophys Acta 1994; 1226:73–82.

57. Bowling AC, Mutisya EM, Walker LC, et al. Age-dependent impairment of mitochondrial function in primate brain. J Neurochem 1993; 60:1964–1967.

58. Yen T-C, Chen Y-S, King K-L, Yeh S-H, Wei Y-H. Liver mitochondrial respiratory function declines with age. Biochem Biophys Res Commun 1989; 165:994–1003.

59. Trounce I, Byrne E, Marzuki S. Declines in skeletal muscle mitochondrial respiratory chain function: possible factor in aging. Lancet 1986; i:637–639.

60. Singer TP, Ramsay RR, Ackrell BAC. Deficiencies of NADH and succinate dehydrogenases in degenerative diseases and myopathies. Biochim Biophys Acta 1995; 1271:211–219.

61. Zhang Y, Marcillat O, Giulivi C, Ernster L, Davies KJA. The oxidative inactivation of mitochondrial electron transport chain components and ATPase. J Biol Chem 1990; 265:16330–16336.

62. Nagley P, Mackay IR, Baumer A, et al. Mitochondrial DNA mutation associated with aging and degenerative disease. In: Fabris N, Harman D, Knook DL, Steinhagen-Thiessen E, Zs.-Nagy I, eds. Physiopathology Processes of Aging. New York: New York Academy of Sciences, 1992:92–102.

63. Benzi G, Curti D, Pastori O, et al. Sequential damage in mitochondrial complexes by peroxidative stress. Neurochem Res 1991; 16:1295–1302.

64. Ho Y-S, Magnenat J-L, Bronsen RT, et al. Mice deficient in cellular glutathione

peroxidase develop normally and show no increased sensitivity to hyperoxia. J Biol Chem 1997; 272:16644–16651.

65. Esworthy RS, Ho Y-S, and Chu F-F. The GPX1 gene encodes mitochondrial glutathione peroxidase in the mouse liver. Arch Biochem Biophys 1997; 340:59–63.

66. Spector A, Ma W, Wang R-R, Yang Y, Ho Y-S. The contribution of GSH peroxidase-1, catalase and GSH to the degradation of H_2O_2 by the mouse lens. Exp Eye Res 1997; 64:477–485.

67. Shichi H, Demar JC. Non-selenium glutathione peroxidase without glutathione s-transferase activity from bovine ciliary body. Exp Eye Res 1990; 50:513–520.

68. Zhang L, Maiorino M, Roveri A, Ursini F. Phospholipid hydroperoxide glutathione peroxidase: specific activity in tissues of rats of different age and comparison with other glutathione peroxidases. Biochem Biophys Acta 1989; 1006:140–143.

69. Lawrence RA, Burk RF. Glutathione peroxidase activity in selenium-deficient rat liver. Biochem Biophys Res Commun 1976; 71:952–958.

70. Oldfield JE. The two faces of selenium. J Nutr 1987; 117:2002–2008.

71. Behne D, Hofer-Bosse T. Effects of a low selenium status on the distribution and retention of selenium in the rat. J Nutr 1984; 114:1289–1296.

72. Behne D, Hilmert H, Scheid S, Gessner H, Elger W. Evidence for specific target tissues and new biologically important selenoproteins. Biochim Biophys Acta 1988; 966:12–21.

73. Weitzel F, Ursini F, Wendel A. Phospholipid hydroperoxide glutathione peroxidase in various mouse organs during selenium deficiency and repletion. Biochim Biophys Acta 1990; 1036:88–94.

74. Sunde RA, Dyer JA, Moran TV, Everson JK, Suglimoto M. Phospholipid hydroperoxidase glutathione peroxidase: full-length pig blastocyst cDNA sequence and regulation by selenium status. Biochem Biophys Res Commun 1993; 193:905–911.

75. Burk RF, Hill KE, Read R, Bellow T. Response of rat selenoprotein P to selenium administration and fate of its selenium. Am J Physiol 1991; 261:E26–E30.

76. Hill KE, McCollum GW, Boeglin ME, Burk RF. Thioredoxin reductase activity is decreased by selenium deficiency. Biochem Biophys Res Commun 1997; 234:293–295.

77. Jones DP, Eklow L, Thor H, Orrenius S. Metabolism of hydrogen peroxide in isolated hepatocytes: relative contributions of catalase and glutathione peroxidase in decomposition of endogenously generated H_2O_2 Arch Biochem Biophys 1981; 210:505–516.

78. Burk RF, Nishiki K, Lawrence RA, Chance B. Peroxide removal by selenium-dependent and selenium-independent glutathione peroxidases in hemoglobin-free perfused rat liver. J Biol Chem 1978; 253:43–46.

79. Reiter R, Wendel A. Selenium and drug metabolism—I: multiple modulations of mouse liver enzymes. Biochem Pharmacol 1983; 32:3063–3067.

80. Simmons TW, Jamall IS. Significance of alterations in hepatic antioxidant enzymes. Biochem J 1988; 251:913–917.

81. Baliga R, Baliga M, Shah SV. Effect of selenium-deficient diet in experimental glomerular disease. Am J Physiol 1992; 263:F56–F61.

82. Cheng W-H, Ho Y-S, Ross DA, et al. Cellular glutathione peroxidase knockout

mice express normal levels of selenium-dependent plasma and phospholipid hydroperoxide glutathione peroxidases in various tissues. J Nutr 1997; 127:1445–1450.

83. Hong YH, Li C-H, Burgess JR, et al. The role of selenium-dependent and selenium-independent glutathione peroxidases in the formation of prostaglandin F_2. J Biol Chem 1989; 264:13793–13800.

84. Reiter R, Wendel A. Selenium and drug metabolism—II: independence of glutathione peroxidase and reversibility of hepatic enzyme modulations in deficient mice. Biochem Pharmacol 1984; 33:1923–1928.

85. Munz B, Frank S, Hubner G, Olsen E, Werner S. A novel type of glutathione peroxidase: expression and regulation during wound repair. Biochem J 1991; 326: 579–585.

86. Kwon SJ, Park J-W, Choi W-K, Kim K. Inhibition of metal-catalyzed oxidation systems by a yeast protector protein in the presence of thioredoxin. Biochem Biophys Res Commun 1994; 201:8–15.

87. Levine RL, Mosoni L, Berlett BS, Stadtman ER. Methione residues as endogenous antioxidants in proteins. Proc Natl Acad Sci USA 1996; 93:15036–15040.

88. Ching W-H, Ho Y-S, Valentine BA, Ross DA, Combs J, Lei XG. Cellular glutathione peroxidase is the mediator of body selenium to protect against paraquat lethality in transgenic mice. J Nutr 1998; 128:1020–1076.

89. Beck M, Esworthy RS, Ho Y-S, Chu F-F. Glutathione peroxidase protects mice from viral-induced myocarditis. FASEB J 1998.

90. Martensson J, Meister A. Mitochondrial damage in muscle occurs after marked depletion of glutathione and is prevented by giving glutathione monoester. Proc Natl Acad Sci USA 1989; 86:471–475.

91. Martensson J, Jain A, Frayer W, Meister A. Glutathione metabolism in the lung: inhibition of its synthesis leads to lamellar body and mitochondrial defects. Proc Natl Acad Sci USA 1989; 86:5296–5300.

92. Jain A, Martensson J, Stole E, Auld PAM, Meister A. Glutathione deficiency leads to mitochondrial damage in brain. Proc Natl Acad Sci USA 1991; 88:1913–1917.

93. Nohl H, Hegner D. Evidence for the existence of catalase in the matrix space of rat-heart mitochondria. FEBS Lett 1978; 89:126–130.

94. Radi R, Turrens JF, Chang LY, et al. Detection of catalase in rats heart mitochondria. J Biol Chem 1991; 266:22028–22034.

95. Feinstein RN, Faulhaber JT, Howard JB. Sensitivity of acatalasemic mice to acute and chronic irradiation and related conditions. Radiat Res 1968; 35:341–349.

96. Goldfischer S, Roheim P, Edelstein D. Hypolipidemia in a mutant strain of "acatalasemic" mice. Science 1971; 173:65–66.

97. Ishii K, Zhen L-X, Wang D-H, et al. Prevention of mammary tumorigenesis in acatalasemic mice by vitamin E supplementation. Jpn J Cancer Res 1996; 87:680–684.

98. Feinstein RN, Fry RJM, Staffeldt EF. Carcinogenic and antitumor effects of aminotriazole on acatalasemic and normal catalase mice. J Natl Cancer Inst 1978; 60: 1113–1116.

99. Kang YJ, Chen Y, Epstein PN. Suppression of doxorubicin cardiotoxicity by overexpression of catalase in the heart of transgenic mice. J Biol Chem 1996; 271: 12610–12616.

100. Chen Z, Siu B, Ho Y-S, et al. Overexpression of MnSOD protects myocardial ischemia/reperfusion injury in transgenic mice. Circ Res. In press.
101. Wispe JR, Warner BB, Clark JC, et al. Human Mn-superoxide dismutase in pulmonary epithelial cells of transgenic mice confers protection from oxygen injury. J Biol Chem 1992; 267:23937–23941.
102. St. Clair DK, Oberley TD, Ho Y-S. Overproduction of human Mn-superoxide dismutase modulates paraquat-mediated toxicity in mammalian cells. FEBS Lett 1991; 293:199–203.
103. Kinouch H, Epstien CJ, Mizui T, et al. Attenuation of focal cerebral ischemic injury in transgenic mice overexpressing CuZn superoxide dismutase. Proc Natl Acad Sci USA 1991; 88:11158–11162.
104. White CW, Avraham KB, Shanley PF, Groner Y. Transgenic mice with expression of elevated levels of copper-zinc superoxide dismutase in the lungs are resistant to pulmonary oxygen toxicity. J Clin Invest 1991; 87:2162–2168.
105. Mirochintchenko O, Palnitkar U, Philbert M, Inouye M. Thermosensitive phenotype of transgenic mice overproducing human glutathione peroxidases. Proc Natl Acad Sci USA 1995; 92:8120–8124.
106. Yoshida T, Watenabe M, Engelman DT, et al. Transgenic mice overexpressing glutathione peroxidase are resistant to myocardial ischemia reperfusion injury. J Mol Cell Cardiol 1996; 28:1759–1767.
107. Lu Y-P, Lou Y-R, Yen P, et al. Enhanced skin carcinogenesis in transgenic mice with high expression of glutathione peroxidase or both glutathione peroxidase and superoxide dismutase. Cancer Res 1997; 57:1468–1474.
108. Yoshida T, Maulik N, Engelman RM, et al. Glutathione peroxidase knockout mice are more susceptible to myocardial ischemia reperfusion. Circulation. In press.

9

Mitochondria and Apoptosis

Bernard M. Babior and Roberta A. Gottlieb
The Scripps Research Institute, La Jolla, California

Christoph Richter
Eidgenössische Technische Hochschule, Zürich, Switzerland

I. INTRODUCTION

Apoptosis refers to a lethal program that is activated in cells when they are induced to commit suicide. When this program is activated, a highly characteristic series of events takes place. These events include the activation of the caspases, a family of cysteine proteases that cleave their substrates on the carboxyl side of an aspartate residue; the cleavage of the caspase targets, which include poly(ADP-ribose) polymerase, actin, the chromosomal protein lamin B, a subunit of DNA-dependent protein kinase (1), and a subunit of the DNA replication complex C, among others; the cleavage of the genome into fragments whose lengths are integral multiples of ≈ 200 bp, which is the amount of DNA found in a nucleosome; the transfer of phosphatidylserine from the inner to the outer plasma membrane leaflet; the activation of tissue transglutaminase and the cross-linking of actin; acidification of the cytosol by ≈ 0.5 pH units; and finally the disintegration of the cell by the formation of blebs that break off and float away, to be disposed of by mononuclear phagocytes which recognize phosphatidylserine and certain adhesion proteins that appear on the surface of the blebs. The net effect is the destruction of the cell without loss of its contents into the extracellular environment.

The participation of mitochondria in the events of apoptosis was first recognized when it was shown that Bcl-2, an antiapoptotic protein first identified in certain malignancies of the lymphatic system, was located in the mitochondria. Since that discovery, many laboratories have begun to study mitochondria and apoptosis. Work from these laboratories has yielded important information about the behavior of mitochondria in apoptotic cells and the surprising role of certain mitochondrial proteins in the apoptotic process. This work has also led to a number of controversies whose resolution is still in the future.

II. MITOCHONDRIA

Mitochondria consist of four major elements: the matrix, which contains among other things the Krebs cycle enzymes; the inner membrane, an extensively folded structure that encloses the matrix and contains a series of proton-transporting electron carriers together with the ATP-synthesizing enzyme; the outer membrane, which surrounds the entire mitochondrion; and the intermembrane space, which lies between the outer and inner membranes.

The major function of the mitochondria is the production of chemical energy in the form of ATP. It accomplishes this task by generating an electrochemical potential across its inner membrane, using the energy stored in this potential to carry out the uphill synthesis of ATP from ADP and inorganic phosphate. The potential is generated in the course of electron transfer from substrate (either NADH or succinate produced by the Krebs cycle enzymes) to oxygen. From NADH, an electron passes successively through complex I (NADH-Q reductase), coenzyme Q, complex III (coenzyme Q–cytochrome-c reductase), cytochrome c, and finally complex IV (cytochrome oxidase), which transfers the electron to oxygen. As they pass the electron on to the next component, each of the three complexes pumps protons from the matrix into the intermembrane space, generating a potential across the inner membrane that consists partly of an outside-to-inside proton gradient and partly of an electrical potential that is negative inside. From succinate, electrons pass from complex II (succinate dehydrogenase) to coenzyme Q, but complex II does not pump protons. It can be seen that the mitochondrion is actually a tiny fuel cell, capturing part of the energy of a chemical oxidation in the form of an electrochemical potential.

The ATP-synthesizing component is known as the mitochondrial ATPase, or complex V. It has a proton channel in the inner membrane and an ATP-synthesizing head that extends into the matrix. ADP and P_i bind to the active site of the enzyme, where they are held very tightly and forced together to form ATP. (At the active site, the equilibrium constant for the reaction ADP + P_i = ATP + H_2O is about 1.) The newly formed ATP cannot be released, however, without a conformational change on the part of the ATPase. This conformational change takes place when a proton passes through the proton channel from the intermembrane space to the matrix of the mitochondrion. Thus the gradient generated by the continuous pumping of protons out of the matrix guarantees a steady supply of ATP as the protons move down their gradient to return to the matrix.

III. MITOCHONDRIA AND APOPTOSIS

After being known for decades only as the major source of biochemical energy for the cell, the discovery that mitochondria are active participants in the apoptosis program was a considerable surprise. In the past several years a large amount of information has been developed regarding mitochondria and apoptosis, but current knowledge is patchy and a great deal more will have to be learned before it will become possible to develop a comprehensive view, even at an elementary level, of what mitochondria do during apoptosis. The following discussion will deal with these patches of knowledge which have not yet begun to coalesce.

A. Mitochondrial Changes in Apoptosis

In addition to genome destruction, cell fragmentation, and other features of apoptosis that were described above, changes occur in the mitochondria of apoptotic cells. Oxygen uptake and ATP production ceases and the outer membrane becomes leaky (2,3). Later the mitochondria swell greatly and lose calcium into the cytoplasm while the inner membrane depolarizes, a series of events that have been postulated to be caused by the opening of a "permeability transition pore," described as a protein-lined megapore through which is released, among other things, a factor called AIF which causes genome degradation in isolated nuclei. A number of proteins and molecular events characteristic of mitochondria have been shown to be related to apoptosis.

B. Bcl-2 and Its Relatives

Bcl-2 is the prototype of a large family of proteins that are involved in apoptosis. Some, like Bcl-2, protect against apoptosis, while others promote apoptosis. Several of the proteins in this family have been identified in mitochondria: Bcl-2, Bcl-xL (4), and Mcl-1 (5), which oppose apoptosis, and Bax and Bcl-xS (4), which promote apoptosis. There are probably other proteins of the very large Bcl-2 family that are localized in the mitochondria, but studies concerning their localization are as yet unavailable.

Of all the members of this family, Bcl-2 is the one that has been most extensively studied. The initial description of Bcl-2 as an antiapoptotic protein reported that the molecule was located in the inner mitochondrial membrane (6–8), but several subsequent reports claimed that it was in the outer membrane

(8–11). It probably resides in both membranes, perhaps at the contact zones where the outer and inner mitochondrial membranes touch (12,13). It inserts into the mitochondrial membrane through a string of hydrophobic amino acids that lie within 22 residues of the C-terminus, with the remainder of the protein projecting into the cytosol (10). There are also alternatively spliced forms of some of these proteins that reside in the cytosol. One such alternatively spliced form is Bcl-2β, a 22 kDa variant of Bcl-2 (7,14), and another is Bcl-x ΔTM (4). Both are missing their transmembrane domains, and both are located in the cytosol; only Bcl-x ΔTM retains its activity as an antiapoptotic protein.

Bcl-2 and other proteins of the Bcl-2 family interact with each other through domains that show homologies across the family and are therefore known as BH (Bcl-2 homology) domains. The 239-residue Bcl-2 molecule itself contains four such domains, not surprisingly named BH1 (residues 136–155), BH2 (residues 187–202), BH3 (residues 93–107), and BH4 (residues 10–30). These domains are necessary for protein-protein interactions. Bcl-2 binds to itself, dimerizing head-to-tail through interactions involving BH1 and BH2 on one polypeptide and BH4 on the other (15–17); elimination of any of these domains abolishes homodimer formation and antiapoptotic activity (15,18–20).

Bcl-2 also binds to apoptosis-inducing proteins of its own family. Of these, Bax is the prototype. Dimerization between Bcl-2 and Bax takes place by a tail-to-tail interaction that requires the BH1 and BH2 domains of Bcl-2 and the BH3 domain of Bax (15,17,18,20–22). The BH4 domain of Bcl-2 is not required for the formation of a Bcl-2/Bax dimer, but in its absence Bcl-2 loses its antiapoptotic function (15,17,19). Bax, like Bcl-2, forms a homodimer, but the only domain required for homodimer formation is BH3 (21). In fact, of the four BH domains that characterize the Bcl-2 family, only the BH3 domain is found in a death-dealing cousin named BID (20).

In addition to the interactions described above, Bcl-2 associates with proteins outside its own immediate family. These proteins include the kinase Raf-1 (23); carnitine palmitoyltransferase I, a mitochondrial membrane protein (24); R-Ras (25); the protein phosphatase calcineurin (26), and probably other proteins as well. The functional significance of these associations is a mystery.

Bcl-2 protects cells against death caused by many but not all apoptosis-inducing agents. For example, cells treated with DNA-damaging agents are saved from death by overexpression of Bcl-2. Similarly, Bcl-2 can prevent death in cells exposed to hypoxia or ischemia-reperfusion (27). On the other hand, Bcl-2 has only a variable effect on cells driven into apoptosis by Fas ligand or tumor necrosis factor, which bind to receptors that initiate the apoptosis program. In opposing apoptosis, Bcl-2 delays or prevents all the cellular manifestations described above. In particular, it prevents the mitochondrial alterations characteristic of apoptosis. The mitochondrial transmembrane potential is preserved, the permeability transition pore remains closed, and O_2^- generation caused by the leakage of electrons from reduced coenzyme Q to oxygen is prevented (28,29).

In addition, Bcl-2 delays the fall in the mitochondrial transmembrane potential caused by CN⁻, rotenone, or antimycin A, although it has no effect on the associated decline in cytoplasmic ATP (30). Of great interest is the observation that the effects of Bcl-2 family proteins differ in mitochondria-containing and mitochondria-deficient strains of *Saccharomyces cerevisiae* (grande and petite strains, respectively) (31). Grande strains are killed as expected when they express the death protein Bax, and are rescued by the coexpression of Bcl-2. In contrast, petite strains remain alive when they express Bax, although they stop growing. Furthermore, a Bcl-2 mutant that lacks the transmembrane domain and therefore remains in the cytosol can rescue petite strains but not grande strains. These tantalizing findings are saying something important about the relationship between Bcl-2, mitochondria, and apoptosis, but to date the message has not been decoded. This is more or less true of all the information that has been obtained on the Bcl-2 family. Many members of this family have been cloned and sequenced, their intracellular locations have been determined, and a great number of their protein-protein interactions have been dissected and analyzed. Yet their mechanism of action is as obscure today as it was when they were first discovered.

C. Cytochrome *c*

Since apoptosis affects so many cell functions, it was not surprising to find that mitochondrial function too is altered during apoptosis. Oxygen uptake was found to fall sharply in apoptotic cells, a fall that was caused by a derangement in the interaction between cytochrome *c* and cytochrome oxidase (2). This was not a surprising finding, considering the wide variety of changes that occur in apoptotic cells. It was truly astonishing, however, when Wang and associates (32) discovered that cytochrome *c* participated in the activation of the caspases. This discovery was made in the course of studies to identify the components necessary for caspase activation in a crude cell-free system. These components were called "Apafs" (*a*poptotic *p*rotease *a*ctivation *f*actors), and Wang's group found that there were three such components, which they designated Apaf-1, -2, and -3. When Apaf-2 was purified, it proved to be cytochrome *c*, a remarkable and surprising finding. A second component, Apaf-1, was a 130 kDa protein containing a domain that was highly homologous to ced-4, a *C. elegans* protein that binds to both ced-3 (a caspase) and ced-9 (a Bcl-2 homolog) (33,34) and participates in apoptosis during the worm's development; an N-terminal domain homologous to ced-3; and a long C-terminal domain containing 12 WD40 motifs, whose prototypes are found in the propeller-shaped β-subunit of trimeric G proteins (35). The presence of Walker A and B boxes in Apaf-1 suggests that the protein is able to bind nucleotides. Most recently, Apaf-3 was cloned, and found to be identical to caspase-9. The complete system comprises Apaf-1, cytochrome *c*, caspase-9, and ATP or deoxy-ATP, and its function is to activate caspase-3.

How does cytochrome *c*, situated as it is in the mitochondrial intermembrane space and separated from the cytosol by the outer membrane, gain access to the rest of the components of the Apaf system? A number of studies from several laboratories have suggested that early in the course of apoptosis, the cyto-chrome leaks from the mitochondria into the cytosol, where it completes the caspase-3–activating Apaf system (32,36,37). We (R.A.G. and B.M.B.), how-ever, found that cytochrome c was not released from mitochondria isolated from apoptotic Jurkat cells, as long as the cells were lysed by cavitation, though cyto-chrome c was liberated from mitochondria obtained from homogenized cells (2,3,38). The outer membranes of the apoptotic mitochondria became leaky, how-ever, if the pH of the buffer was less than 7.0, a value typical of apoptotic cyto-plasm. Whether released or not, however, something happens to cytochrome c during apoptosis: first, because the cytochrome is not released from nonapoptotic mitochondria regardless of how the cells are disrupted; and second, because in anti-Fas-treated Jurkat cells, cytochrome-c function is rapidly lost, with a conse-quent failure in mitochondrial oxygen uptake. Whatever it is that happens to cytochrome c during apoptosis, it is prevented by Bcl-2 and Bcl-x_L, two antiapop-totic members of the Bcl-2 family (5,39,40). Bcl-x_L exerts this effect even if the cytochrome is microinjected into the cell (39).

The elimination of cytochrome c function results from the action of CIFA (cytochrome c-inactivating factor of apoptosis), an as yet uncharacterized protein that appears in the cytoplasm of cells undergoing apoptosis. It is possible that caspase-3 is activated, not by the leakage of cytochrome c from the mitochondria to the cytosol, but through the leakage of pro-caspase-3 together with the other components of the Apaf system into the intermembrane space. Alternatively, Apaf-mediated activation of caspase-3 may be delayed until the cytochrome be-gins to leak from the mitochondria, a late event in the case of anti-Fas-treated Jurkat cells.

The role of cytochrome c in apoptosis may not be restricted to caspase activation. This is suggested by certain observations made in *S. cerevisiae*. The expression of Bax in these yeast causes them to undergo apoptosis. During the course of this event, cytochrome c is released from their mitochondria. Caspases, however, are not activated. What if anything cytochrome c is doing to promote apoptosis under these circumstances is not clear.

D. Calcium and the Permeability Transition Pore

1. Apoptosis and the Permeability Transition Pore

A large number of studies from one laboratory have provided evidence for the permeability transition pore as the proximate cause of apoptosis. These studies have indicated that other events characteristic of apoptosis, including for example chromatin condensation and the degradation of the genome into nucleosomal fragments (DNA fragments that are integral multiples of 200 bp in size), are preceded by changes suggesting that the mitochondrial permeability transition pore has been opened. These changes include loss of the potential across the mitochondrial inner membrane; the onset of production of O_2^- and other oxidants, an event thought to reflect a leakage of electrons out of the mitochondrial electron transport change at the level of ubiquinone; a sharp fall in reduced glutathione; an increase in cytosolic calcium (41,42); and the release of AIF, a 50 kDa protein with caspase-like properties (29). The foregoing events, plus the translocation of NF-kappa B to the nucleus, the transfer of phosphatidylserine to the outer leaflet of the plasma membrane, cytoplasmic vacuolization, and the destruction of the genome, all occur in thymocyte apoptosis induced by glucocorticoids or DNA-damaging agents, and all are blocked by bongkrekic acid, an inhibitor of the adenine nucleotide translocase that prevents the opening of the permeability transition pore (42). Similar events, also prevented by bongkrekic acid or cyclosporin (an inhibitor of the opening of the permeability transition pore, among other things), occur in other systems undergoing apoptosis (29,43,44). Furthermore, this group has reported that the opening of the permeability transition pore precedes other changes of apoptosis, including nuclear changes and caspase-3 activation (29).

The views of this laboratory, however, are not universally held. Other groups have found that the fall in mitochondrial transmembrane potential follows other cellular changes of apoptosis (for example, Ref. 2), and the whole question of whether the permeability transition pore really exists or is merely a late consequence of mitochondrial damage remains open. This issue is discussed in the following section.

2. Calcium and Apoptosis

Attention was first drawn to Ca^{2+}-induced cell death many years ago (45). Excessive intracellular Ca^{2+} is thought to contribute to a final common pathway of cytotoxic events leading to formation of reactive oxygen species (ROS), necrosis, or apoptosis. These events include overaction of protein kinase C, Ca^{2+}/calmodulin-dependent protein kinase II, phospholipases, proteases, protein phosphatases, xanthine oxidase, endonucleases, and nitric oxide synthase. Although the exact role of Ca^{2+} in cell killing is unclear, a disturbance of mitochondrial Ca^{2+} handling can be fatal.

Ca^{2+} uptake by mitochondria was discovered more than 30 years ago (46,47). Already at that time it became evident that Ca^{2+} can damage mitochondria (for a review of the early findings, see reference 48).

Routes of Ca^{2+} Transport by Mitochondria. Mitochondria take up and release Ca^{2+} via different routes. This is the basis of the well-known cyclic uptake and release of Ca^{2+}. The normal Ca^{2+} "cycling" across the inner mitochondrial membrane requires little energy (49). However, when the Ca^{2+} release pathway is stimulated, for example, by reactive oxygen species, "cycling" may become excessive and lead to loss of the mitochondrial membrane potential ($\Delta\psi$), general leakiness of the inner mitochondrial membrane, inhibition of ATP synthesis, mitochondrial damage, and cell death (reviewed in Ref. 50). Consistent with this concept, cyclosporine A (CSA), an inhibitor of the prooxidant-induced Ca^{2+} release from mitochondria (see also below), protects against loss of cell viability induced by prooxidants (51) or by nitric oxide (52), and favorably alters liver mitochondrial functions in the postischemic phase at the organ level (53).

The importance of mitochondria as short-term modulators of cytosolic Ca^{2+} under physiological conditions was, until recently, considered minor. However, there is now compelling evidence (reviewed in Refs. 54 and 55) that during physiological cell stimulation mitochondrial Ca^{2+} transport directly participates in the modulation and maintenance of cellular Ca^{2+} homeostasis. Thus mitochondria are of central importance for physiological Ca^{2+} handling. They act as a reservoir for Ca^{2+}, provide much of the ATP used by Ca^{2+}-ATPases, and regulate with Ca^{2+} the activity of intramitochondrial dehydrogenases as well as nucleic acid and protein synthesis (reviewed in Ref. 56).

3.   Prooxidant-Induced, NAD⁺-Linked Ca²⁺ Release: Studies with Isolated Mitochondria

Early studies. Ca^{2+} release from mitochondria was first associated with the oxidation of their pyridine nucleotides by Lehninger's group (57) who showed that enzymatic oxidation of NAD(P)H by acetoacetate or oxaloacetate promotes calcium release, whereas reduction of NAD(P)⁺ by β-hydroxybutyrate in the presence of rotenone prevents it. Shortly thereafter we reported (58,59) that hydroperoxides such as *t*-butylhydroperoxide or H_2O_2 promote mitochondrial pyridine nucleotide oxidation followed by hydrolysis of NAD⁺ to ADPribose and nicotinamide, and by Ca^{2+} release. Since then, many prooxidants were identified who stimulate Ca^{2+} release from intact liver, heart, brain, and kidney mitochondria secondary to pyridine nucleotide oxidation and hydrolysis (reviewed in Ref. 60).

Ca²⁺ Release is Specific and Occurs from Intact Mitochondria.   The specificity of the NAD(P)⁺-linked Ca^{2+} efflux was initially questioned, and it was claimed that Ca^{2+} release stimulated by prooxidants is preceded by mitochondrial swelling and $\Delta\psi$ collapse due to nonspecific increase in membrane permeability. However, several lines of evidence clearly show that nonspecific permeability changes can be dissociated from pyridine nucleotide redox changes and Ca^{2+}

fluxes, and are the consequence of Ca^{2+} "cycling" (summarized in Refs. 50 and 61; see also the following paragraphs).

4. Mitochondrial "Pore" Formation or "Permeability Transition"

It was suggested that Ca^{2+} release induced by hydroperoxides and/or inorganic phosphate is accompanied by opening of a nonspecific "pore" (62,64). Operation of the putative pore, also termed "permeability transition" (65), is supposedly indicated by general-leakiness of the inner mitochondrial membrane as measured by mitochondrial swelling, K^+, and protein release from and sucrose entry into mitochondria. Another important feature of the postulated pore is that it is supposedly closed by ethylene glycol bis(β-aminoethyl ether)-N,N,N',N'-tetraactic acid (EGTA) (62).

It is experimentally proven that neither the unstimulated nor the hydroperoxide-dependent Ca^{2+} release from mitochondria require pore formation (60,66,67): hydroperoxides stimulate Ca^{2+} release from mitochondria in the presence of EGTA, and Ca^{2+} release is neither accompanied by sucrose entry into nor K^+ release from nor swelling of mitochondria, nor $\Delta\psi$ collapse. The Ca^{2+} uptake inhibitor ruthenium red, when added to Ca^{2+}-loaded mitochondria, also prevents the unspecific solute fluxes, yet the unstimulated as well as the hydroperoxide-dependent Ca^{2+}-release occur.

In the absence of EGTA or ruthenium red, that is, when Ca^{2+} "cycling" is allowed, the prooxidant-dependent Ca^{2+} release is accompanied by $\Delta\psi$ collapse (68), sucrose entry into, K^+ release from, and swelling of mitochondria (62–64). Whether this reflects formation of a pore or the initial, reversible phase of the notorious mitochondrial damage due to continuous Ca^{2+} "cycling" remains to be established.

The term "permeability transition" implies the reversible opening of the inner membrane of individual mitochondria. Neither this nor a component of the pore has so far been documented. This concept lacks experimental support and should be replaced by the simple and experimentally verified concept of mitochondrial damage.

Cyclosporine as a Tool for the Study of Mitochondrial Ca^{2+}. CSA and its derivatives are useful for the study of mitochondrial Ca^{2+} handling. A number of them were tested as inhibitors of the matrix-located peptidyl-prolyl *cis-trans* isomerase, pyridine nucleotide hydrolysis, and Ca^{2+} release (69). There is an impressive positive correlation between the extent of inhibition of these three parameters by the different cyclosporine derivatives. This strongly suggests that the prooxidant-dependent Ca^{2+} release engages peptidyl-prolyl *cis-trans* isomerase and further documents the specificity of this pathway. CSA protects mitochondria in an in vitro model of hypoxia/reperfusion injury against loss of complex I activity by preventing pyridine nucleotide hydrolysis and Ca^{2+} cycling (70), protects liver mitochondrial functions at the organ level (53), and prevents under certain conditions apoptosis (see below).

Mitochondrial "Pore" or "Permeability Transition" in Apoptosis. $\Delta\psi$ can be measured in intact cells using a variety of lipophilic cations. The most useful probe for $\Delta\psi$ is JC-1, whereas other commonly used dyes have various drawbacks (71). A collapse of $\Delta\psi$ is taken by many researchers as the sign for the "mitochondrial permeability transition." In light of the studies with isolated mitochondria (see above), it is clear that a decrease in $\Delta\psi$ shown in cells most likely reflects damage of mitochondria and not a "permeability transition." In addition, if in cells the $\Delta\psi$ decrease is inhibitable with CSA, it can be safely concluded that the preservation of $\Delta\psi$ was due to CSA's ability to inhibit the specific, NAD-linked release from intact mitochondria, and that the collapse of $\Delta\psi$ was the consequence of mitochondrial Ca^{2+} "cycling."

5. Prevention of Mitochondrial Ca^{2+} "Cycling" Prevents Apoptosis

Ca^{2+} "cycling" can, in principle, be prevented by either preventing Ca^{2+} release from or Ca^{2+} reuptake by mitochondria. In two models of apoptosis prevention of mitochondrial Ca^{2+} cycling resulted in prevention of apoptosis. In a long-term model, apoptosis was induced in mouse fibrosarcoid cells by tumor necrosis factor-α, which causes ROS formation in mitochondria. Apoptosis was abrogated by ruthenium red (72), the inhibitor of Ca^{2+} reuptake. In a short-term model, apoptosis was induced in freshly isolated rat hepatocytes by nitric oxide and its congeners, which cause Ca^{2+} release from mitochondria. Apoptosis was largely prevented by CSA, and also by trapping the released Ca^{2+} in the cytosol with a Ca^{2+} chelator (52). It should be noted here that the use of CSA or Ca^{2+} chelators as inhibitors of apoptosis may not be successful in long-term models of apoptosis, because the former eventually leads to a detrimental Ca^{2+} overload in mitochondria, and the latter perturbs the cytosolic Ca^{2+} homeostasis.

REFERENCES

1. McConnell KR, Dynan WS, Hardin JA. The DNA-dependent protein kinase catalytic subunit (p460) is cleaved during Fas-mediated apoptosis in Jurkat cells. J Immunol 1997; 158:2083–2089.
2. Krippner A, Matsuno-Yagi A, Gottlieb RA, Babior BM. Loss of function of cytochrome c in Jurkat cells undergoing Fas-mediated apoptosis. J Biol Chem 1996; 271:21629–21636.
3. Adachi S, Cross AR, Babior BM, Gottlieb RA. Bcl-2 and the outer mitochondrial

membrane in the inactivation of cytochrome c during Fas-mediated apoptosis. J Biol Chem 1997; 272:21878–21882.

4. Fang W, Rivard JJ, Mueller DL, Behrens TW. Cloning and molecular characterization of mouse Bcl-x in B and T lymphocytes. J Immunol 1994; 153:4388–4398.

5. Yang J, Liu X, Bhalla K, et al. Prevention of apoptosis by Bcl-2: release of cytochrome c from mitochondria blocked. Science 1997; 275:1129–1132.

6. Hockenbery D, Nunez G, Milliman C, Schreiber RD, Korsmeyer SJ. Bcl-2 is an inner mitochondrial membrane protein that blocks programmed cell death. Nature 1990; 348:334–336.

7. Tanaka S, Saito K, Reed JC. Structure-function analysis of the Bcl-2 oncoprotein: addition of a heterologous transmembrane domain to portions of the Bcl-2β protein restores function as a regulator of cell survival. J Biol Chem 1993; 268:10920–10926.

8. Akao Y, Otsuki Y, Kataoka S, Ito Y, Tsujimoto Y. Multiple subcellular localization of Bcl-2: detection in nuclear outer membrane, endoplasmic reticulum membrane, and mitochondrial membranes. Cancer Res 1994; 54:2468–2471.

9. Lithgow T, van Driel R, Bertram JF, Strasser A. The protein product of the oncogene Bcl-2 is a component of the nuclear envelope, the endoplasmic reticulum, and the outer mitochondrial membrane. Cell Growth Differ 1994; 5:411–417.

10. Nguyen M, Millar DG, Yong VW, Korsmeyer SJ, Shore GC. Targeting of Bcl-2 to the mitochondrial outer membrane by a COOH-terminal signal anchor sequence. J Biol Chem 1993; 268:25265–25268.

11. Monaghan P, Robertson D, Amos TA, et al. Ultrastructural localization of Bcl-2 protein. J Histochem Cytochem 1992; 40:1819–1825.

12. Krajewski S, Tanaka S, Takayama S, et al. Investigation of the subcellular distribution of the Bcl-2 oncoprotein: residence in the nuclear envelope, endoplasmic reticulum, and outer mitochondrial membranes. Cancer Res 1993; 53:4701–4714.

13. de Jong D, Prins FA, Mason DY, et al. Subcellular localization of the Bcl-2 protein in malignant and normal lymphoid cells. Cancer Res 1994; 54:256–260.

14. Borner C, Martinou I, Mattmann C, et al. The protein Bcl-2α does not require membrane attachment, but two conserved domains to suppress apoptosis. J Cell Biol 1994; 126:1059–1068.

15. Reed JC, Zha H, Aime-Sempe C, Takayama S, Wang HG. Structure-function analysis of Bcl-2 family proteins. Regulators of programmed cell death. Adv Exp Med Biol 1996; 406:99–112.

16. Sato T, Hanada M, Bodrug S, et al. Interactions among members of the Bcl-2 protein family analyzed with a yeast two-hybrid system. Proc Natl Acad Sci USA 1994; 91:9238–9242.

17. Hanada M, Aime-Sempe C, Sato T, Reed JC. Structure-function analysis of Bcl-2 protein. Identification of conserved domains important for homodimerization with Bcl-2 and heterodimerization with Bax. J Biol Chem 1995; 270:11962–11969.

18. Yin XM, Oltvai ZN, Korsmeyer SJ. BH1 and BH2 domains of Bcl-2 are required for inhibition of apoptosis and heterodimerization with Bax. Nature 1994; 369:321–323.

19. Hunter JJ, Bond BL, Parslow TG. Functional dissection of the human Bcl2 protein: sequence requirements for inhibition of apoptosis. Mol Cell Biol 1996; 16:877–883.

20. Wang K, Yin XM, Chao DT, Milliman CL, Korsmeyer SJ. BID: a novel BH3 domain-only death agonist. Genes Dev 1996; 10:2859–2869.

21. Zha H, Aime-Sempe C, Sato T, Reed JC. Proapoptotic protein Bax heterodimerizes with Bcl-2 and homodimerizes with Bax via a novel domain (BH3) distinct from BH1 and BH2. J Biol Chem 1996; 271:7440–7444.

22. Ottilie S, Diaz JL, Horne W, et al. Dimerization properties of human BAD. Identification of a BH-3 domain and analysis of its binding to mutant Bcl-2 and Bcl-x$_L$ proteins. J Biol Chem 1997; 272:30866–30872.

23. Wang H-G, Miyashita T, Takayama S, et al. Apoptosis regulation by interaction of Bcl-2 protein and Raf-1 kinase. Oncogene 1994; 9:2751–2756.

24. Paumen MB, Ishida Y, Han H, et al. Direct interaction of the mitochondrial membrane protein carnitine palmitoyltransferase I with Bcl-2. Biochem Biophys Res Commun 1997; 231:523–525.

25. McCormick F. Ras-related proteins in signal transduction and growth control. Mol Reprod Dev 1995; 42:500–506.

26. Shibasaki F, Kondo E, Akagi T, McKeon F. Suppression of signalling through transcription factor NF-AT by interactions between calcineurin and Bcl-2. Nature 1997; 386:728–731.

27. Myers KM, Fiskum G, Liu Y, et al. Bcl-2 protects neural cells from cyanide/aglycemia-induced lipid oxidation, mitochondrial injury, and loss of viability. J Neurochem 1995; 65:2432–2440.

28. Zamzami N, Marchetti P, Castedo M, et al. Sequential reduction of mitochondrial transmembrane potential and generation of reactive oxygen species in early programmed cell death. J Exp Med 1995; 182:367–377.

29. Susin SA, Zamzami N, Castedo M, et al. The central executioner of apoptosis: multiple connections between protease activation and mitochondria in Fas/APO-1/CD95- and ceramide-induced apoptosis. J Exp Med 1997; 186:25–37.

30. Shimizu S, Eguchi Y, Kamiike W, Matsuda H, Tsujimoto Y. Bcl-2 expression prevents activation of the ICE protease cascade. Oncogene 1996; 12:2251–2257.

31. Greenhalf W, Stephan C, Chaudhuri B. Role of mitochondria and C-terminal membrane anchor of Bcl-2 in Bax induced growth arrest and mortality in *Saccharomyces cerevisiae*. FEBS Lett 1996; 380:169–175.

32. Liu X, Kim CN, Yang J, Jemmerson R, Wang X. Induction of apoptotic program in cell-free extracts: requirement for dATP and cytochrome c. Cell 1996; 86:147–157.

33. Chinnaiyan AM, O'Rourke K, Lane BR, Dixit VM. Interaction of CED-4 with CED-3 and CED-9: a molecular framework for cell death. Science 1997; 275:1122–1126.

34. Wu D, Wallen HD, Nunez G. Interaction and regulation of subcellular localization of CED-4 by CED-9. Science 1997; 275:1126–1129.

35. Zou H, Henzel WJ, Liu X, Lutschg A, Wang X. Apaf-1, a human protein homologous to *C. elegans* CED-4, participates in cytochrome c-dependent activation of caspase-3. Cell 1997; 90:405–413.

36. Kluck RM, Bossy-Wetzel E, Green DR, Newmeyer DD. The release of cytochrome c from mitochondria: a primary site for Bcl-2 regulation of apoptosis. Science 1997; 275:1132–1136.

37. Kluck RM, Martin SJ, Hoffman BM, et al. Cytochrome c activation of CPP32-like

proteolysis plays a critical role in a *Xenopus* cell-free apoptosis system. EMBO J 1997; 16:4639–4649.

38. Adachi S, Gottlieb RA, Babior BM. Lack of release of cytochrome c from mitochondria into cytosol early in the course of Fas-mediated apoptosis of Jurkat cells. J Biol Chem, in press.
39. Li F, Srinivasan A, Wang Y, et al. Cell-specific induction of apoptosis by microinjection of cytochrome c. J Biol Chem 1997; 272:30299–30305.
40. Rosse T, Olivier R, Monney L, et al. Bcl-2 prolongs cell survival after Bax-induced release of cytochrome c. Nature 1998; 391:496.
41. Hirsch T, Marzo I, Kroemer G. Role of the mitochondrial permeability transition pore in apoptosis. Biosci Rep 1997; 17:67–76.
42. Macho A, Hirsch T, Marzo I, et al. Glutathione depletion is an early and calcium elevation is a late event of thymocyte apoptosis. J Immunol 1997; 158:4612–4619.
43. Zamzami N, Susin SA, Marchetti P, et al. Mitochondrial control of nuclear apoptosis. J Exp Med 1996; 183:1533–1544.
44. Marchetti P, Hirsch T, Zamzami N, et al. Mitochondrial permeability transition triggers lymphocyte apoptosis. J Immunol 1996; 157:4830–4836.
45. Schanne FA, Kane A, Young E, Farber J. Calcium dependence of toxic cell death: a final common pathway. Science 1979; 206:699–700.
46. Vasington FD, Murphy PA. Active binding of calcium by mitochondria. Fed Proc 1961; 20:146.
47. DeLuca HF, Engström GW. Calcium uptake by rat kidney mitochondria. Proc Natl Acad Sci USA 1961; 47:1744–1750.
48. Lehninger AL, Carafoli E, Rossi CS. Energy-linked ion movements in mitochondrial systems. Adv Enzymol 1967; 29:259–320.
49. Carafoli E. Intracellular calcium homeostasis. Annu Rev Biochem 1987; 56:395–433.
50. Richter C, Kass GEN. Oxidative stress in mitochondria: its relationship to cellular Ca^{2+} homeostasis, cell death, proliferation, and differentiation. Chem Biol Interact 1991; 77:1–23.
51. Kass GEN, Juedes M, Orrenius S. Cyclosporine A protects hepatocytes against prooxidant-induced cell killing. A study on the role of mitochondrial Ca^{2+} cycling in cytotoxicity. Biochem Pharmacol 1992; 44:1995–2003.
52. Richter C, Gogvadze V, Schlapbach R, Schweizer M, Schlegel J. Nitric oxide kills hepatocytes by mobilizing mitochondrial calcium. Biochem Biophys Res Commun 1994; 205:1143–1150.
53. Kurokawa T, Kobayashi H, Harada A, et al. Beneficial effects of cyclosporine on postischemic liver injury in rats. Transplantation 1992; 53:308–311.
54. Rizzuto R, Bastianutto C, Brini M, Murgia M, Pozzan T. Mitochondrial Ca^{2+} homeostasis in intact cells. J Cell Biol 1994; 126:1183–1194.
55. Hajnoczky G, Robb-Gaspers LD, Seitz M, Thomas AP. Decoding of cytosolic calcium oscillations in the mitochondria. Cell 1995; 82:415–424.
56. Richter C. Control of the prooxidant-dependent calcium release from intact liver mitochondria. Redox Rep 1996; 2:217–221.
57. Lehninger AL, Vercesi A, Bababunmi EA. Regulation of Ca^{2+} release from mitochondria by the oxidation-reduction state of pyridine nucleotides. Proc Natl Acad Sci USA 1978; 75:1690–1694.

58. Lotscher HR, Winterhalter KH, Carafoli E, Richter C. Hydroperoxides can modulate the redox state of mitochondrial pyridine nucleotides and the calcium balance in rat liver mitochondria. Proc Natl Acad Sci USA 1979; 76:4340–4344.

59. Lotscher HR, Winterhalter KH, Carafoli E, Richter C. Hydroperoxide-induced loss of pyridine nucleotides and release of calcium from rat liver mitochondria. J Biol Chem 1980; 255:12579–12583.

60. Richter C, Schlegel J. Mitochondrial Ca^{2+} release induced by prooxidants. Toxicol Lett 1993; 67:119–127.

61. Richter C. Mitochondrial calcium transport. In: Neuberger A, Van Deenen LLM, eds. New Comprehensive Biochemistry. Amsterdam: Elsevier, 1992:349–358.

62. Al-Nasser I, Crompton M. The reversible Ca^{2+}-induced permeabilization of rat liver mitochondria. Biochem J 1986; 239:19–29.

63. Crompton M, Ellinger H, Costi A. Inhibition by cyclosporin A of a Ca^{2+}-dependent pore in heart mitochondria activated by inorganic phosphate and oxidative stress. Biochem J 1988; 255:357–360.

64. Crompton M, Costi A. A heart mitochondrial Ca^{2+}-dependent pore of possible relevance to reperfusion-induced injury. Biochem J 1990; 266:33–39.

65. Broekemeier KM, Dempsey ME, Pfeiffer DR. Cyclosporine A is a potent inhibitor of the inner membrane permeability transition in liver mitochondria. J Biol Chem 1989; 264:7826–7830.

66. Schlegel J, Schweizer M, Richter C. "Pore" formation is not required for the hydroperoxide-induced Ca^{2+} release from rat liver mitochondria. Biochem J 1992; 285: 65–69.

67. Richter C, Schlegel J, Schweizer M. The prooxidant-induced Ca^{2+} release from mitochondria: specific versus nonspecific pathways. Ann N Y Acad Sci 1992; 663:262–268.

68. Lotscher HR, Winterhalter KH, Carafoli E, Richter C. The energy-state of mitochondria during the transport of Ca^{2+}. Eur J Biochem 1980; 110:211–216.

69. Schweizer M, Schlegel J, Baumgartner D, Richter C. Sensitivity of mitochondrial peptidyl-prolyl cis-trans isomerase, pyridine nucleotide hydrolysis, and Ca^{2+} release to cyclosporine A and related compounds. Biochem Pharmacol 1993; 45:641–646.

70. Gogvadze V, Richter C. Cyclosporine A protects mitochondria in an in vitro model of hypoxia/reperfusion injury. FEBS Lett 1993; 333:334–338.

71. Salvioli S, Ardizzoni A, Franceschi C, Cossarizza A. JC-1 but not DiOC6(3) or rhodamine 123, is a reliable fluorescent probe to assess delta-psi changes in intact cells: implications for studies on mitochondrial functionality during apoptosis. FEBS Lett 1997; 411:77–82.

72. Hennet T, Richter C, Peterhans E. Tumour necrosis factor-α induces superoxide anion generation in mitochondria of L929 cells. Biochem J 1993; 289:587–592.

10

Role of Mitochondria in Oxidative Stress Associated with Aging and Apoptosis: Studies in Intact Cells

Juan Sastre, Federico V. Pallardó, José García de la Asunción, Miguel A. Asensi, Juan M. Esteve, and José Viña-Ribes
University of Valencia, Valencia, Spain

I. THE MITOCHONDRIAL THEORY OF AGING

One of the most relevant theories raised to explain aging is the free radical theory of aging, which was first proposed by Harman more than 40 years ago (1). According to this theory, oxygen-derived free radicals are responsible for age-associated impairment at the cellular and tissue levels (1).

The free radical theory of aging assumes that cellular antioxidant systems are not able to cope with the oxygen free radicals generated continuously throughout cell life. Thus cellular aging would be associated with a "chronic" oxidative stress, which was defined by Sies (2) as a disturbance in the balance between prooxidants and antioxidants, in favor of the former.

Experimental evidence supports the free radical theory of aging, especially the extension of mean life span obtained by increasing antioxidant defense. Thus administration of antioxidants can increase the life span of flies (3,4). Orr and Sohal (5) have recently found that simultaneous overexpression of copper-zinc superoxide dismutase (Cu, Zn-SOD) and catalase genes in transgenic *Drosophila* extends their mean and maximum life span. Furthermore, these transgenic flies exhibited a delayed loss of physical performance and a lower amount of protein oxidative damage (5). Barja and coworkers (6) have also found that a simultaneous induction of SOD, glutathione reductase, GSH, and ascorbate, following complete inhibition of catalase activity, increases the mean life span in frogs.

Reactive oxygen species (ROS), a term used for oxygen free radicals and peroxides, are generated continuously in aerobic cells and especially in the mitochondrial respiratory chain (7,8). Indeed, 1–2% of all oxygen used by mammalian mitochondria in state 4 does not form water but oxygen activated species (7,8).

On this basis, Miquel et al. (9) proposed the mitochondrial theory of cell aging. This theory suggests that senescence is a by-product of oxyradical attack to the mitochondrial genome of fixed postmitotic cells. According to Miquel and Fleming (10), it is essential that cells contain "differentiated" mitochondria for cellular aging to occur. Such mitochondria use high levels of O_2 due to the high energy requirements of postmitotic cells, thereby releasing oxygen radicals which exceed cellular homeostatic protection (10).

Rubner (11) was the first to point out the inverse relationship between the rate of oxygen consumption and the maximum life span. Much later, Harman (12) suggested that mitochondria might be the biological clock in aging since the rate of oxygen consumption determines the rate of accumulation of products resulting from mitochondrial damage produced by free radical reactions. Rubner's theory explains the differences in maximum life-span potential among numerous, but not all species. Exceptions to this theory are birds and primates (13). These groups exhibit at the same time high oxygen consumption and high longevity. The explanation for this paradox is that mitochondrial production of ROS is lower in pigeons than in rats (13–15). Thus mitochondria from birds use oxygen more efficiently and exhibit less free radical leakage through the respiratory chain. Studying up to seven species, Sohal and coworkers (16,17) have found that mitochondria from shorter-lived species produce relatively higher amounts of ROS than those from longer-lived species. Hence the rate of ROS production, and not merely the rate of oxygen consumption, appears to determine the maximum life-span potential.

ROS are generated by complexes I, II, and III of the mitochondrial respiratory chain (18), but not by cytochrome oxidase, which seems to exhibit SOD activity (19,20). The processes forming ROS are nonenzymatic chemical reactions whose rate increases with the oxygen concentration and electronegativity—or degree of reduction—of autooxidizable redox carriers of complexes II and III (21,22). Thus mitochondrial ROS production is markedly decreased in state 4 to state 3 transition when using complex II–linked substrates such as succinate (7). However, when using complex I–linked substrates such as pyruvate and malate, mitochondrial ROS production is maintained at comparable rates in states 3 and 4 (23). Since mitochondria in cells are considered to be in a state between states 3 and 4 (24), ROS generation by complex I may be of major importance for mitochondrial aging.

At present, numerous studies have shown that oxidative damage to mitochondrial DNA, proteins, and lipids occurs with aging (25,26). This oxidative stress may be responsible for age-associated deficits in mitochondrial function as well as changes in mitochondrial morphology, as we will discuss later in this chapter. The role of old mitochondria in cell aging has also been outlined by the degeneration induced in cells microinjected with mitochondria isolated from fibroblasts of old rats (27).

Several studies have observed a decline in activities of complexes I, II, and especially IV (28–30). Moreover, the respiratory activity of isolated mitochondria decreases with age in liver, skeletal muscle, and brain (31–33). Age-related changes in mitochondrial membrane potential and transport systems have also been reported (26,34–36).

An increased generation of oxygen free radicals may be responsible for the decline in the activity of mitochondrial membrane proteins, such as metabolite carriers and respiratory chain complexes. In fact, it is known that exposure of mitochondria to free radicals causes impairment of the mitochondrial inner membrane proteins (37). In addition, studies in isolated mitochondria have shown that acute oxidative stress causes an inhibition of mitochondrial respiration (27).

II. MITOCHONDRIA ARE DAMAGED WITHIN INTACT CELLS OF OLD ANIMALS: THE METABOLIC APPROACH

All findings mentioned above support the hypothesis that mitochondrial damage plays a key role in the aging process. However, most mitochondrial changes were found in experiments using isolated mitochondria. Thus some of these effects could be due to an increased susceptibility of old mitochondria to the stress caused by the isolation procedure. Moreover, when intact cells were not used, the mitochondrial-cytosolic interactions were also ignored. Thus the use of isolated mitochondria might be misleading for aging studies. Hence whole cells, such as isolated hepatocytes, should be considered an excellent model for studies on mitochondrial aging (26).

Using isolated hepatocytes, we measured the rate of biochemical pathways which critically depend on mitochondrial function (26). Table 1 shows that gluconeogenesis from lactate plus pyruvate, but not from glycerol or fructose, fell with aging in isolated hepatocytes. Gluconeogenesis from lactate involves mitochondria, whereas it does not from glycerol or fructose. Figure 1 shows that the lower rate of gluconeogenesis from lactate plus pyruvate is due to an impaired transport of malate across mitochondrial membrane using the dicarboxylate carrier (26). Furthermore, posttranscriptional modifications appear to be involved in the age-related impairment of such carriers, since its gene expression does not change

Table 1   Rate of Gluconeogenesis in Hepatocytes Isolated from Young and Old Rats

Additions	3-month-old	22-month-old	32-month-old
Krebs-Henseleit	0.39 ± 0.09 (9)	0.33 ± 0.11 (10)	0.34 ± 0.10 (7)
L(10 mM) + P(1 mM)	3.5 ± 0.3 (7)	2.5 ± 0.5 (10)*	2.2 ± 0.2 (6)*
Fructose (10 mM)	10.4 ± 1.4 (5)	9.5 ± 1.5 (8)	9.5 ± 0.6 (3)
Glycerol (10 mM)	2.3 ± 0.6 (4)	2.0 ± 0.5 (9)	1.8 ± 0.9 (4)

Abbreviations used: L = lactate, P = pyruvate.
The rate of glucose synthesis is expressed as μmol/min/g of dry weight.
Results are expressed as means ± SD for the number of experiments in parentheses.
Statistical difference is indicated as follows: *$P < .01$ versus hepatocytes from 3-month-old rats.

with age (26). Oxidized proteins accumulate with cellular aging and oxidative posttransductional modifications have been reported as the main cause for the loss of some enzyme activities such as that of malic enzyme (38). These findings support the hypothesis that an increased generation of oxygen free radicals may be responsible for the decline in the activity of mitochondrial membrane proteins, such as metabolite carriers. In fact, it is known that exposure of mitochondria to free radicals causes impairment of the mitochondrial inner-membrane proteins (37).

    Ketogenesis from oleate, which depends on mitochondrial performance, also decreased in hepatocytes from old animals (26). Hansford (34) reported an

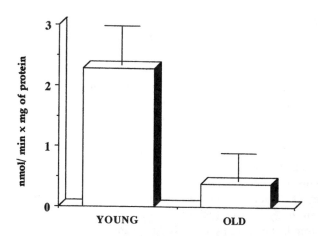

Figure 1   Rate of malate efflux from mitochondria isolated from livers of young and old rats. Data are expressed as mean values of three individual experiments. Statistical significance is expressed as $P < .01$.

age-related decrease in the activity of acylcarnytine-carnitine translocation. Since the activity of the carnitine acyltransferase system controls the rate of oxidation of fatty acids, it may be responsible for the age-related decrease in ketogenesis that we observed.

Age-related decreases in the activities of other mitochondrial anion carrier proteins, such as the phosphate carrier and ATP/ADP translocation in liver mitochondria (35,36), and Ca^{2+}, adenine nucleotide, and pyruvate carriers in heart mitochondria (39–42), have also been reported.

Nevertheless, the fact that some mitochondrial membrane carriers are impaired upon aging does not necessarily imply that all mitochondrial functions are affected by aging. For instance, the rate of urea synthesis in hepatocytes does not change with age (26).

Hence metabolic studies using whole cells show that aging affects mitochondrial function by an impairment of specific processes which involve the mitochondrial inner membrane, such as malate transport. Posttranscriptional modifications appear to be involved in the age-related loss of these carriers. These changes are likely due to the chronic oxidative stress associated with mitochondrial aging. Metabolic pathways which do not involve mitochondria are not affected. All these findings point out the role of mitochondria as key targets of the cellular changes which occur with aging.

III. MITOCHONDRIA ARE DAMAGED WITHIN INTACT CELLS OF OLD ANIMALS: THE FLOW CYTOMETRIC APPROACH

Another way of studying mitochondrial aging within intact cells of old animals is by flow cytometry, which allows a noninvasive analysis of individual cells. Thus the effect of aging on liver mitochondrial membrane potential has been studied by flow cytometry (26,43). Figure 2 shows that aging of the liver causes a significant decrease in mitochondrial membrane potential (26,43). This may reduce the energy supply in old hepatocytes since the mitochondrial membrane potential is the driving force for ATP synthesis. Furthermore, this impairment in mitochondrial function may affect mitochondrial protein synthesis (44) and might be involved in the age-related decrease in the level of mitochondrial transcripts (45).

In agreement with the age-related decrease in mitochondrial membrane potential, changes in respiratory activity have also been reported (31,36). Moreover, we have found an impairment in energization of brain and liver mitochondria under respiratory state 4 (46).

Studies in isolated mitochondria have shown that acute oxidative stress causes an inhibition of mitochondrial respiration (27), which affects the mito-

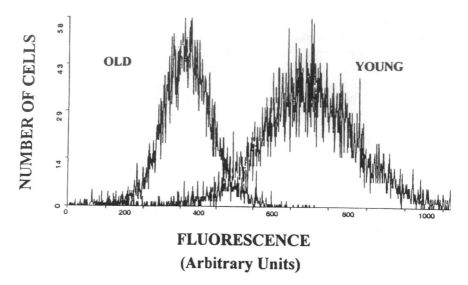

FLUORESCENCE
(Arbitrary Units)

Figure 2   Representative histogram of mitochondrial membrane potential, estimated by flow cytometry using rhodamine 123.

chondrial membrane potential. Moreover, hyperoxia reduces the mitochondrial membrane potential in microvascular cells (47). Hence the oxidative stress associated with aging may be responsible, at least in part, for the age-related impairment in mitochondrial membrane potential and respiratory activity. Indeed, intracellular peroxide levels increase with age in whole cells (26,43), which correlates with parallel changes in peroxide generation by isolated mitochondria (26,48; see Fig. 3). It is likely that the accumulation of peroxides in whole cells upon aging comes from the continuous peroxide generation by mitochondria throughout the cell life, although we cannot rule out that other structures, such as peroxisomes, may also have a role.

On the other hand, mitochondrial morphology is important because changes in mitochondrial ultrastructure modulate mitochondrial function (49). Indeed, volume-dependent regulation of matrix protein packing modulates metabolite diffusion and, in turn, mitochondrial metabolism (49). Enlargement, matrix vacuolization, and altered cristae have been evidenced in mitochondria from old animals by flow cytometry and electron microscopy (26,46,50,51). Moreover, structural complexity of brain and liver mitochondria increased upon aging (46). Alterations of mitochondrial crests which occur in old mitochondria may be responsible for the age-related impairment in mitochondrial membrane potential that we have found.

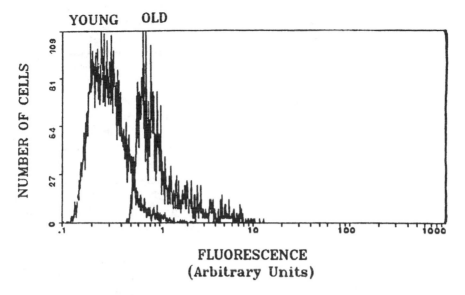

Figure 3 Representative histogram of peroxide generation by liver mitochondria, estimated by flow cytometry using dihydrorhodamine 123.

In spite of changes in mitochondrial size, the number of mitochondria per cell and the mitochondrial inner-membrane mass did not change with age in liver (26). The age-related increase in mitochondrial size can be explained by an increase in mitochondrial matrix volume, without changes in inner mitochondrial membrane mass. In fact, the electron microscopy studies showed a higher degree of vacuolization in liver mitochondria from old rats than in controls.

It is well known that acute oxidative stress causes mitochondrial swelling (37). Thus age-associated chronic oxidative stress may be the cause, at least in part, of mitochondrial swelling. Furthermore, a correlation between changes in mitochondrial morphology and function seems to occur upon aging.

IV. OXIDATION OF MITOCHONDRIAL GLUTATHIONE IN AGING

The involvement of glutathione metabolism in aging has been known since the work of Pinto and Bartley (52) in the late 1960s. Later, different authors found age-related glutathione oxidation in several animal models (4,53,54) and even in humans (55). Glutathione oxidation in aging can be caused by an increased production of oxidative species, a decreased antioxidant capacity, or both.

In general terms, aging is associated with a decrease in the activities of enzymes which catalyze reactions tending to reduce GSSG, such as glucose-6 phosphate dehydrogenase or glutathione reductase, rather than increasing the activity of those enzymes which favor oxidation of glutathione, such as glutathione peroxidase or transferase (4).

So far, glutathione compartmentation during aging has not received sufficient attention. Mitochondria cannot synthesize GSH because they do not have γ-glutamylcysteine synthetase or glutathione synthetase activities (56). Thus mitochondria obtain GSH by transport from the cytosol. We have recently found that glutathione oxidation increases in mitochondria from liver, kidney, and brain of old rats (57). It is worth noting that this increase was much higher in mitochondria than in the whole tissue. These results support the idea that mitochondria are a major source of free radicals in aging, as suggested by different authors (9,58), and emphasize the relevance of mitochondria as primary targets of damage associated with aging (9).

The glutathione system protects against oxidative damage to lipids, proteins, and DNA. The occurrence of oxidative injury to these cell components proves that the antioxidant action of GSH and related enzymes in aged cells is not completely effective. It would be of great importance to find ways of improving the function of the glutathione system by increasing GSH levels and/or the activities of glutathione peroxidase and reductase.

V. AGE-ASSOCIATED DAMAGE TO MITOCHONDRIAL DNA CORRELATES WITH OXIDATION OF MITOCHONDRIAL GLUTATHIONE

Mitochondrial DNA (mtDNA) is especially susceptible to oxidative damage and mutation because it lacks protective histones or effective repair systems (59). Indeed, levels of oxidative damage to mtDNA are several times higher than those in nuclear DNA, and mtDNA mutates several times more frequently than nuclear DNA (25,60–62).

As already mentioned, a great deal of experimental evidence shows that age-associated mitochondrial damage is due to oxidative stress (25,30). Oxidative DNA damage represents another index of oxidative stress and serves to confirm the involvement of free radical damage in aging. In fact, oxidative lesions in mtDNA accumulate with age in human and rodent tissues (57,63–65). This oxidative damage to mtDNA may affect transcription of mitochondrial genes (66). Indeed, an age-related decrease in the levels of mitochondrial transcripts in some rat tissues and in *Drosophila* have been reported (45,67).

Ames et al. (64) calculated that oxygen free radicals are responsible for approximately 10,000 DNA base modifications per cell per day. DNA repairing enzymes are able to remove most of these oxidative lesions, but not all of them (68). Thus unrepaired oxidative lesions in DNA, such as 8-oxo-7,8-dihydro-2'-deoxyguanosine, accumulate with age (68). As mentioned previously, most of this damage occurs in mtDNA, not in nuclear DNA (62).

The age-associated increase in the level of common deletions produced spontaneously in the absence of inherited cases seems to be very low (<0.1%) and may not be significant (25). Nevertheless, these deletions may represent only a small portion of the numerous deletions and point mutations which could accumulate with age (69). Indeed, several studies have found increased deletions, point mutations, and aberrant forms in mtDNA of postmitotic tissues upon aging (70–73). Furthermore, since mtDNA has no introns, any mutation affects a DNA coding sequence (59). Thus it was suggested that mitochondrial DNA mutations may be important contributors to aging and neurodegenerative diseases (10,62,74).

Oxidative damage to proteins and DNA should not be considered separately because they can potentiate each other. Thus accumulation of inactive forms of DNA repairing enzymes might enhance the accumulation of DNA oxidative damage and vice versa. Moreover, a loss of repairing enzymes leads to an increased spontaneous mutation rate when oxidative lesions to guanine residues are present. Therefore oxidative lesions in DNA exhibit mutagenic potential. On these bases, oxidative damage to DNA appears to be not only involved in cell aging, but also in the pathogenesis of associated diseases, such as cancer.

Mitochondrial glutathione plays a key role in the protection against oxidative damage associated with aging. Indeed, Figure 4 shows that oxidative damage to mtDNA which occurs upon aging is directly related to oxidation of mitochondrial glutathione (57). Similarly, glutathione oxidation may also be correlated with the oxidative damage to mitochondrial lipids and proteins related to aging. A change in the glutathione redox status would indicate that mitochondrial antioxidant systems cannot cope with the oxidant species generated throughout cell life. Therefore glutathione oxidation may occur prior to oxidative damage to other mitochondrial components, and it might be an early event in oxidative stress associated with mitochondrial aging. This points out the importance of maintaining a reduced glutathione status to protect cells against oxidative damage of important molecules such as DNA.

VI. ANTIOXIDANTS PREVENT AGE-ASSOCIATED OXIDATIVE DAMAGE TO MITOCHONDRIA

The free radical theory of aging proposed by Harman (1) is especially attractive because it provides a rationale for intervention, that is, antioxidant administration may slow the aging process. Indeed, certain impairments associated with aging

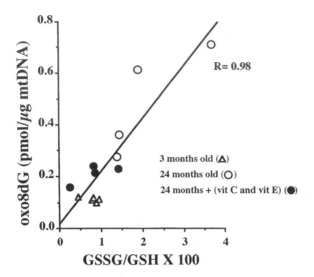

Figure 4 Relationship between values of mitochondrial GSSG/GSH ratio and levels of oxo-8-deoxyguanosine in mitochondrial DNA from livers of rats. Line of regression and correlation coefficient (r) are shown. 3-month-old (Δ), 24-month-old (○), and 24-month-old rats fed with supplemented diet (vitamins C and E) (●).

can be prevented by antioxidant administration (3,4,75). In 1979, Miquel and Economos (3) were the first to show that administration of thiazolidine carboxylate increases the vitality and mean life span of mice. Later, Furukawa et al. (75) reported that oral administration of glutathione protects against the age-associated decline in immune responsiveness. More recently, we found that administration of thiol-containing antioxidants protects against age-associated glutathione depletion in mouse tissues as well as partially preventing the age-related decline in neuromuscular coordination (4). These antioxidants also increased the mean life span of *Drosophila* (4). In addition, we found that some antioxidants, such as thiazolidine carboxylate derivatives or vitamins C and E, protect against mitochondrial glutathione oxidation and mtDNA oxidative damage associated with aging (57). Recently we found that EGb 761, a standardized *Ginkgo biloba* extract with antioxidant properties, prevents age-related changes in mitochondrial function and morphology by protecting against age-associated oxidative damage to mitochondria (46). These protective effects are summarized in Figure 5. Hence the beneficial effects of antioxidant treatment on physiological performance could be associated with the prevention of age-related oxidative stress.

The importance of an adequate dietary intake of antioxidants is corrobo-

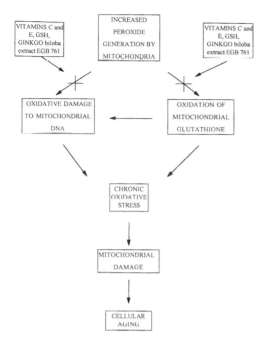

Figure 5 Protective effects of vitamins C and E as well as the *Ginkgo biloba* extract EGb 761 on the age-associated oxidative damage to mitochondria.

rated by the fact that a decline in antioxidant defense, such as that obtained when dietary ascorbate is insufficient, is associated with oxidative stress evidenced by increased oxidative damage to DNA (64,76).

When using thiol-containing antioxidants such as GSH or N-acetyl cysteine, the dose must be chosen carefully to enhance the cellular antioxidant defense without increasing intracellular cysteine levels, which may be prooxidant. Free cysteine undergoes spontaneous autooxidation and thus it may cause cell damage (77,78). Treatment of rats with high doses of N-acetyl cysteine, administered iP, causes a decrease in hepatic GSH levels (79). Cysteine accumulates intracellularly because it is not used in glutathione synthesis when this is inhibited by physiological levels of GSH. Therefore the dose of thiol antioxidants must be chosen carefully to prevent an excessive increase in intracellular cysteine levels.

Recent research shows that vitamin C, which can be prooxidant in vitro in the presence of heavy metals, acts as an antioxidant in plasma even in the presence of high levels of free iron (80).

The use of antioxidant supplementation in aging still requires further research, especially epidemiological studies in humans. Recent epidemiological studies suggest that dietary supplementation with antioxidants should not be applied to certain groups. In fact, it was found that administration of β-carotene to smokers increases the incidence of cancer in these patients (81). On the other hand, vitamin E administration may increase the acute response of neutrophils during exercise in older patients (82). Therefore, administration of antioxidants to patients must be done carefully and should not be recommended to the overall population.

On the other hand, it should be emphasized that it is possible to prolong the life span of some species without antioxidant supplementation, that is, by dietary or caloric restriction (83). It is well known that dietary or caloric restriction can increase the mean life span of animals (83). Furthermore, calorie-restricted rats exhibit less oxidative damage and a delay in the age-dependent decline of some antioxidant defenses (83–85). A restriction of only certain nutrients may be enough to extend life span. Indeed, Richie et al. (86) reported that methionine restriction increases longevity in rats.

In conclusion, administration of some, but not all antioxidants may prevent the oxidative stress and the physiological impairment associated with aging. Nevertheless, further studies on dietary supplementation with antioxidants need to be carried out to establish the adequate doses that provide beneficial action.

VII. CELL DEATH: APOPTOSIS IS A COMMON FEATURE IN ALL CELLS

The maintenance of tissue homeostasis involves the removal of superfluous or damaged cells. Failure to accomplish these goals will induce malformations during development or cancer. This process is often referred to as "programmed cell death" (PCD), because it is believed that cells activate an intrinsic death program contributing to their own death. It has also been termed "apoptosis" which was originally related to the morphological description of this universal mechanism. Now these terms are used as synonyms.

Apoptosis contrasts with necrosis (for a review see Refs. 87–90). In necrosis cell death is induced by osmotic, physical, or chemical damage. These agents cause an early disruption of external and internal membranes, releasing denatured proteins into the cellular space and inducing an inflammatory response in the vicinity of the dying cell. By contrast, in apoptosis cells undergo nuclear condensation and shrinkage as major morphological features. A ladder-type fragmentation affecting nuclear DNA, but not mtDNA, is typical in apoptosis. Recent studies, however, criticize the concept that alterations of the nucleus are obligatory events in PCD (90).

VIII. ROLE OF FREE RADICALS IN APOPTOSIS: OXIDATIVE STRESS IN APOPTOTIC CELLS

It is well known that cellular redox status modulates various aspects of cellular function. Recent reports show that reactive oxygen species play a major role in apoptosis (91,92). Apoptosis can be induced by exogenous sources of reactive oxygen species, such as t-butyl hydroperoxide, in several cell lines (91). A decrease in GSH concentration is a common feature in most apoptotic cells. Van den Dobbelsteen et al. (93) recently reported that apoptosis in Jurkat cells causes an efflux, not an oxidation, of cellular glutathione. However, we have found a clear increase of oxidized glutathione in fibroblasts 24 h after induction of apoptosis (94). GSSG concentration rose from 0.025 ± 0.005 nmol/mg of protein in control cells to 0.12 ± 0.06 nmols/mg of protein in apoptotic fibroblasts.

Glutathione depletion increases the percentage of apoptotic cells in a given population. Thus antioxidants have been shown to protect against apoptosis in different experimental models (92,95). Similarly, inhibition of antioxidant pathways, such as glutathione synthesis by dl-buthionine sulfoximine, provokes apoptosis in different cell lines. It was suggested (96) that proto-oncogene Bcl-2, which is an inhibitor of apoptosis, exerts an antioxidant action, mainly in mitochondria. However, several authors have shown that apoptosis can be induced under anaerobic conditions (97,98). As suggested by Jacobson and Raff (97), it is possible that ROS may play a role in different signal transduction pathways that induce apoptosis but do no play any role during the effector phase of programmed cell death.

IX. ROLE OF MITOCHONDRIA IN APOPTOSIS

Recent experiments have shown that a change in mitochondrial activity is another common feature in apoptosis (90,99). Zamzami et al. (100) have shown that apoptosis is closely related to mitochondrial impairment.

A recent report by Kroemer et al. (90) shows that the permeability transition (PT) of mitochondria specifically increases during apoptosis. Permeability transition involves the opening of the so-called PT pores, which are identical to mitochondrial megachannels and which are located at the inner/outer membrane contact sites. In a very interesting experiment, Kroemer and coworkers (100) showed that isolated mitochondria can induce nuclear DNA digestion in a cell-free system when PT is activated. In addition, inhibition of PT blocks apoptosis. Thus mitochondrial PT appears to be a critical step in apoptosis (100).

Mitochondria are an important source of reactive oxygen species. Leakage of electrons along the mitochondrial electron transport chain causes the formation of $O_2{}^-$ anion radicals. We have found a 40% decrease in mitochondrial membrane potential from apoptotic fibroblasts and a 45% increase in the peroxide content of apoptotic fibroblasts (see Fig. 6). The effect of mitochondrial-induced oxidative stress during apoptosis requires further studies.

Thus new evidence suggests that at least two independent cytoplasmic pathways may induce apoptosis: one which requires the presence of mitochondria

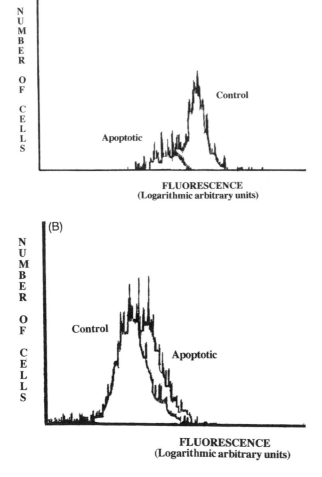

Figure 6   (**A**) A representative histogram of mitochondrial membrane potential from fibroblasts, using rhodamine 123 as a fluorescent probe. (**B**) A representative histogram of peroxide levels from fibroblasts, which were estimated using dihydroethidium.

and the other which directly involves the action of specific proteases. This model fits with the contradictory data available on apoptosis research.

X. APOPTOSIS DURING THE AGING PROCESS

There are very few reports on the effect of aging on apoptosis, but it has been proposed that the efficiency of apoptosis may correlate with the rate of aging. Experimental studies on rats have suggested that apoptotic cell death provides protective mechanisms by removing senescent or damaged cells which might undergo neoplastic transformation (101).

Livers from old rats show a higher in situ rate of apoptosis. Using the dietary restriction model, which is known to retard aging, it has been shown that tumor incidence in liver may be related to the intrinsic rate of apoptosis (102). Diet-restricted animals exhibit a higher rate of apoptosis. These authors concluded that increased apoptotic activity in livers of old animals is a cellular mechanism of defense against neoplasic degeneration.

XI. CONCLUSIONS

Free radicals are involved in aging. The role of mitochondria in the generation of oxidative stress associated with aging was postulated some 20 years ago. Several studies have shown, using isolated mitochondria, that mitochondrial function is impaired in aging. We have reviewed here experiments showing for the first time the impairment of mitochondrial function within intact cells of old animals. This has been achieved using both a metabolic approach—the study of specific metabolic pathways that involve both cytosol and mitochondria—and a flow cytometric approach. Mitochondria from old animals have a lower membrane potential and produce more peroxides than those from young ones. Specific transport systems, such as the one for malate, are impaired with aging. Other important mitochondrial functions such as oxidative phosphorylation are also affected. These changes may be due to oxidation of key molecules such as mtDNA, which is affected in aging. Mitochondria are also affected in apoptotic cells. Some common features of apoptotic cells and of cells from old animals include increased mitochondrial peroxide production, oxidation of glutathione, and oxidation of mtDNA. However, a relationship between aging and apoptosis has not been established.

Antioxidants protect against the oxidative stress associated with aging. However, not all antioxidants are efficient under all circumstances. Indeed, some antioxidants may have a prooxidant effect in vivo. Thus, although antioxidants have a beneficial effect in protecting against aging, care should be taken to test each specific antioxidant and its potential protective effect against age-associated impairment in molecular, metabolic, and physiological functions.

REFERENCES

1. Harman D. Aging: a theory based on free radical and radiation chemistry. J Gerontol 1956; 11:298–300.
2. Sies H. Biochemistry of oxidative stress. Angew Chem 1986; 25:1058–1071.
3. Miquel J, Economos AC. Favorable effects of the antioxidants sodium and magnesium thiazolidine carboxylate on the vitality and life span of *Drosophila* and mice. Exp Geront 1979; 14:279–285.
4. Viña J, Sastre J, Anton V, et al. Effect of aging on glutathione metabolism. Protection by antioxidants. In: Emerit I, Chance B, eds. Free Radicals and Aging. Basel: Birkhauser Verlag, 1992:136–144.
5. Orr WC, Sohal RS. Extension of life-span by overexpression of superoxide dismutase and catalase in *Drosophila melanogaster*. Science 1994; 263:1128–1130.
6. López-Torres M, Pérez-Campo R, Rojas C, Cadenas S, Barja G. Maximum life span in vertebrates: correlation with liver antioxidant enzymes, glutathione system, ascorbate, urate sensitivity to peroxidation, true malondialdehyde, in vivo H_2O_2 and basal and maximum aerobic capacity. Mech Ageing Dev 1993; 70:177–199.
7. Boveris A, Oshino N, Chance B. The cellular production of hydrogen peroxide. Biochem J 1972; 128:617–630.
8. Chance B, Sies H, Boveris A. Hydroperoxide metabolism in mammalian organs. Physiol Rev 1979; 59:527–604.
9. Miquel J, Economos AC, Fleming J, Johnson JE Jr. Mitochondrial role in cell aging. Exp Gerontol 1980; 15:579–591.
10. Miquel J, Fleming JE. Theoretical and experimental support for an "oxygen radical-mitochondrial injury" hypothesis of cell aging. In: Johnson JE Jr, Walford R, Harman D, Miquel J, eds. Free Radicals, Aging and Degenerative Diseases. New York: Alan R. Liss, 1986:51–74.
11. Rubner M. Das problem der Lebensdauer und seine Beziehumgen zu Wachstum und Ernährung (Oldenburg R, ed.) München, 1908.
12. Harman D. The biological clock: the mitochondria. J Am Geriatr Soc 1972; 20: 145–147.
13. Barja G, Cadenas S, Rojas C, Pérez-Campo R, López-Torres M. Low mitochondrial free radical production per unit O_2 consumption can explain the simultaneous presence of high longevity and high aerobic metabolic rate in birds. Free Radic Res 1994; 21:317–328.
14. Ku HH, Sohal RS. Comparison of mitochondrial prooxidant and antioxidant defenses between rat and pigeon: possible basis of variation in longevity and metabolic potential. Mech Ageing Dev 1993; 72:67–76.
15. Pamplona R, Prat J, Cadenas S, et al. Low fatty acid unsaturation protects against lipid peroxidation in liver mitochondria from long lived species: the pigeon and the human case. Mech Ageing Dev 1996; 86:53–66.

16. Sohal RS. Hydrogen peroxide production by mitochondria may be a biomarker of aging. Mech Ageing Dev 1991; 60:189–198.

17. Ku H, Brunk UT, Sohal RS. Relationship between mitochondrial superoxide and hydroperoxide production and longevity of mammalian species. Free Radic Biol Med 1993; 15:621–627.

18. Cadenas E, Boveris A, Ragan CI, Stoppani AOM. Production of superoxide radicals and hydrogen peroxide by NADH-ubiquinone reductase and ubiquinol-cytochrome c reductase from beef-heart mitochondria. Arch Biochem Biophys 1977; 180:248–257.

19. Markossian KA, Poghossian AA, Paitian NA, Nalbandyan RM. Superoxide dismutase activity of cytochrome oxidase. Biochem Biophys Res Commun 1978; 81:1336–1343.

20. Papa S, Guerrieri F, Capitanio N. A possible role of slips in cytochrome c oxidase in the antioxygen defense system of the cell. Biosci Rep 1997; 17:23–31.

21. Boveris A, Chance B. The mitochondrial generation of hydrogen peroxide. Biochem J 1973; 134:707–716.

22. Skulachev VP. Role of uncoupled and non-coupled oxidations in maintenance of safely low levels of oxygen and its one-electron reductants. Q Rev Biophys 1996; 29:169–202.

23. Herrero A, Barja G. ADP-regulation of mitochondrial free radical production is different with complex I or complex II-linked substrates: implications for the exercise paradox and brain hypermetabolism. J Bioenerg Biomembr. In press.

24. Brand MD, Murphy MP. Control of electron flux through the respiratory chain in mitochondria and cells. Biol Rev 1987; 62:141–193.

25. Shigenaga MK, Hagen TM, Ames BN. Oxidative damage and mitochondrial decay in aging. Proc Natl Acad Sci USA 1994; 91:10771–10778.

26. Sastre J, Pallardó FV, Plá R, et al. Aging of the liver: age associated mitochondrial damage in intact hepatocytes. Hepatology 1996; 24:1199–1205.

27. Corbisier P, Raes M, Michiels C, et al. Respiratory activity of isolated rat liver mitochondria following in vitro exposure to oxygen species: a threshold study. Mech Ageing Dev 1990; 51:249–263.

28. Di Monte DA, Sandy MS, DeLanney LE, et al. Age-dependent changes in mitochondrial energy production in striatum and cerebellum of the monkey brain. Neurodegeneration 1993; 2:93–99.

29. Bowling AC, Mutisya EM, Walker LC, et al. Age-dependent impairment of mitochondrial function in primate brain. J Neurochem 1993; 60:1964–1967.

30. Benzi G, Moretti A. Age- and peroxidative stress-related modifications of the cerebral enzymatic activities linked to mitochondria and the glutathione system. Free Radic Biol Med 1995; 19:77–101.

31. Yen TC, Chen YS, King KL, Yeh SH, Wei YH. Liver mitochondrial respiratory functions decline with age. Biochem Biophys Res Commun 1989; 165:994–1003.

32. Trounce I, Byrne E, Marzuki S. Decline in skeletal muscle mitochondrial chain function: possible factor in ageing. Lancet 1989; 25 March:637–639.

33. Beal MF, Hyman BT, Koroshetz W. Do defects in mitochondrial energy metabolism underlie the pathology of neurodegenerative diseases? Trends in Neurosci 1993; 16:125–131.

34. Hansford RG. Lipid oxidation by heart mitochondria from young adult and senescent rats. Biochem J 1978; 170:285–295.
35. Paradies G, Ruggiero FM. Effect of aging on the activity of the phosphate carrier and on the lipid composition in rat liver mitochondria. Arch Biochem Biophys 1991; 284:332–337.
36. Tummino PJ, Gafni A. A comparative study of succinate-supported respiration and ATP/ADP translocation in liver mitochondria from adult and old rats. Mech Ageing Dev 1991; 59:177–188.
37. Takeyama N, Matsuo N, Tanaka T. Oxidative damage to mitochondria is mediated by the Ca^{2+}-dependent inner membrane permeability transition. Biochem J 1993; 294:719–725.
38. Gordillo EA, Ayala F, Lobato M, Bautista J, Machado A. Possible involvement of histidine residues in the loss of enzymatic activity of rat liver malic enzyme during aging. J Biol Chem 1988; 208:8053–8056.
39. Nohl H, Kramer R. Molecular basis of age-dependent changes in the activity of adenine nucleotide translocase. Mech Ageing Dev 1980; 14:137–144.
40. Hansford RG, Castro F. Effect of senescence on Ca^{++}-ion transport by heart mitochondria. Mech Ageing Dev 1982; 19:5–13.
41. Kim JH, Woldgiorgis G, Elson CE, Shrago E. Age-related changes in perspiration coupled to phosphorylation. I. Hepatic mitochondria. Mech Ageing Dev 1988; 46: 263–277.
42. Paradies G, Ruggiero FM. Age-related changes in the activity of the pyruvate carrier and in the lipid composition in rat-heart mitochondria. Biochim Biophys Acta 1990; 1016:207–212.
43. Hagen TM, Yowe DL, Bartholomew JC, et al. Mitochondrial decay in hepatocytes from old rats: membrane potential declines, heterogeneity and oxidants increase. Proc Natl Acad Sci USA 1997; 94:3064–3069.
44. Chen LB. Mitochondrial membrane potential in living cells. Annu Rev Cell Biol 1988; 4:155–181.
45. Calleja M, Peña P, Ugalde C, et al. Mitochondrial DNA remains intact during *Drosophila* aging, but the levels of mitochondrial transcripts are significantly reduced. J Biol Chem 1993; 268:18891–18897.
46. Sastre J, Millán A, García de la Asunción J, et al. A *Ginkgo biloba* extract (EGb 761) prevents mitochondrial aging by protecting against oxidative stress. Free Radic Biol Med. In press.
47. D'Amore PA, Sweet E. Effects of hyperoxia on microvascular cells in vitro. In Vitro Cell Dev Biol 1987; 23:123–128.
48. Sohal RS, Arnold LA, Sohal BH. Age-related changes in antioxidant enzymes and prooxidant generation in tissues of the rat with special reference to parameters in two insect species. Free Radic Biol Med 1990; 9:495–500.
49. Scalettar BA, Abney JR, Hackenbrock CR. Dynamics, structure and function are coupled in the mitochondrial matrix. Proc Natl Acad Sci USA 1991; 88:8057–8061.
50. Wilson PD, Franks LM. The effect of age on mitochondrial ultrastructure and enzymes. Adv Exp Med Biol 1975; 53:171–183.
51. De la Cruz J, Burón I, Roncero I. Morphological and functional studies during aging at mitochondrial level. Action of drugs. Int J Biochem 1990; 22:729–735.

52. Pinto RE, Bartley W. A negative correlation between oxygen uptake and glutathione oxidation in rat liver homogenates. Biochem J 1969; 114:5–9.
53. Hazelton GA, Lang CA. Glutathione contents of tissues in the aging mouse. Biochem J 1980; 188:25–30.
54. Allen RG, Sohal RS. Role of glutathione in the aging and development of insects. In: Collatz KG, Sohal RS, eds. Insect Aging. Berlin: Springer Verlag, 1986; 168–181.
55. Goldschmidt L. Seasonal variations in red cell glutathione levels with aging in mental patients and normal controls. Proc Soc Exp Biol Med 1970; 133:555–559.
56. Meister A. Glutathione deficiency produced by inhibition of its synthesis, and its reversal: applications in research and therapy. Pharmacol Ther 1991; 51:155–194.
57. García de la Asunción J, Millán A, Plá R, et al. Mitochondrial glutathione oxidation correlates with age-associated oxidative damage to mitochondrial DNA. FASEB J 1996; 10:333–338.
58. Sohal RS, Dubey A. Mitochondrial oxidative damage, hydrogen peroxide release, and aging. Free Radic Biol Med 1994; 16:621–626.
59. Johns DR. Mitochondrial DNA and disease. N Engl J Med 1995; 333:638–644.
60. Brown WM, George M, Wilson AC. Rapid evolution of animal mitochondrial DNA. Proc Natl Acad Sci USA 1979; 76:1967–1971.
61. Giles RE, Blanc H, Cann HM, Wallace DC. Maternal inheritance of human mitochondrial DNA. Proc Natl Acad Sci USA 1980; 77:6715–6719.
62. Richter C, Park JW, Ames B. Normal oxidative damage to mitochondrial and nuclear DNA is extensive. Proc Natl Acad Sci USA 1988; 85:6465–6467.
63. Hayakawa M, Torii K, Sugiyama S, Tanaka M, Ozawa T. Age-associated accumulation of 8-hydroxydeoxyguanosine in mitochondrial DNA of human diaphragm. Biochem Biophys Res Commun 1991; 179:1023–1029.
64. Ames BN, Shigenaga M, Hagen TM. Oxidants, antioxidants and the degenerative diseases of aging. Proc Natl Acad Sci USA 1993; 90:7915–7922.
65. Mecocci P, MacGarvey U, Kaufman AE, et al. Oxidative damage to mitochondrial DNA shows marked age-dependent increases in human brain. Ann Neurol 1993; 34:609–616.
66. Kristal BA, Chen J, Yu BP. Sensitivity of mitochondrial transcription to different free radical species. Free Radic Biol Med 1994; 16:323–329.
67. Gadaleta MN, Petruzzella V, Renis M, Fracasso F, Cantatore P. Reduced transcription of mitochondrial DNA in the senescent rat. Tissue dependence and effect of L-carnitine. Eur J Biochem 1990; 187:501–506.
68. Ames BN, Shigenaga MK. Oxidants are major contributor to aging. Ann NY Acad Sci 1992; 663:85–96.
69. Arnheim N, Cortopassi G. Deleterious mitochondrial DNA mutations accumulate in aging human tissues. Mutat Res 1992; 275:157–167.
70. Cortopassi GA, Arnheim N. Detection of a specific mitochondrial DNA deletion in tissues of older humans. Nucleic Acids Res 1990; 18:6927–6933.
71. Linnane AW, Baumer A, Maxwell RJ, et al. Mitochondrial gene mutation: the ageing process and degenerative diseases. Biochem Int 1990; 22:1067–1076.
72. Münscher C, Muller-Hocker J, Kadenbach B. Human aging is associated with vari-

ous point mutations in tRNA genes of mitochondrial DNA. Biol Chem Hoppe Seyler 1993; 374:1099–1104.

73. Wallace DC. Mitochondrial DNA sequence variation in human evolution and disease. Proc Natl Acad Sci USA 1994; 91:8739–8746.

74. Linnane A, Marzuki S, Ozawa T, Tanaka M. Mitochondrial DNA as an important contributor to ageing and degenerative diseases. Lancet 1989; i:642–645.

75. Furukawa T, Meydani SN, Blumberg JB. Reversal of age-associated decline in immune responsiveness by dietary glutathione supplementation in mice. Mech Ageing Dev 1987; 38:107–117.

76. Fraga CG, Motchnick PA, Shigenaga MK, et al. Ascorbic acid protects against endogenous oxidative DNA damage in human sperm. Proc Natl Acad Sci USA 1991; 88:11003–11006.

77. Viña J, Hems R, Krebs HA. Maintenance of glutathione content of isolated hepatocytes. Biochem J 1978; 170:627–630.

78. Viña J, Sáez GT, Wiggins D, et al. The effect of cysteine oxidation on isolated hepatocytes. Biochem J 1983; 212:39–44.

79. Viña J, Romero FJ, Estrela JM, Viña JR. Effect of acetaminophen (paracetamol) and its antagonists on glutathione (GSH) content in rat liver. Biochem Pharmacol 1980; 29:1968–1970.

80. Berger TM, Polidori MC, Dabbagh A, et al. Antioxidant activity of vitamin C in iron overloaded human plasma. J Biol Chem 1997; 272:15656–15660.

81. Omenn GS, Goodman GE, Thornquist MD, et al. Effects of a combination of beta carotene and vitamin A on lung cancer and cardiovascular disease. N Engl J Med 1996; 334:1150–1155.

82. Cannon JG, Orencole SF, Fielding RA, et al. Acute phase response in exercise-interaction of age and vitamin-E on neutrophils and muscle enzyme release. Am J Physiol 1990; 259:R1214–R1219.

83. Weindruch R. In: Armstrong D, ed. Free Radicals in Molecular Biology Aging and Diseases. New York: Raven Press, 1984: 181–202.

84. Weraarchakul N, Strong R, Wood WG, Richardson A. The effect of aging and dietary restriction on DNA repair. Exp Cell Res 1989; 181:197–204.

85. Koizumi A, Weindruch R, Walford RL. Influences of dietary restriction and age on liver enzyme activities and lipid peroxidation in mice. J Nutr 1987; 117:361–367.

86. Richie JP, Leutzinger Y, Partharsarathy S, et al. Methionine restriction increases blood glutathione and longevity in F344 rats. FASEB J 1994; 8:1302–1307.

87. Buja LM, Eigenbrodt ML, Eigenbrodt EH. Apoptosis and necrosis: basic types and mechanisms of cell death. Arch Pathol Lab Med 1993; 117:1208–1214.

88. Schwartzman RA, Cidlowski JA. Apoptosis: the biochemistry and molecular biology of programmed cell death. Endocr Rev 1993; 14:133–151.

89. Vaux DL. Toward an understanding of the molecular mechanisms of physiological cell death. Proc Natl Acad Sci USA 1993; 90:786–789.

90. Kroemer G, Petit P, Zamzami N, Vaysiere JL, Mignotte B. The biochemistry of programmed cell death. FASEB J 1995; 9:1277–1287.

91. Hockenbery DM, Oltvai ZN, Yin XM, Milliman CL, Korsmeyer SJ. Bcl-2 functions in an antioxidant pathway to prevent apoptosis. Cell 1993; 75:241–251.

92. Ratan RR, Murphy TH, Baraban JM. Oxidative stress induces apoptosis in embryonic cortical neurons. J Neurochem 1994; 62:376–379.

93. Van den Dobbelsteen DJ, Stefan C, Nobel Y, et al. Rapid and specific efflux of reduced glutathione during apoptosis induced by anti-Fas/APO-1 antibody. J Biol Chem 1996; 271:15420–15427.

94. Pallardó FV, Mompó J, Esteve JM, et al. Glutathione oxidation increases in apoptotic fibroblasts. Role of apoptosis in aging. 8th Biennial Meeting of the International Society for Free Radical Research, Barcelona, Spain, Oct 1–5, 1996.

95. Schulze-Osthoff K, Kramer PH, Dröge W. Divergent signalling via APO-1Fas and TNF receptor, two homologous molecules involved in physiological cell death. EMBO J 1994; 13:4587–4596.

96. Kane DJT, Sarafian A, Anton R, et al. Bcl-2 inhibition of neural death: decreased generation of reactive oxygen species. Science 1993; 262:1274–1277.

97. Jacobson MD, Raff MC. Programmed cell death and Bcl-2 protection by very low oxygen. Nature 1995; 374:814–816.

98. Muschelm RJ, Bernhard EJ, Gaza L, McKenna WG, Koch CJ. Induction of apoptosis at different oxygen tensions: evidence that oxygen radicals do not mediate apoptotic signaling. Cancer Res 1995; 55:995–998.

99. Deckwerth TL, Johnson EM Jr. Neurotrophic factor deprivation-induced death. Ann NY Acad Sci 1993; 679:121–131.

100. Zamzami NS, Susin A, Marchetti P, et al. Mitochondrial control of nuclear apoptosis. J Exp Med 1996; 183:1533–1544.

101. Monti D, Troiano L, Tropea F, et al. Apoptosis—programmed cell death: a role in the aging process? Am J Clin Nutr 1992; 55:1208S–1214S.

102. Muskhelishvili L, Hart RW, Turturro A, James SJ. Age-related changes in the intrinsic rate of apoptosis in livers of diet-restricted and ad libitum-fed B6C3F1 mice. Am J Pathol 1995; 147:20–24.

11

The Role of Iron and Mitochondria in Aging

Patrick B. Walter, Kenneth B. Beckman, and Bruce N. Ames
University of California, Berkeley, California

I. INTRODUCTION

The free radical theory of aging, first proposed by Harman in 1956 (1), is attracting more interest recently. While we do not intend to discuss this theory here, its major tenet—that the accumulation of damage due to oxidants contributes to aging—cannot be overlooked, because at the center of this concept are mitochondria.

Iron can induce the formation of potentially reactive oxygen species (ROS) by the reduction of H_2O_2 to $HO^\bullet$ via the Fenton reaction. The presence of iron therefore potentiates oxidative stress (reviewed in Refs. 2 and 3). ROS are produced by normal metabolism, for example, O_2^- is produced by the mitochondrial respiratory chain (4). With aging, mitochondrial function decays and mitochondrial O_2^- production may increase, and both of these phenomena have been hypothesized to be a factor in aging (reviewed in Ref. 5). Mitochondria take up iron as part of their normal metabolic role (6), and increased levels of iron have been found to damage mitochondrial macromolecules and function (7–9). Tissue iron levels increase with advancing age (10), and elevated mitochondrial iron levels have been found in normal, aged human subcortical brain (11). There are specific cellular and mitochondrial changes in morphology caused by iron overload (8,12) that are similar to those seen in aging (13). Iron is also found to be increased in many of the degenerative diseases of aging such as Parkinson's disease (14–16), Alzheimer's disease (17), cancer (18), and heart disease (19,20). It is also clear that oxidative damage, and in particular oxidative damage to mitochondria, could be a major factor in aging (21). We now argue that accumulated iron is causing some of the loss of mitochondrial function with age, and that this becomes evident in some of the degenerative diseases of aging.

II. BACKGROUND: RELEVANT IRON BIOCHEMISTRY AND MITOCHONDRIAL PHYSIOLOGY

A. Iron Potentiates Oxidant Stress

For the majority of organisms on earth, life without oxygen is not possible. However, oxygen also is a double-edged sword. Gershman et al. (22) first recognized that the toxic effects of oxygen are based on the formation of ROS. Biologically relevant ROS include superoxide (O_2^-), hydrogen peroxide (H_2O_2), singlet oxygen ($^1O_2$), and hydroxyl radical (HO$^\bullet$) (23,24).

Much of ROS toxicity is due to reduced iron (3). The potential for iron to cause damage to organic molecules was first noted by Fenton in 1894 (25) when he found that the oxidation of polycarboxylic acids by H_2O_2 was strongly promoted by ferrous iron. Forty years later Haber and Weiss suggested that the hydroxyl radical (HO) is formed by the reduction of H_2O_2 by reduced iron. It is now generally agreed that in biological systems HO is one of the most damaging radicals (23).

Why do cells possess a metal with such potential to cause damage? In the biological milieu, complex molecules that contain iron readily gain and lose electrons. This is because iron is a transition metal able to redox cycle between Fe(II) and Fe(III). This property makes iron exceedingly useful when coordinated in redox enzymes, but Fe(II) can also reduce oxygen or other oxygen intermediates, giving rise to reactive organic and oxygen-based free radicals. Thus the availability of transition metal ions such as iron is a limiting condition for oxygen toxicity. The concentration of iron is, therefore, under tight control (see below).

B. Metabolism Results in an Accumulation of Iron with Age

For mammals, iron's biological availability depends on ingestion in the diet. Once absorbed, tissue iron accumulation occurs via both transferrin-mediated and non-transferrin-mediated mechanisms. Cellular uptake of transferrin-bound iron depends on the number of membrane transferrin receptors (TRs). Increases in intracellular iron down-regulate the number of TRs. This is accomplished by posttranscriptional regulation via the iron responsive element binding protein (IRE-BP, also known as iron regulatory protein 1 or IRP-1) (cytosolic aconitase) (reviewed in Refs. 26 and 27). Iron can interact with IRE-BP, reducing its binding affinity for a stem loop structure on the TR-mRNA. Without bound IRE-BP, the stability of TR-mRNA is greatly reduced, leading to a reduction in receptor number (28). Conversely, when cytosolic iron concentration is low, the IRE-BP binds to the stem loop structure, which is a 3′ regulatory untranslated region in TR-mRNA. The bound IRE-BP increases TR-mRNA stability, thus leading to increased translation of TRs, greater receptor number, and the potential increase of TR binding of iron-transferrin complex.

The iron-transferrin complex is endocytotically taken up by cells and iron is released from transferrin by the combined effects of low pH and an unknown chelator present in the endosomes (29,30). After release, iron is incorporated into the chelatable iron pool ("low molecular weight iron pool"), then into cytosolic ferritin and mitochondria. The exact means by which these transfers are accomplished are not known (31). Recent work by Pollack and coworkers has shed light on this problem. They have found (in the reticulocyte) that the major low molecular weight ligand for iron is ATP. Iron ATP, together with a 13,800 MW polypeptide, forms a high molecular weight polynuclear aggregate which comprises at least part of the nonheme, nonferritin iron pool, and which may deliver iron to mitochondrial receptors (31). There may be two receptors on mitochondria for Fe(III), one of which would acquire iron AMP hydrolyzed near the mitochondria from iron ATP (31). In addition to ATP, other intracellular ligands may exist. It is important that iron is ligated and quickly transported into ferritin and/or mitochondria, because as discussed above, free iron can induce oxidation to important cellular targets. While discussion of potential iron-induced tissue damage is important, one must not forget that too little iron is also deleterious to normal human function (32,33).

Three main factors control iron balance and metabolism: iron intake and bioavailability, storage, and loss. Iron metabolism is unusual; unlike other ions, whose concentration in the body is controlled at the level of loss (for example, by excretion in the urine), iron is controlled primarily by its intake from the gastrointestinal tract. Iron absorption is up-regulated in iron deficiency and ferropenic anemia (34), and is also efficiently down-regulated as anemia disappears and iron stores increase (34–36). Moreover, the body's ability to store iron in its two main storage compounds (ferritin and hemosiderin) is flexible, as revealed by diseases of iron overload, where the liver parenchymal cell iron content may increase by 100-fold (7). [There is also evidence to show that mitochondria also store iron for eventual heme synthesis (37).] At the same time, the body does not have as much control over iron loss. The primary [0.6 mg/day (can be variable)] cause of iron loss is a combination of desquamated mucosal cells, loss of small amounts of blood, and bile excretion into the feces (38). Smaller amounts of iron (0.2–0.3 mg/day) are lost in a combination of desquamated skin cells, hair, and sweat, while urinary losses are minor (<0.1 mg/day) (38). Premenopausal women incur a further loss with menstruation [0.4–0.5 mg/day (30–40 ml blood/cycle)]. Thus males differ from females in iron loss, with men averaging 1 mg/day and premenopausal women 1.3 mg/day. Postmenopausal women do not have as great an iron loss [and suffer from increased risk of heart disease, cancer (39), and Alzheimer's disease (40)].

The accumulation of iron is most likely the result of systems evolved to conserve a scarce element. In order to prevent iron deficiency, the body retains iron as it becomes metabolically available. At the same time, the storage/transport proteins ferritin and transferrin appear to have evolved to avert oxidative damage by binding iron in its benign Fe(III) state. In summary, iron is not easily lost from the body, and this metabolic dead end forces the body to slowly accumulate iron over time (38).

C. Mitochondria Play a Role in Aging

Mitochondria have many important cellular functions, including ATP production, heme and cholesterol synthesis, and calcium regulation (reviewed in Ref. 6). Furthermore they are thought to be the most important intracellular source of oxidants (reviewed in Refs. 21 and 41), although it remains controversial as to exactly how much O_2^- they form (reviewed in Ref. 5). The main sites at which O_2^- is generated are ubiquinone and NADH dehydrogenase. Electrons are transferred one at a time to form ubisemiquinone, which can then react with oxygen to form O_2^-. It has been suggested that during aging mitochondrial macromolecules and membranes can be damaged by mitochondrially generated oxidants, and that isolated mitochondria from aged animals produce more H_2O_2 than their young counterparts (although some controversy exists) (reviewed in Refs. 5, 13, and 21).

III. AGE-RELATED CHANGES TO MITOCHONDRIA INVOLVING IRON

A. Accumulated Iron in Aging Has Potential to Damage Mitochondria

In many aging and stress-related models it has been shown that oxidative stress is due to iron-induced formation of reactive radicals (3), which may cause peroxidation, loss of integrity, and dysfunction of the mitochondrial membrane (reviewed in reference 42). Such damage to membranes is an important factor in the molecular basis of iron overload disease (7,12,43–45), and has also been cited in support for the free radical theory of aging (5). Furthermore, there is in vivo evidence that accumulated iron may be a culprit in the increased oxidative stress seen in aging (46). For example, iron's participation in aging is revealed by experiments that have shown that iron overload of *Drosophila* decreases life span (46), while iron deprivation extends life span (47). It has been documented that redox active iron can damage proteins (8,48,49), lipids (50,51), and DNA (8,52–54) in mitochondria, and furthermore, iron has been found to accumulate in the mitochondria of postmitotic tissue, such as the subcortical brain, with age (11,55). There are three possible ways in which such iron could contribute to the mitochondrial decay seen in aging: (1) It may increase the production of mitochondrial O_2^- and H_2O_2 (56–58), presumably by Fe(II) reduction of O_2. (2) It may catalyze oxidative damage, by Fenton chemistry. (3) It may sensitize the mitochondrial macromolecules to oxidation by site-specific iron binding. The concept of site-specific iron binding is important because Fe(III) bound to DNA, proteins, or lipids can undergo cyclic reduction and reoxidation. This idea is distinguishable from free iron in that bound iron is not as dynamically diffusible, and partially explains the idea of specific multihit attack on the same macromolecule. This concept has been applied experimentally to analyze H_2O_2 toxicity (59), and has been reviewed (60,61). The following sections discuss the potential involvement of the above mechanisms in damage to lipids, proteins, and DNA.

B. Damage to Mitochondrial Lipids Associated with Aging

It is generally agreed that iron-induced formation of hydroperoxides is an initiating event in lipid peroxidation (2). Lipid peroxidation leads to a great diversity of aldehydes when lipid hydroperoxides break down in biological systems (reviewed in Refs. 62–65). Some of these aldehydes are highly reactive and propagate further free radical events. Therefore lipid peroxidation is an important index of oxidative stress.

With aging there is a change in cell and mitochondrial membranes which may be due to membrane peroxidation (reviewed in Ref. 21). Mitochondria of aged animals are characterized by the loss of membrane structural integrity, essential to functions such as ATP production (42). During aging, mitochondrial membrane fluidity decreases and enzymes associated with membrane lipids exhibit reduced activities. Because the activities of most such enzymes are regulated by the physicochemical state of the lipid environment of the membrane, it seems likely that their impaired function is related to decreased membrane fluidity. Mitochondrial membrane fluidity is altered by lipid peroxidation products, implicating reactive products of lipid peroxidation in age-associated decreases in mitochondrial membrane fluidity (66).

There are similarities in the peroxidative damage detected in several models of iron overload and aging. Initial work with aged mitochondria was done as early as 1977, when Player et al. (67,68) showed that lipids of old rat liver mitochondria were more sensitive than young when challenged in vitro by Fe(III) plus NADPH and ADP. This age-associated increase in sensitivity to peroxidation of mitochondrial lipids could be attributed to one of four possibilities; (1) an increase in the degree of polyunsaturation; (2) a decreased number of membrane-bound SH groups (42); (3) an increase in O_2^- production by mitochondria with age (first shown in Ref. 56; reviewed in Ref. 5); or (4) an increased iron content of aged mitochondria (69,70). The increased sensitivity to peroxidation of aged lipids is likely a combination of these.

Other models of lipid peroxidation also show a similarity between iron-induced and age-accumulated damage. Incubation of freshly isolated rat hepatocytes with 100 μM iron nitrilotriacetic acid (FeNTA) caused an increase in lipid peroxidation products as measured by conjugated dienes (71). Furthermore, FeNTA treatment also induced the loss of mitochondrial membrane potential, a drop in ATP content, and an increase in cytosolic Ca^{2+}. Similarly, lipid peroxidation and a loss of membrane potential are also evident in hepatocyte mitochondria from old rats (21,72). Similarities between aging and iron overload are also seen in vivo when rats were overloaded with dietary carbonyl iron. As in recent aging studies, in which an age-related decline in mitochondrial respiratory control ratio was also found (72), the iron carbonyl–supplemented diet increased the levels of mitochondrial conjugated dienes and impaired the mitochondrial respiratory control ratio (73).

Last, similarities also exist in the apparently increased rate of H_2O_2 production by aged and iron overloaded mitochondria. Since it was reported by Nohl and Hegner (56) that aged mitochondria produce more H_2O_2 than young mitochondria, others have reported similar findings (57,58). Recently, however, there has appeared data that challenges this conclusion. It was shown that mitochondrial production of H_2O_2 depends on the concentration of substrates and O_2 used. No difference was found [between aged (24 months) and young (6 months) mitochondria] when using physiological substrate concentrations (74). The steady-state level of mitochondrial H_2O_2 in vivo (which could be argued to be the most important factor) is also influenced by the activity of mitochondrial glutathione peroxidase. In support of the idea that the steady-state level of H_2O_2 increases with age, Sohal's group has found that one of the correlates of longevity of different species is the activity of glutathione peroxidase (75). Furthermore, Sohal et al. (76) and others (77,78) have found (although not consistently) that glutathione peroxidase activity is reduced in aging. With all of these studies in mind, there is possibly an elevation in the steady-state level of H_2O_2 in aging, due either to an increased rate of H_2O_2 formation or a decline in glutathione peroxidase activity.

Similarly, the exposure of mitochondria to iron elevates their rate of H_2O_2 production. Data in support of this comes from mitochondria isolated from the livers of ethanol fed rats; when challenged with Fe(III) ATP and NADH or NADPH they produced more H_2O_2 and $HO^\bullet$ than nonchallenged mitochondria (79). Kukielka et al. (79) suggest that NADH acts via the mitochondrial outer-membrane NADH reductase to catalyze an iron-dependent production of H_2O_2. Thus iron may have the ability to enhance the rate of H_2O_2 formation both from the outside and the inside of the mitochondrial membrane.

We also have preliminary evidence to show that in parallel with the aging increase in iron content, there is an increase in MDA (malondialdehyde) in the mitochondria from the livers of aged (27 months) Fisher 334 rats when compared to young rats (4 months) (J. Liu et al., unpublished results). Further support for iron-induced lipid peroxidation to mitochondrial membranes during aging comes from the degenerative diseases of aging. In three of these diseases (described in more detail in section IV)—Parkinson's, Alzheimer's, and Friedrich's ataxia—there is an increase in mitochondrial iron that leads to lipid peroxidation and losses in mitochondrial respiratory capacity (80–82).

Clearly, iron has the ability to catalyze the reduction of O_2 to O_2^- and to reduce H_2O_2 to $HO^\bullet$, and because we accumulate iron in aging, its role in age-related changes may be an important factor to consider. However, even if there is an enhanced tissue iron content as we age, the body has (as outlined in Sec. II) very efficient pathways for controlling the iron that may become available to catalyze these reactions (the so-called free iron). How does free iron become available in mitochondria? The mechanism could involve the release of iron from iron-sulfur clusters by O_2^--mediated oxidation, as has been observed for aconitase (83).

C. Damage to Mitochondrial Proteins Associated with Aging

When isolated mitochondria from normal rats are exposed to excess Fe(III) (500 μM) they show an increase in protein carbonyl content (8), and similarly, the level of mitochondrial protein oxidative damage has also been shown to increase with age (76,84,85). Sohal and coworkers have studied whether such damage is random or whether some proteins are relatively more susceptible. High molecular weight proteins (such as aconitase) were found to be relatively more oxidatively damaged during aging, and this was interpreted to indicate that protein oxidative damage during aging is not random (86,87). The mechanism for this protein damage most likely involves the one described in detail by Stadtman (49), which involves only a few amino acid residues that are modified and relatively little peptide bond cleavage when proteins are exposed to metal-catalyzed oxidations. Evidence indicates that Fenton reactions occur at metal-binding sites on the protein to produce $HO^\bullet$ which attack the side chains of amino acid residues at the metal-binding sites. Protein carbonyl derivatives are among the modifications that are formed on some amino acid residues (48,49).

D.   Damage to Mitochondrial DNA Associated with Aging

There has been a steady interest (8,52,54,88–94) in mitochondrial DNA
(mtDNA) damage induced by iron since Lim and Neims (92) first published that
mtDNA is damaged when isolated mitochondria are exposed to iron-bleomycin.
Since that time a great deal of work has established that the metal plays a significant role in oxidative DNA damage in both nuclear and mtDNA. Both in vivo
and in vitro, the binding of iron by DNA mediates strand scission by H_2O_2 via
Fenton chemistry ($HO^{\bullet}$ generation) (60). The use of a variety of free radical
scavengers has demonstrated that freely diffusible $HO^{\bullet}$ are less important than
those generated by iron tightly bound to the double helix (95). DNA cleavage
by $Fe(II)/H_2O_2$ is somewhat sequence specific as well, suggesting that specific
iron binding sites exist (60). Until recently, the accepted wisdom held that the
role of superoxide ($O_2^{-\bullet}$) in DNA oxidation was its ability to reduce Fe(III) to
Fe(II). However, as mentioned above, the role of $O_2^{-\bullet}$ in reducing free Fe(III)
has been challenged by experiments suggesting that the principal role of $O_2^{-\bullet}$ is
to release iron from protein-bound iron-sulfur clusters (96). Besides $O_2^{-\bullet}$, there
are other reductants (such as NADH) which effectively reduce Fe(III) to Fe(II),
and which may be more relevant as reductants of free or DNA-bound iron than
is $O_2^{-\bullet}$ (96).

Most experimental approaches used to study mtDNA damage induced by
iron have used electrophoresis with or without Southern blotting (8,52,54,89–
93). Hruszkewycz (90,91) investigated the combination of Fe(II) and NADPH
exposed to isolated mitochondria and found the resulting mtDNA damage correlated to lipid peroxidation. Yaffee et al. (54) used both electrophoresis and high-
resolution scanning electron microscopy to find that there was a dose-dependent
increase in relaxation of supercoiled mtDNA when isolated mitochondria were
exposed to Fe(III) gluconate. Tissue culture studies have shown that exposure
of cultured rat HTC cells (a tumor cell line) to 100 µM iron for as little as 3 h
resulted in conversion of supercoiled mtDNA to open circular and linear forms,
evidence of the formation of single-strand breaks (52). In the same experiments,
mtDNA in freshly prepared rat hepatocytes was completely resistant to iron-
mediated damage. Differences between the transformed and primary cells may
have been due to dramatically lower levels of mitochondrial antioxidant defenses
in the cancer cell line, including the virtual absence of Mn-SOD activity (52).

Regarding aging, Harman (97) was first to suggest that one of the possible
sites of free radical attack is mtDNA. This has also been proposed by Miquel et
al. (98), who suggested that aging results from changes in mtDNA of differentiated cells. There are two basic ways that damage to mtDNA may lead to aging.
The first was proposed by Richter (94), suggesting that mtDNA fragments escape
from mitochondria and incorporate in a time-dependent manner in nuclear DNA,

which would progressively change the nuclear information content and thereby cause aging. In the second, mtDNA alterations would cause a reduced capacity in oxidative phosphorylation (reviewed in Refs. 99–101). Recent results showed that brain cells of senescent rats have a decreased content of the D-loop portion of mtDNA (102). Cross-links between proteins and DNA may also play a role. That such a covalent mtDNA modification by proteins may occur during aging is supported by the report of Asano et al. (103), that the buoyant density banding of mtDNA of old rats is broad, but becomes similar to that of mtDNA of young rats after treatment with proteinase K. Furthermore, age-dependent accumulation of point mutations in human mtDNA have been detected (104) and age-associated deletions in mtDNA have also been extensively reported (reviewed in Ref. 13).

With respect to possible base modification, evidence that mtDNA is particularly prone to oxidation was stimulated by a report that the amount of the oxidative DNA adduct 8-oxo-2′-deoxyguanosine (oxo^8-dG) is more than 10-fold higher in mtDNA than in nuclear DNA (53). This report and another (91) have also revealed that iron exposed to isolated mitochondria induced increased formation of oxo^8-dG in mtDNA. In comparison, oxo^8-dG has also been found to be elevated in mtDNA from aged tissues (105–107). However, these and other reported measurements of oxo^8-dG in mtDNA have resulted in a range of estimates that spans four orders of magnitude (108). As a result of the huge disparities between these estimates, as well as other concerns about the reliability of methods for measuring oxo^8-dG (109), more investigations into the analytical measurement of oxo^8-dG in mtDNA must be made before one can conclude that mtDNA is, in fact, more heavily oxidized in vivo than is nuclear DNA. An alternative and sensitive method for measuring damage in DNA is to assess its ability to serve as a template for amplification by PCR; nicks and other lesions which block thermostable polymerases decrease the efficiency of amplification (110). When used to compare damage to nuclear and mitochondrial genes in cultured cells exposed to exogenous H_2O_2, the method revealed that mtDNA was considerably more sensitive to oxidative damage than nuclear DNA, and that the damage was also more persistent in the mitochondrial genome than the nuclear genome (111). Even if mtDNA is more extensively oxidized, or more prone to oxidation, than nuclear DNA, this is not necessarily due to iron. For example, mtDNA has been shown to be more sensitive than nuclear DNA to nonoxidative mutagens as well (112).

Last, indirect evidence for the involvement of iron in mtDNA damage has recently come from studies of the disease Friedrich's ataxia (FRDA). As is discussed below, this autosomal recessive degenerative disorder, long associated with mitochondrial dysfunction, has recently been linked to mitochondrial iron accumulation. The existence of a yeast homologue of the FRDA gene product frataxin, called YFH1, has permitted focused studies of frataxin's likely physiological role. Yeast cells in which YFH1 has been deleted accumulate iron in their mitochondria to 10 times control values and show a high frequency of formation of petite mutants, cells in which there has been a complete loss of mitochondrial DNA. A possible explanation for this observation is that mtDNA deletion in YFH1 mutant strains is a primary effect of iron-mediated damage to the mitochondrial genome (113). Alternatively, it has been proposed that the abundance of mitochondrial iron due to the loss of the *YFH1* gene product selects for petite mutants, since respiration-deficient cells would be less likely to generate the mitochondrial free radicals whose toxicity is amplified by the presence of excess iron (114). Of course, these two hypotheses are not mutually exclusive.

E. Morphological Changes of Mitochondria with Aging

Changes in the size and number of mitochondria with age have been noted. The phenomenon was first described in 1968 (115); in human hepatocytes from donors between 21 and 79 years of age there was an increase in area and circumference but a decrease in the number of mitochondria. Similar morphological changes with age have been revealed in studies of synaptic mitochondria of young and old rats, in primary cultures of human skin fibroblasts from young and old humans (116,117), and in fruit flies between 1 and 72 days of age (118). Also observed was a decrease in supranuclear mitochondria by 50% in the old flies compared to young flies. Moreover, the volume of the old fly mitochondria doubled as a function of age (118).

There seems to be a general trend: mitochondria from old organisms are decreased in number and this is balanced by an increase in their size. This could be due to a decreased duplicative capacity (reviewed in Ref. 13). Interestingly, similar morphological findings have now been found with animals fed a diet supplemented with 3.5% iron fumarate. Fine-needle biopsy samples from rat liver were visualized by cryotransmission electron microscopy to investigate mitochondrial morphology (8). It was found that hepatic mitochondria from animals on the iron-supplemented diet had an increased volume and surface density and also a larger diameter than mitochondria from rats on a normal diet. They also found that there were always fewer mitochondria in the hepatocytes of iron-loaded rats than control rats, but because serial sectioning was lacking, definitive proof of this observation must await further investigation. Thus at this point it seems that both aging and iron overload share a common morphological change. The increased mitochondrial volume may indicate that the organelle is responding to stress by increasing the potential area for respiration. The mechanism may be similar in both cases, and one could speculate that because of the similarities in macromolecular damage discussed above, iron-mediated oxidation could be responsible in both cases.

During aging there is an accumulation of lipofuscin in postmitotic tissues such as the nervous system, muscle, and retinal pigment epithelium, while a similar pigment, named ceroid, is found in cells with mitotic activity such as hepatocytes, smooth muscle cells, and a variety of other cells. Lipofuscin represents one of the most visible subcellular modifications in aging (119). By definition, lipofuscin is a yellow to brown, granular, lipid pigment that accumulates in lysosomes with age. The accumulation rate has been reported in humans to be about 0.6% of the volume of myocardial cells per decade (120). They are the oxidation product of membrane lipids and autophagocytosed organelles (in particular mitochondria). Lysosomes are known to digest the autophagocytosed organelles, and iron has been found in the degradation products. Thus there is potentially redox active iron in the lysosome during the digestion (121). Inside the lysosome there is a complex mixture of oxidation reactions occurring involving mitochondrial iron, which is thought to lead to the generation of cross-linked biocomplexes that are not degradable by the lysosomal enzymes (120).

Parallel with age-associated increases in lipofuscin, there is also sometimes an increase in hemosiderin. Hemosiderin is an intracellular storage form of iron that has a characteristic electron microscopic appearance. The internal makeup consists of granules containing an ill-defined complex of ferrihydrites, polysaccharides, and proteins, and data indicate that the ferrihydrite structure is similar to that of the ferritin core (122). A concentrated deposit of hemosiderin is sometimes called a siderosome, and the process of formation is often referred to as siderosis. During aging in the spleen, for instance, it has been reported that there is an increase of hemosiderin in certain areas; furthermore, vascular siderosis in the globus pallidus of the brain is considered a regional hallmark of aging (discussed more below). Thus, as we age, hemosiderin granules accumulate in some tissues. Moreover, as a storage granule full of iron hemosiderin provides a continual source of iron for the cell, it is possible that this is a source of redox active iron for the mitochondria.

In Parkinson's and Alzheimer's disease there is an abnormally high level of brain tissue iron that has been implicated in the production of oxidants (11,16,17,123,124; see Sec. IV). Iron-containing cytoplasmic inclusions accumulate in striatal and periventricular astrocytes of the aging rodent and human brain (11). These inclusions have been found to be distinct from the aging pigment lipofuscin, have an affinity for Gomori's chrome alum hematoxylin stain (11), and have nonenzymatic peroxidase activity believed to be mediated by iron (55). The inclusions contain iron-laden autophagic mitochondria (69,70), and peroxidase activity of the inclusions is sufficient to oxidize dopamine and other catechols to orthosemiquinone radicals with neurotoxic activity (11). Thus the age-associated increase in iron-containing mitochondria may render the aging central nervous system vulnerable to free radical–related neurodegeneration. In summary, it appears that all these aging pigments interact with lysosomes, increasing the cytosolic reactive oxidants and thus contributing to cellular damage and the potential to age.

IV. CONSEQUENCES OF ACCUMULATED IRON IN AGE-ASSOCIATED DISEASE

A. Parkinson's Disease

Parkinson's disease (PD) is a slowly progressive neurodegenerative disorder affecting more men than women. The disease results from degeneration of the basal ganglia. Specifically affected are dopaminergic cells of the substantia nigra, the caudate nucleus, and the putamen. There are many publications that report the substantia nigra and basal ganglia have elevated iron levels (14,125,126). This redox active iron has been implicated as a major generator of reactive oxygen species in PD (11,127) by two possible mechanisms: (1) H_2O_2, generated by the oxidation of dopamine via monoamine oxidase in neurons and astroglia, may be reduced by Fe(II) to HO$^{\bullet}$; (2) as discussed in Section III.E there is an increase in cytoplasmic inclusions containing autophagic iron-laden mitochondria, and this Fe(II) may have a nonenzymatic peroxidase activity capable of oxidizing catecholamines to neurotoxic orthosemiquinone radicals with neurotoxic capabilities (55). These oxidants have been suggested to induce complex I deficiency of mitochondria (128). Interestingly, it has also been found that parallel to the increase in iron there is an increase in the so-called common 5 kb deletion of mtDNA in PD (129).

Also, the iron content of mitochondria in animal and tissue culture models of PD are now known to increase (55). The cell death found in PD may well be induced, if not at least exacerbated, by the high iron content found in the mitochondria. This hypothesis may have support from Mizuno et al. (130), who have revealed that in MPTP-treated monkeys there is a marked increase in iron of the substantia nigra. Although this model shows striking similarities to PD, whether or not the iron increase is secondary to neuronal degradation awaits further investigation.

B. Alzheimer's Disease

Alzheimer's Disease (AD) is a common progressive neurodegenerative disease resulting in a loss of memory. Approximately 5% of AD cases are familial (with autosomal dominant transmission of early-onset disease) and three nuclear genes have been identified as the cause. The other 95% of AD are sporadic, without nuclear genetic associations. Yet first-degree relatives of affected individuals are at greater risk for AD than others (131), and furthermore, the lack of family history is a negative risk factor (132). This is suggestive of an unidentified genetic contribution to AD. Interestingly the risk for AD increases when a maternal relative has the disease (133). Because mtDNA are almost entirely maternally inherited, it is attractive to think that sporadic AD is a disease induced by mtDNA mutations. In fact, supportive data to this hypothesis have been reported: deficiencies in cytochrome oxidase activity have been identified in brain and in platelets of AD patients (134). Clearly there is strong evidence to show that defects in energy metabolism are involved in the pathogenesis of AD (reviewed in Refs. 135–137). Interestingly there was a recent attempt to show that there were mutations in mtDNA genes for cytochrome oxidase that segregated with sporadic late-onset AD (138). However, these mutations have been revealed to be most likely a PCR amplification of nucleus-embedded mtDNA pseudogenes (139). Thus it remains to be found what causes these reduced activities of cytochrome oxidase in mitochondria of AD patients and what role they play in the disease.

The hallmark lesions of Alzheimer's disease include the formation of abnormal proteins known as neurofibrillary tangles and neuritic plaques. The nature of these abnormal proteins and the location of the gene for producing the precursor protein has been identified. Alzheimer's is also characterized by profound deficits in neurotransmitters, particularly acetylcholine, which has been linked with memory function. The important scientific issue concerning AD revolves around the question of why particular classes of nerve cells are vulnerable and subject to cell death. Many researchers are actively pursuing an answer to this question in studies examining the potential effects of genetic factors, toxins, infectious agents, metabolic abnormalities, and a combination of these factors. Relevant to this chapter, it has been found that damage due to oxidants has been demonstrated in susceptible neuronal cell populations in cases of AD (17). Furthermore, the neurofibrillary tangles and neuritic plaques described above have been revealed to bind redox active iron (17). This plaque-associated iron is able to catalyze H_2O_2-dependent damage (17). Thus iron deposition could be an important factor in the oxidative stress of AD.

C. Ischemia-Reperfusion in Heart Disease

Ischemic heart disease is the leading cause of death in the United States (140) and in all Western cultures (141) and advanced age is a major risk factor (140). As with other diseases of aging, heart disease afflicts men more than women (142). It has been suggested that this difference exists due to regular iron loss by premenopausal women through menstruation (142). Ischemic heart disease is the result of insufficient coronary blood flow due to blockage (sometimes due to plaque formation via the process of atherosclerosis) of one of the coronary arteries that supply the heart (141). This is commonly called a heart attack or myocardial infarction. Death may result from this occlusion by ischemic necrosis or by fibrillation of the heart, but also may occur slowly over a period of weeks to years as a result of progressive weakening of the heart muscle due to the ischemic damage. The best way to limit infarct size and improve survival in patients with heart infarcts is to restore the flow in the obstructed artery as quickly as possible and thus the blood flow to the threatened myocardium. However, if restoration of blood flow is the sole method for saving ischemic tissue, oxidative damage may occur during reoxidation, contributing to ischemia-reperfusion injury. As discussed above, isolated mitochondria from heart tissue of aged animals have been reported to have increased H_2O_2 production compared to young animals. Thus the myocardium of senescent animals may be more sensitive to ischemia-reperfusion damage (143).

If ischemia-reperfusion damage is the mechanism by which an infarct inflicts its final fatal blow, what is the mechanism by which this takes place? Biochemical data show that mitochondrial production of radicals coupled with iron-induced HO• formation can lead to lipid peroxidation and permanent damage to ischemic rabbit hearts upon reperfusion (144) (reviewed in Refs. 145 and 146). Furthermore, recent evidence also provides compelling data that mitochondria are a subcellular target of reperfusion damage and a site of age-associated increases in sensitivity to this injury (143). Also, animal experiments have indicated that iron can promote ischemic myocardial injury (147). These findings have prompted the human studies of iron content and its effects on heart disease. At this point the evidence is not conclusive. First, increased serum ferritin was correlated with increased incidence of heart disease in Finland (19,20,142,148–154). Furthermore, lowering serum ferritin concentration by 44% using phlebotomy increased the oxidative resistance of serum very-low-density lipoprotein/low-density lipoprotein to challenge by oxidants (150). However, some studies into increased serum ferritin have found no or even inverse associations to heart disease (155,156). Recent studies using dietary iron and heart disease as variables have found a correlation between them, especially in aged men and postmenopausal women (157). These observations show that body iron stores are probably involved in heart disease in some way, and coupled with the biochemical data known for mitochondria it seems likely that aged mitochondria and high iron levels may interact to produce more damage after an ischemic episode, thus lowering the chance of recovery. Work in this area is being pursued by many laboratories.

D. Friedrich's Ataxia

In the last year, rapid advances in the study of a crippling disorder have provided a new model of mitochondrial iron overload. Friedrich's ataxia (FRDA), the most common inherited ataxia in humans, presents as a progressive loss of coordination of the limbs, and is transmitted as an autosomal recessive trait (158). In addition to neuronal dysfunction in the central nervous system and the periphery, symptoms of FRDA include diabetes and cardiomyopathy (often the cause of death). The fact that most of the affected cell types share a requirement for a high rate of mitochondrial respiration implicated mitochondria as a potential target of the disease (159). The fact that another inherited ataxia (called AVED for ataxia with isolated vitamin E deficiency), which results from an inability to transport vitamin E in low-density lipoprotein, shares similar symptoms as FRDA, strongly suggested that the etiology of FRDA involves oxidative stress (160).

Physical mapping and positional cloning of the gene responsible for FRDA identified a small (210 amino acid) protein, called frataxin, which is deficient in the disease (161). Interestingly, virtually all of the FRDA individuals analyzed had a lower expression of frataxin; null individuals have not been identified, presumably because this results in a lethal phenotype. The genetic lesions responsible for the disease in the large majority of cases are an expansion of a GAA triplet in the first intron of the frataxin gene which causes a decrease in steady-state frataxin mRNA expression levels. Strong inverse correlations between the lengths of the repeat expansion and the age of onset, as well as positive correlations between the length of the expansions and the rate of disease progression, have buttressed the genetic evidence implicating low levels of frataxin expression as the cause of FRDA (162).

Comparative analysis of the FRDA sequence to published databases has revealed homologues in eight different species of γ-purple bacteria (including *E. coli*), as well as eukaryotes such as yeast (YFH1) and *C. elegans* (163). The prokaryotic homologues share similarity only to the C-terminal end of the eukaryotic consensus sequence, whose N-terminal end comprises a predicated mitochondrial targeting sequence. Since purple bacteria are thought to be the living descendants of the prokaryotic endosymbiont which became modern-day mitochondria, the C-terminal homology of the bacterial genes and the addition of an N-terminal leader peptide in eukaryotes fits nicely into evolutionary models (163).

The yeast homologue of FRDA, called YFH1 (yeast frataxin homologue 1), is localized to mitochondria, as shown by fluorescence microscopy of fusions between Yfh1 protein and green fluorescent protein (113,114,164). [Mitochondrial localization has also been demonstrated in human cells in culture (115,165).] Deletion of the yeast gene (*Dyfh1*) results in a 10-fold increase in mitochondrial iron content (and a threefold overall increase in total cellular iron) (114). YFH1 deletion mutants grow slowly on fermentable carbon sources and are unable to grow on nonfermentable nutrients requiring mitochondrial respiration (113,114,164). Moreover, *Dyfh1* cells are extremely sensitive to H_2O_2 exposure (114) and frequently give rise to petite mutants lacking mtDNA (so-called r^0 cells) (113,114,164).

The mouse frataxin homologue has also been cloned and the murine protein shown to be localized to mitochondria; high-level expression of the murine Frda is found in the tissues most affected in FRDA in humans (164). Last, in human endomyocardial biopsies from patients with FRDA, the loss of activities of a number of mitochondrial iron-sulfur enzymes (such as aconitase) was observed (166). Since it has been well established that iron-sulfur clusters are exquisitely sensitive to destructive inactivation by oxygen free radicals (167), these data also support the conclusion that FRDA is a disease of mitochondrial iron overload and oxidative stress. Further elucidation of the mechanisms of FRDA pathophysiology should provide insights into the role of iron in normal and abnormal mitochondrial function.

V. CONCLUSIONS

Given the damaging Fenton chemistry of iron and the evidence that iron accumulates in tissues with age, it is surprising that the role of iron has received so little attention. Recently, however, with the discovery of the gene responsible for Friedrich's ataxia, and the correlation of neurodegeneration with mitochondrial iron overload and dysfunction, attention has now focused on the role of iron in mitochondria. The suspicion that iron plays a role in aging has existed for years, based on the fact that women (who live in a state of iron deficiency for most of their lives and have lower iron stores than men) have a longer life span and have less age-associated disease than men. Recent experiments showing life-span extension in *Drosophila* by iron deprivation and life-span reduction by iron overload implicate iron in aging mechanisms. It is clear that during aging there is an accumulation of oxidant-induced cellular damage, and that mitochondrial function also decays. Experimental evidence now reveals that increased iron in mitochondria during aging may contribute to neurodegeneration. It is also intriguing that mitochondria from iron-overload tissues and mitochondria from aged tissues share similar changes in mitochondrial morphology. However, despite all of this supporting evidence, few experiments have directly tested the hypothesis that iron is either necessary or sufficient for age-related disease. Future work should explore this interesting question.

ACKNOWLEDGMENTS

We thank Fernando Viteri, Mitchell Knutson, Roni Kohn, Hal Helbock, and Vladimir Gogvadze for their critical reading and/or helpful discussion of the manuscript.

REFERENCES

1. Harman, D. Aging: a theory based on free radical and radiation chemistry. J Gerontol 1956; 2:298–300.
2. Halliwell B, Gutteridge JM. Role of free radicals and catalytic metal ions in human disease: an overview. Methods Enzymol 1990; 186:1–85.
3. Floyd RA, Carney JM. The role of metal ions in oxidative processes and aging. Toxicol Ind Health 1993; 9:197–214.
4. Loschen G, Azzi A, Richter C, Flohé L. Superoxide radicals as precursors of mitochondrial hydrogen peroxide. FEBS Lett 1974; 42:68 72.
5. Beckman KB, Ames BN. The free radical theory of aging matures. Physiol Rev 1998; 78:547–581.
6. Tyler D. The Mitochondrion in Health and Disease. Cambridge, England: VCH Publishers, 1992.
7. Britton RS, Ramm GA, Olynyk J, et al. Pathophysiology of iron toxicity. Adv Exp Med Biol 1994; 356:239–253.
8. Walter P. Alterations in mitochondrial DNA metabolism caused by iron and reactive oxygen species. Ph.D. dissertation. Swiss Federal Institute of Technology (ETH), Zurich.
9. Hermes-Lima M. How do Ca^{2+} and 5-aminolevulinic acid-derived oxyradicals promote injury to isolated mitochondria? Free Radic Biol Med 1995; 19:381–390.
10. Yip R. Age related changes in iron metabolism. In: Brock JH, Halliday JW, Pippard MJ, Powell LW, ed. Iron Metabolism in Health and Disease. London: WB Saunders, 1994:427–448.
11. Schipper HM, and Cisse S. Mitochondrial constituents of corpora amylacea and autofluorescent astrocytic inclusions in senescent human brain. Glia 1995; 14:55–64.
12. Bacon BR, Park CH, Brittenham GM, O'Neill R, Tavill AS. Hepatic mitochondrial oxidative metabolism in rats with chronic dietary iron overload. Hepatology 1985; 5:789–797.
13. Ozawa T. Genetic and functional changes in mitochondria associated with aging. Physiol Rev 1997; 77:425–464.
14. Gorell J, Ordidge R, Brown G, et al. Increased iron-related MRI contrast in the substantia nigra in Parkinson's disease. Neurology 1995; 45:1138–1143.
15. Gerlach M, Ben SD, Riederer P, Youdim MB. Altered brain metabolism of iron as a cause of neurodegenerative diseases? J Neurochem 1994; 63:793–807.
16. Youdim MB, Riederer P. The role of iron in senescence of dopaminergic neurons in Parkinson's disease. J Neural Transm Suppl 1993; 40:57–67.
17. Smith MA, Harris PLR, Sayre LM, Perry G. Iron accumulated in Alzheimer's disease is a source of redox-generated free radicals. Proc Natl Acad Sci USA 1997; 94:9866–9868.
18. Toyokuni S. Iron-induced carcinogenesis: the role of redox regulation. Free Radic Biol Med 1996; 20:553–566.
19. Sullivan JL. Stored iron and ischemic heart disease—empirical support for a new paradigm. Circulation 1992; 86:1036–1037.

20. Salonen JT, Nyyssonen K, Korpela H, et al. High stored iron levels are associated with excess risk of myocardial infarction in eastern Finnish men. Circulation 1992; 86:803–811.

21. Shigenaga MK, Hagen TM, Ames BN. Oxidative damage and mitochondrial decay in aging. Proc Natl Acad Sci USA 1994; 91:10771–10778.

22. Gershman R, Gilbert DL, Sylvanus WN, Dwyer P, Fenn WO. Oxygen poisoning and X-irradiation; mechanism in common. Science 1954; 119:623–626.

23. Halliwell B. Free radicals, antioxidants, and human disease: curiosity, cause, or consequence? Lancet 1994; 344:721–724.

24. Koppenol WH. Oxyradical reactions: from bond-dissociation energies to reduction potentials. FEBS Lett 1990; 264:165–167.

25. Fenton HJH. Oxidation of tartaric acid in the presence of iron. J Chem Soc 1894; 65:899–910.

26. Munro HN, Eisenstein RS. Translational control: the ferritin story. Curr Opin Cell Biol 1989; 1:1154–1159.

27. Munro HN. Iron regulation of ferritin gene expression. J Cell Biochem 1990; 44: 107–115.

28. Lesnefsky EJ. Tissue iron overload and mechanisms of iron-catalyzed oxidative injury. Adv Exp Med Biol 1994; 366:129–146.

29. Crichton RR. Inorganic Biochemistry of Iron Metabolism. Chichester, England: Ellis Horwood.

30. Crichton RR, Ward RJ. Iron metabolism—new perspectives in view. Biochemistry 1992; 31:11255–11264.

31. Pollack S. Intracellular iron. Adv Exp Med Biol 1994; 356:165–171.

32. Masini A, Salvioli G, Cremonesi P, et al. Dietary iron deficiency in the rat. 1. Abnormalities in energy metabolism of the hepatic tissue. Biochim Biophys Acta 1994; 1188:46–52.

33. Masini A, Trenti T, Caramazza I, et al. Dietary iron deficiency in the rat. 2. Recovery from energy metabolism derangement of the hepatic tissue by iron therapy. Biochim Biophys Acta 1994; 1188:53–57.

34. Finch C. Regulators of iron balance in humans. Blood 1994; 84:1697–1702.

35. Hallberg L, Hulten L, Gramatkovsky E. Iron absorption from the whole diet in men: how effective is the regulation of iron absorption? Am J Clin Nutr 1997; 66: 347–356.

36. Hulten L, Gramatkovsky E, Gleerup A, Hallberg L. Iron absorption from the whole diet. Relation to meal composition, iron requirements and iron stores. Eur J Clin Nutr 1995; 49:794–808.

37. Tangeras A, Flatmark T. Mitochondrial iron not bound in heme and iron-sulfur centers and its availability for heme synthesis in vitro. Biochim Biophys Acta 1985; 843:199–207.

38. Yip R, Dallman PR. Iron. In: Ziegler EE, Filer LJ, eds. Present knowledge in nutrition. Washington, DC: ILSI Press, 1996:277–292.

39. van Asperen IA, Feskens EJ, Bowles CH, Kromhout D. Body iron stores and mortality due to cancer and ischaemic heart disease: a 17-year follow-up study of elderly men and women. Int J Epidemiol 1995; 24:665–670.

40. Keller JN, Germeyer A, Begley JG, Mattson MP. 17Beta-estradiol attenuates oxida-

tive impairment of synaptic Na$^+$/K$^+$-ATPase activity, glucose transport, and gluta-mate transport induced by amyloid beta-peptide and iron. J Neurosci Res 1997; 50:522–530.

41. Chance B, Sies H, Boveris A. Hydroperoxide metabolism in mammalian organs. Physiol Rev 1979; 59:527–605.

42. Bindoli A. Lipid peroxidation in mitochondria. Free Radic Biol Med 1988; 5:247–261.

43. Bacon BR, Oneill R, Britton RS. Hepatic mitochondrial energy production in rats with chronic iron overload. Gastroenterology 1993; 105:1134–1140.

44. Britton RS, O'Neill R, Bacon BR. Chronic dietary iron overload in rats results in impaired calcium sequestration by hepatic mitochondria and microsomes. Gastro-enterology 1991; 101:806–811.

45. Park CH, Bacon B, Brittenham GM, Tavill AS. Pathology of dietary carbonyl iron overload in rats. Lab Invest 1987; 57:555–563.

46. Sohal RS, Allen RG, Farmer KJ, Newton RK. Iron induces oxidative stress and may alter the rate of aging in the housefly, *Musca domestica*. Mech Ageing Dev 1985; 32:33–38.

47. Massie HR, Aiello VR, Williams TR. Inhibition of iron absorption prolongs the life span of *Drosophila*. Mech Ageing Dev 1993;67:227–237.

48. Stadtman ER, Starke-Reed PE, Oliver CN, Carney JM, Floyd RA. Protein modifi-cation in aging. EXS 1992; 62:64–72.

49. Stadtman ER. Metal ion-catalyzed oxidation of proteins: biochemical mechanism and biological consequences. Free Radic Biol Med 1990; 9:315–325.

50. Britton RS. Metal-induced hepatotoxicity. Semin Liver Dis 1996; 16:3–12.

51. Hermes-Lima M, Castilho RF, Meinicke AR, Vercesi AE. Characteristics of Fe(II)-ATP complex-induced damage to the rat liver mitochondrial membrane. Mol Cell Biochem 1995; 145:53–60.

52. Itoh H, Shioda T, Matsura T, et al. Iron ion induces mitochondrial DNA dam-age in HTC rat hepatoma cell culture—role of antioxidants in mitochondrial DNA protection from oxidative stresses. Arch Biochem Biophys 1994; 313:120–125.

53. Richter C, Park JW, Ames BN. Normal oxidative damage to mitochondrial and nuclear DNA is extensive. Proc Natl Acad Sci USA 1988; 85:6465–6467.

54. Yaffee M, Walter P, Richter C, Müller M. Direct observation of iron-induced con-formational changes of mitochondrial DNA by high resolution field-emission in-lens scanning electron microscopy. Proc Natl Acad Sci USA 1996; 93:5341–5346.

55. Wang XD, Manganaro F, Schipper HM. A cellular stress model for the sequestra-tion of redox-active glial iron in the aging and degenerating nervous system. J Neurochem 1995; 64:1868–1877.

56. Nohl H, Hegner D. Do mitochondria produce oxygen radicals in vivo? Eur J Bio-chem 1978; 82:563–567.

57. Sohal RS, Ku HH, Agarwal S, Forster MJ, Lal H. Oxidative damage, mitochondrial oxidant generation and antioxidant defenses during aging and in response to food restriction in the mouse. Mech Ageing Dev 1994; 74:121–133.

58. Sohal RS, Dubey A. Mitochondrial oxidative damage, hydrogen peroxide release, and aging. Free Radic Biol Med 1994; 16:621–626.

59. Luo YZ, Han ZX, Chin SM, Linn S. Three chemically distinct types of oxidants formed by iron-mediated Fenton reactions in the presence of DNA. Proc Natl Acad Sci USA 1994; 91:12438–12442.

60. Henle ES, Linn S. Formation, prevention, and repair of DNA damage by iron/hydrogen peroxide. J Biol Chem 1997; 272:19095–19098.

61. Chevion M. A site-specific mechanism for free radical induced biological damage: the essential role of redox-active transition metals. Free Radic Biol Med 1988; 5: 27–37.

62. Cadenas E, Sies H. Oxidative stress: excited oxygen species and enzyme activity. Adv Enzyme Regul 1985; 23:217–237.

63. Esterbauer H. Estimation of peroxidative damage. A critical review. Pathol Biol (Paris) 1996; 44:25–28.

64. Esterbauer H. Cytotoxicity and genotoxicity of lipid-oxidation products. Am J Clin Nutr 1993; 57:779S–785S; discussion 785S–786S.

65. Esterbauer H, Schaur RJ, Zollner H. Chemistry and biochemistry of 4-hydroxy-nonenal, malonaldehyde and related aldehydes. Free Radic Biol Med 1991; 11:81–128.

66. Chen JJ, Yu BP. Alterations in mitochondrial membrane fluidity by lipid peroxidation products. Free Radic Biol Med 1994; 17:411–418.

67. Player TJ, Mills DJ, Horton AA. NADPH-dependent lipid peroxidation in mitochondria from livers of young and old rats and from rat hepatoma D30 [proceedings]. Biochem Soc Trans 1977; 5:1506–1508.

68. Player TJ, Mills DJ, Horton AA. Age-dependent changes in rat liver microsomal and mitochondrial NADPH-dependent lipid peroxidation. Biochem Biophys Res Commun 1977; 78:1397–1402.

69. Brawer JR, Reichard G, Small L, Schipper HM. The origin and composition of peroxidase-positive granules in cysteamine-treated astrocytes in culture. Brain Res 1994; 633:9–20.

70. Brawer JR, Stein R, Small L, Cisse S, Schipper HM. Composition of Gomori-positive inclusions in astrocytes of the hypothalamic arcuate nucleus. Ant Rec 1994; 240:407–415.

71. Carini R, Parola M, Dianzani MU, Albano E. Mitochondrial damage and its role in causing hepatocyte injury during stimulation of lipid peroxidation by iron nitriloacetate. Arch Biochem Biophys 1992; 297:110–118.

72. Hagen TM, Yowe DL, Bartholomew JC, et al. Mitochondrial decay in hepatocytes from old rats: membrane potential declines, heterogeneity and oxidants increase. Proc Natl Acad Sci USA 1997; 94:3064–3069.

73. Tector AJ, Olynyk JK, Britton RS, et al. Hepatic mitochondrial oxidative metabolism and lipid peroxidation in iron-loaded rats fed ethanol. J Lab Clin Med 1995; 126:597–602.

74. Hansford RG, Hogue BA, Mildaziene V. Dependence of H_2O_2 formation by rat heart mitochondria on substrate availability and donor age. J Bioenerg Biomembr 1997; 29:89–95.

75. Sohal RS, Ku HH, Agarwal S. Biochemical correlates of longevity in two closely related rodent species. Biochem Biophys Res Commun 1993; 196:7–11.

76. Sohal RS, Sohal BH, Orr WC. Mitochondrial superoxide and hydrogen peroxide

generation, protein oxidative damage, and longevity in different species of flies. Free Radic Biol Med 1995; 19:499–504.

77. Azhar S, Cao L, Reaven E. Alteration of the adrenal antioxidant defense system during aging in rats. J Clin Invest 1995; 96:1414–1424.

78. Ji LL, Dillon D, Wu E. Myocardial aging: antioxidant enzyme systems and related biochemical properties. Am J Physiol 1991; 261:R386–R392.

79. Kukielka E, Dicker E, Cederbaum AI. Increased production of reactive oxygen species by rat liver mitochondria after chronic ethanol treatment. Arch Biochem Biophys 1994; 309:377–386.

80. Jenner P. Oxidative stress as a cause of Parkinson's disease. Acta Neurol Scand Suppl 1991; 136:6–15.

81. Benzi G, Moretti A. Are reactive oxygen species involved in Alzheimer's disease? Neurobiol Aging 1995; 16:661–674.

82. Foury F, Cazzalini O. Deletion of the yeast homologue of the human gene associated with Friedreich's ataxia elicits iron accumulation in mitochondria. FEBS Lett 1997; 411:373–377.

83. Liochev SI, Fridovich I. How does superoxide dismutase protect against tumor necrosis factor: a hypothesis informed by effect of superoxide on "free" iron. Free Radic Biol Med 1997; 23:668–671.

84. Agarwal S, Sohal RS. Differential oxidative damage to mitochondrial proteins during aging. Mech Ageing Dev 1995; 85:55–63.

85. Yan LJ, Levine RL, Sohal RS. Oxidative damage during aging targets mitochondrial aconitase. Proc Natl Acad Sci USA 1997; 94:11168–11172.

86. Agarwal S, Sohal RS. Aging and protein oxidative damage. Mech Ageing Dev 1994; 75:11–19.

87. Agarwal S, Sohal RS. Aging and proteolysis of oxidized proteins. Arch Biochem Biophys 1994; 309:24–28.

88. Forsmark-Andree P, Ernster L. Evidence for a protective effect of endogenous ubiquinol against oxidative damage to mitochondrial protein and DNA during lipid peroxidation. Mol Aspects Med 1994; 15:s73–s81.

89. Higuchi Y, Linn S. Purification of all forms of HeLa cell mitochondrial DNA and assessment of damage to it caused by hydrogen peroxide treatment of mitochondria or cells. J Biol Chem 1995; 270:7950–7956.

90. Hruszkewycz AM. Evidence for mitochondrial DNA damage by lipid peroxidation. Biochem Biophys Res Commun 1988; 153:191–197.

91. Hruszkewycz AM. Lipid peroxidation and mtDNA degeneration—a hypothesis. Mutat Res 1992; 275:243–248.

92. Lim LO, Neims AH. Mitochondrial DNA damage by bleomycin. Biochem Pharmacol 1987; 36:2769–2774.

93. Richter C, Gogvadze V, Laffranchi R, et al. Oxidants in mitochondria: from physiology to diseases. Biochim Biophys Acta 1995; 1271:67–74.

94. Richter C. Do mitochondrial DNA fragments promote cancer and aging? FEBS Lett 1988; 241:1–5.

95. Luo Y, Henle ES, Linn S. Oxidative damage to DNA constituents by iron-mediated fenton reactions. The deoxycytidine family. J Biol Chem 1996; 271:21167–21176.

96. Keyer K, Imlay JA. Superoxide accelerates DNA damage by elevating free-iron levels. Proc Natl Acad Sci USA 1996; 93:13635–13640.

97. Harman D. The aging process. Proc Natl Acad Sci USA 1981; 78:7124–7128.

98. Miquel J, Economos A, Fleming J, Johnson JJ. Mitochondrial role in cell aging. Exp Gerontol 1980; 15:575–591.

99. Wallace DC. Mitochondrial genetics: a paradigm for aging and diseases. Science 1992; 256:628–632.

100. Wallace DC. Mitochondrial DNA variation in human evolution, degenerative disease, and aging. Am J Hum Genet 1995; 57:201–223.

101. Wallace D, Shoffner J, Trounce I, et al. Mitochondrial DNA mutations in human degenerative diseases and aging. Biochim Biophys Acta 1995; 1271:141–151.

102. Petruzzella V, Fracasso F, Gadaleta M, Cantatore P. Age-dependent structural variations in rat brain mitochondrial DNA. Ann NY Acad Sci 1992; 673:194–199.

103. Asano K, Amagase S, Matsuura ET, Yamagishi H. Changes in the rat liver mitochondrial DNA upon aging. Mech Ageing Dev 1991; 60:275–284.

104. Lee CM, Weindruch R, Aiken JM. Age-associated alterations of the mitochondrial genome. Free Radic Biol Med 1997; 22:1259–1269.

105. Hayakawa M, Hattori K, Sugiyama S, Ozawa T. Age-associated oxygen damage and mutations in mitochondrial DNA in human hearts. Biochem Biophys Res Commun 1992; 189:979–985.

106. Hayakawa M, Sugiyama S, Hattori K, Takasawa M, Ozawa T. Age-associated damage in mitochondrial DNA in human hearts. Mol Cell Biochem 1993; 119:95–103.

107. de la Asuncion JG, Millan A, Pla R, et al. Mitochondrial glutathione oxidation correlates with age-associated oxidative damage to mitochondrial DNA. FASEB J 1996; 10:333–338.

108. Beckman KB, Ames BN. Detection and quantification of oxidative adducts of mitochondrial DNA. Methods Enzymol 1996; 264:442–453.

109. Helbock HJ, Beckman KB, Shigenaga MK, et al. DNA oxidation matters: the HPLC-EC assay of 8-oxo-deoxyguanosine and 8-oxo-guanine. Proc Natl Acad Sci USA 1998; 95:288–293.

110. Yakes FM, Chen Y, Van Houten B. PCR-based assays for the detection and quantitation of DNA damage and repair. In: Pfeifer GP, ed. Technologies for Detection of DNA Damage and Mutations. New York: Plenum, 1996:171–184.

111. Yakes FM, Van Houten B. Mitochondrial DNA damage is more extensive and persists longer than nuclear DNA damage in human cells following oxidative stress. Proc Natl Acad Sci USA 1997; 94:514–519.

112. Niranjan BG, Bhat NK, Avadhani NG. Preferential attack of mitochondrial DNA by aflatoxin B1 during hepatocarcinogenesis. Science 1982; 215:73–75.

113. Wilson RB, Roof DM. Respiratory deficiency due to loss of mitochondrial DNA in yeast lacking the frataxin homologue. Nat Genet 1997; 16:352–357.

114. Babcock M, de Silva D, Oaks R, et al. Regulation of mitochondrial iron accumulation by Yfh1p, a putative homolog of frataxin. Science 1997; 276:1709–1712.

115. Tauchi H, Sato T. Age changes in size and number of mitochondria of human hepatic cells. J Gerontol 1968; 23:454–461.

116. Bertoni-Freddari C, Fattoretti P, Casoli T, et al. Morphological plasticity of synaptic mitochondria during aging. Brain Res 1993; 628:193–200.

117. Solmi R, Pallotti F, Rugolo M, et al. Lack of major mitochondrial bioenergetic changes in cultured skin fibroblasts from aged individuals. Biochem Mol Biol Int 1994; 33:477–484.

118. Gartner LP. The fine structural morphology of the midgut of aged *Drosophila*: a morphometric analysis. Exp Gerontol 1987; 22:297–304.

119. Marzabadi MR, Llvaas E. Spermine prevent iron accumulation and depress lipofuscin accumulation in cultured myocardial cells. Free Radic Biol Med 1996; 21:375–381.

120. Yin D. Biochemical basis of lipofuscin, ceroid, and age pigment-like fluorophores. Free Radic Biol Med 1996; 21:871–888.

121. Brunk UT, Jones CB, Sohal RS. A novel hypothesis of lipofuscinogenesis and cellular aging based on interactions between oxidative stress and autophagocytosis. Mutat Res 1992; 275:395–403.

122. Mackle P, Garner CD, Ward RJ, Peters TJ. Iron K-cdgc absorption spectroscopic investigations of the cores of ferritin and haemosiderins. Biochim Biophys Acta 1991; 1115:145–150.

123. Youdim MBH, Benshachar D, Riederer P. The possible role of iron in the etiopathology of Parkinson's disease. Mov Disord 1993; 8:255.

124. Smith MA, Perry G. Free radical damage, iron, and Alzheimer's disease. J Neurol Sci 1995; 134(suppl):92–94.

125. Griffiths PD, Crossman AR. Distribution of iron in the basal ganglia and neocortex in postmortem tissue in Parkinson's disease and Alzheimer's disease. Dementia 1993; 4:61–65.

126. Loeffler DA, Connor JR, Juneau PL, et al. Transferrin and iron in normal, Alzheimer's disease, and Parkinson's disease brain regions. J Neurochem 1995; 65:710–716.

127. Justino L, Welner SA, Tannenbaum GS, Schipper HM. Long-term effects of cysteamine on cognitive and locomotor behavior in rats: relationship to hippocampal glial pathology and somatostatin levels. Brain Res 1997; 761:127–134.

128. Mann VM, Cooper JM, Krige D, et al. Brain, skeletal muscle and platelet homogenate mitochondrial function in Parkinson's disease. Brain 1992; 115:333–342.

129. Mizuno Y, Mochizuki H, Nishi K, et al. Pathogenesis of Parkinson's disease— iron and mitochondrial DNA deletion. In: Riederer P, Youdim MBH, eds. Iron in Central Nervous System Disorders. New York: Springer-Verlag 1993:117–135.

130. Mizuno Y, Ikebe S, Hattori N, et al. Role of mitochondria in the etiology and pathogenesis of Parkinson's disease. Biochim Biophys Acta 1995; 1271:265–274.

131. Silverman JM, Raiford K, Edland S, et al. The Consortium to Establish a Registry for Alzheimer's Disease (CERAD). Part VI. Family history assessment: a multicenter study of first-degree relatives of Alzheimer's disease probands and nondemented spouse controls. Neurology 1994; 44:1253–1259.

132. Payami H, Montee K, Kaye J. Evidence for familial factors that protect against dementia and outweigh the effect of increasing age. Am J Hum Genet 1994; 54:650–657.

133. Edland SD, Silverman JM, Peskind ER, et al. Increased risk of dementia in mothers of Alzheimer's disease cases: evidence for maternal inheritance. Neurology 1996; 47:254–256.

134. Parker WD, Jr, Parks JK. Cytochrome c oxidase in Alzheimer's disease brain: puri-
      fication and characterization. Neurology 1995; 45:482–486.
135. Beal MF. Aging, energy, and oxidative stress in neurodegenerative diseases. Ann
      Neurol 1995; 38:357–366.
136. Schapira AH. Oxidative stress and mitochondrial dysfunction in neurodegeneration.
      Curr Opin Neurol 1996; 9:260–264.
137. Sims NR. Energy metabolism, oxidative stress and neuronal degeneration in Alz-
      heimer's disease. Neurodegeneration 1996; 5:435–440.
138. Davis RE, Miller S, Herrnstadt C, et al. Mutations in mitochondrial cytochrome c
      oxidase genes segregate with late-onset Alzheimer disease. Proc Natl Acad Sci
      USA 1997; 94:4526–4531.
139. Hirano M, Shtilbans A, Mayeux R, et al. Apparent mtDNA heteroplasmy in Alzhei-
      mer's disease patients and in normals due to PCR amplification of nucleus-embed-
      ded mtDNA pseudogenes. Proc Natl Acad Sci USA 1997; 94:14894–14899.
140. Johnson MA, Fischer JG, Bowman BA, Gunter EW. Iron nutriture in elderly indi-
      viduals. FASEB J 1994; 8:609–621.
141. Guyton AC. Textbook of Medical Physiology. Philadelphia: WB Saunders, 1996:
      240–245.
142. Sullivan JL. Iron, hematocrit, and the sex difference in heart disease. Arch Pathol
      Lab Med 1993; 117:966–967.
143. Lucas DT, Szweda LI. Cardiac reperfusion injury: aging, lipid peroxidation, and
      mitochondrial dysfunction. Proc Natl Acad Sci USA 1998; 95:510–514.
144. Lesnefsky EJ, Ye J. Exogenous intracellular, but not extracellular, iron augments
      myocardial reperfusion injury. Am J Physiol 1994; 266:H384–H392.
145. Ferrari R. The role of mitochondria in ischemic heart disease. J Cardiovasc Pharma-
      col 1996; 28(suppl 1):S1–S10.
146. Ferrari R. Importance of oxygen free radicals during ischemia and reperfusion in
      the experimental and clinical setting. Oxygen free radicals and the heart. Am J
      Cardiovasc Pathol 1992; 4:103–114.
147. Farber NE, Vercellotti GM, Jacob HS, Pieper GM, Gross GJ. Evidence for a role
      of iron-catalyzed oxidants in functional and metabolic stunning in the canine heart.
      Circ Res 1988; 63:351–360.
148. Salonen JT, Nyyssonen K, Korpela H, et al. High stored iron levels and the risk
      of myocardial infarction—reply. Circulation 1993; 87:1425–1426.
149. Salonen JT, Nyyssonen K, Salonen R. Body iron stores and the risk of coronary
      heart disease. N Engl J Med 1994; 331:1159.
150. Salonen JT, Korpela H, Nyyssonen K, et al. Lowering of body iron stores by blood
      letting and oxidation resistance of serum lipoproteins: a randomized cross-over trial
      in male smokers. J Intern Med 1995; 237:161–168.
151. Sullivan JL. Vitamin-E and the risk of coronary disease. N Engl J Med 1993; 329:
      1425.
152. Sullivan JL. Effects of dietary iron intake on stored iron, free iron, and coronary
      disease. Circulation 1994; 90:3122–3123.
153. Sullivan JL. Low iron-binding capacity: an independent heart disease risk factor?
      Circulation 1994; 89:2947–2948.

154. Sullivan JL. Hemochromatosis and coronary artery disease. JAMA 1995; 273:25–26.

155. Sempos CT, Looker AC, Gillum RF, Makuc DM. Body iron stores and the risk of coronary heart disease. N Engl J Med 1994; 330:1119–1124.

156. Liao Y, Cooper RS, McGee DL. Iron status and coronary heart disease: negative findings from the NHANES I epidemiologic follow-up study. Am J Epidemiol 1994; 139:704–712.

157. Tzonou A, Lagiou P, Trichopoulou A, Tsoutsos V, Trichopoulos D. Dietary iron and coronary heart disease risk: a study from Greece. Am J Epidemiol 1998; 147: 161–166.

158. Gray JV, Johnson KJ. Waiting for frataxin [news]. Nat Genet 1997; 16:323–325.

159. Harding AE, Holt IJ, Cooper JM, et al. Mitochondrial myopathies: genetic defects. Biochem Soc Trans 1990; 18:519–522.

160. Ouahchi K, Arita M, Kayden H, et al. Ataxia with isolated vitamin E deficiency is caused by mutations in the alpha-tocopherol transfer protein. Nat Genet 1995; 9:141–145.

161. Campuzano V, Montermini L, Molto MD, et al. Friedreich's ataxia: autosomal recessive disease caused by an intronic GAA triplet repeat expansion. Science 1996; 271:1423–1427.

162. Durr A, Cossee M, Agid Y, et al. Clinical and genetic abnormalities in patients with Friedreich's ataxia [see comments]. N Engl J Med 1996; 335:1169–1175.

163. Gibson TJ, Koonin EV, Musco G, Pastore A, Bork P. Friedreich's ataxia protein: phylogenetic evidence for mitochondrial dysfunction. Trends Neurosci 1996; 19: 465–468.

164. Koutnikova H, Campuzano V, Foury F, et al. Studies of human, mouse and yeast homologues indicate a mitochondrial function for frataxin. Nat Genet 1997; 16: 345–351.

165. Priller J, Scherzer CR, Faber PW, MacDonald ME, Young AB. Frataxin gene of Friedreich's ataxia is targeted to mitochondria. Ann Neurol 1997; 42:265–269.

166. Rötig A, de Lonlay P, Chretien D, et al. Aconitase and mitochondrial iron-sulphur protein deficiency in Friedrich ataxia. Nat Genet 1997; 17:215–217.

167. Fridovich I. Superoxide anion radical (O_2^-), superoxide dismutases, and related matters. J Biol Chem 1997; 272:18515–18517.

12

The Mitochondrial-Lysosomal Axis Theory of Cellular Aging

Ulf T. Brunk
University of Linköping, Linköping, Sweden

Alexei Terman
University of Linköping, Linköping, Sweden, and Institute of Gerontology, AMS of Ukraine, Kiev, Ukraine

I. INTRODUCTION

Aging is inevitable and has always fascinated mankind. Still, its basic course and the genetic keys that necessarily must be involved in the regulation of the complicated processes involved are largely unknown. An aged individual is easy to recognize compared to a child or young adult. Wrinkling skin, graying and thinning hair, hardening eye lenses, and changing posture are apparent and well-known manifestations of aging. Morphological studies of aged individuals also reveal some structural disorganization of most organs and tissues, a decreased proportion of functional elements with an increased amount of stroma, and impairment of vascularization and innervation. These age changes chiefly represent alterations of the extracellular matrices due to glucosylation, oxidation, and cross-linkings of various proteins and, although they may be much disturbing to the individual, they certainly do not threaten life.

Manifestations of aging are more variable on the cellular level, depending first on how the cells are renewed. For some cell types, renewal is provided by division and differentiation of stem cells that occurs constantly throughout life. Such cells are the squamous epithelial cells of the skin, epithelial cells along the intestinal canal, and the bone marrow cells which produce erythrocytes, platelets, and leukocytes. Even in very old but otherwise healthy individuals, they seem to carry no structural alterations and function largely in a normal way. Other

229

cells may divide and multiply when necessary, for example, during tissue regeneration and repair processes. Most of the parenchymatous cells which build up organs such as the liver, kidneys, and most endocrine and exocrine glands belong to this group. Also, these cells show only limited and uncharacteristic alterations with age, and healthy older individuals do not need to have problems connected with these types of cells which have a low and retained, mitotic capacity. Needless to say, some endocrine glands stop functioning at the end of the fertile period, and the immune system is known to weaken in old age, but these phenomena may rather be an effect of failure of regulating pituitary and hypothalamic centers than due to a primary loss of function by the endocrine cells themselves. Aging at the cellular level must then be looked for within the third population of cells that build the organism: the postmitotic type of cells. The cells within this group are neurons, cardiac myocytes, skeletal muscle fibers, Sertoli cells, and retinal pigment epithelial cells. Humans, and most other mammals, carry a certain number of them at birth or obtain them in the early neonatal period. They are terminally differentiated throughout the life span, and they do not divide since division is probably incompatible with their highly specialized functions. The renewal of these cell types is solely intracellular, provided by constant degradation of utilized or impaired cellular components, with permanent creation of new and replacing structures. Apparently, intracellular renewal is less perfect than renewal by cell division, and at the end of the expected life span many postmitotic cells start to function less well, even in healthy individuals. The cardiac capacity declines, as do motor, sensory, and mental functions until finally the really aged individual is much reduced in many ways and may pass away without it being possible to exactly define why. In those few cases we have to admit that life has just come to an end. Unfortunately most individuals die long before the end of their theoretical life span because of diseases we still cannot beat or prevent.

Senescent postmitotic cells are easy to recognize by a number of morphological hallmarks when compared to corresponding young ones. Although the degree of these changes varies between different cells, there is an average trend to the enlargement of cell size, to the impairment of functional elements of the cell, and to the appearance of waste products, mainly represented by lipofuscin— the age pigment.

II. AGE-RELATED CHANGES IN POSTMITOTIC CELLS

A. Aging and General Impairment of the Cell

Aging postmitotic cells undergo various alterations of their functional elements. In cell membranes, changes in molecular composition, especially of lipid components and alterations of fluidity (1–3), cause disturbances in transport, excitability, and responsiveness to regulatory stimuli (4–8). Nuclear changes involve chromatin condensation, DNA-protein cross-links, single- and double-strand breaks, and various mutations which lead to alterations of replication and transcription (9–16). In the cytoplasm of senescent cells, mutual arrangement of organelles and organization of cytoskeletal structures is usually not so regular as in young cells (17–20). The most consistent changes, however, occur in the cytoplasmic structures responsible for genetic translation, degradation of macromolecules, and energy production. The decrease in the amount of free- and membrane-associated ribosomes, and alterations in their density and distribution is associated with a decline in protein synthesis (21–28). Mitochondrial changes are, however, together with lipofuscin accumulation within secondary lysosomes, the most prominent ones.

Although some structural changes in postmitotic cells can be classified as compensatory, for example, cell size enlargement, irregularities of nuclear shape, increased number of microvilli, or partial hyperplasia of the endoplasmic reticulum and/or the Golgi complex (29–33), the general outcome of senescent changes is the loss of function, increased susceptibility to pathogenic factors, and finally, cell death.

B. Aging and Mitochondrial Alterations

Mitochondria are usually reduced in numbers, while their size is increased, sometimes enormously (so-called giant mitochondria) (34–36). Swelling and partial destruction of cristae are not infrequent (35,37,38). Mutations of mitochondrial DNA (mtDNA) occur progressively with age, effecting mainly sites coding for complexes I and IV of the respiratory chain (39–44). The ensuing decline in the activity of encoded enzymes inevitably affects energy metabolism because of a decrease in ATP production (41,45–47). Moreover, mitochondria in aged postmitotic cells seem to produce increasing amounts of $O_2^{\cdot-}$ and H_2O_2 due to enhanced loss of electrons from their electron transport chains (48–52).

C. Aging and Accumulation of Lipofuscin

Accumulation of lipofuscin in postmitotic cells is perhaps the most consistent manifestation of aging. In an unstained histologic section, lipofuscin is seen as a brown-yellow pigment that also displays autofluorescence. Ultrastructurally it represents an osmiophilic material surrounded by a typical lysosomal membrane (53–55). Lipofuscin accumulation seems to start very early in life and then goes on about linearly until old age (56–60). Whether or not lipofuscin accumulation is permanently of a linear type, or if it finally becomes exponential, is presently not known. The inverse correlation between the rate of lipofuscin accumulation and longevity (61–63) emphasizes the importance of lipofuscin as a biomarker of aging.

The discovery of lipofuscin, and its relation to aging, is by no means a recent one. In the previous century, scientists such as Hannover (64) and Koneff (65) presented detailed drawings of their light microscopic findings. However, only recently has it been clarified that lipofuscin represents incompletely degraded, oxidized, and polymerized intralysosomal material (54,66–70) composed of various chemical substances, mainly of protein and lipid origin, 30–70% and 20–50%, respectively (reviewed in Refs. 71 and 72). Practically the same properties are displayed by the ceroid pigments, formed in many cell types and under various pathological conditions unrelated to aging, such as malnutrition, lysosomal storage diseases, atherosclerosis, tumors, X irradiation, stress, etc. (73–84).

What chemical structures are responsible for lipofuscin/ceroid autofluorescence is not completely clear. Nevertheless, test tube experiments suggest that reactions between carbonyls (mainly aldehydes resulting from lipid peroxidation reactions) and amino compounds result in the formation of Schiff bases, 1,4-dihydropyridines, or 2-hydroxy-1,2-dihydropyrrol-3-ones that display autofluorescent properties similar to extracts of natural lipofuscin (85–87). A lipofuscin fluorophore with the structure of a Schiff base (a product of retinaldehyde and ethanolamine interaction) has been identified in retinal pigment epithelium (88). The presence of retinol derivatives has been shown also in liver lipofuscin (89).

Solid pigment in intact cells is usually characterized by emission at longer wavelengths, when assayed by microfluorometry, than lipofuscin extracts assayed by ordinary spectrofluorometry. This had caused confusion until it was shown that the difference may be due to a higher concentration of the pigment fluorophores in intact tissue specimens than in extracts (90). It needs to be remembered, however, that a major part of lipofuscin/ceroid is insoluble in both polar and nonpolar solvents (72).

III. OXIDATIVE STRESS IN AGING, MITOCHONDRIAL DAMAGE, AND LIPOFUSCINOGENESIS

A. Oxidative Stress as a Fundamental Factor of Aging

Since age-related alterations start to develop, albeit slowly, in originally young and healthy organisms, the causes of aging must be looked for in nonsenescent postmitotic cells. The aging phenomenon suggests that the intracellular renewal processes are not perfect, even in young healthy cells, that is, utilized or damaged functional elements are not turned over and restored completely, and not all auto-/heterophagocytosed material is degraded properly within the acidic vacuolar apparatus, but partially converted to lipofuscin. Actually there is no comprehensive knowledge on how cell components are damaged, and why cell renewal is not absolutely accurate, even under favorable conditions. However, a large body of data testifies to the importance of oxidative stress in the initiation and progress of senescent changes. As early as 1956, Harman (91) proposed the free radical theory of aging, which tried to explain aging as a result of accumulated damage by oxygen-derived free radicals. This theory was later further developed and modified in numerous studies (reviewed in Refs. 92–94). Oxidative stress inevitably occurs in most tissues throughout life as a result of normal oxygen metabolism, mainly by mitochondria, which is accompanied by the formation of reactive oxygen species (ROS), such as $O_2^{\cdot -}$ anion radicals, singlet oxygen, nitric oxide, and reactive peroxides, including H_2O_2 and aldehydes (95,96). Despite the activity of a host of cellular antioxidant systems, including oxygen radical scavengers, persistent oxidative stress seems to induce the development of age-related changes, and thus favors the validity of the oxidative stress hypothesis of aging (reviewed in Ref. 93).

Although aging involves practically all structures of the postmitotic cells, there are good reasons to believe that the oxidative stress which is mainly induced by mitochondria—and will cause oxidative alterations of the lysosomal compartment and the mitochondria themselves—would play a key role in the progress of senescence.

B. Mechanisms of Oxidative Mitochondrial Damage and Its Contribution to Aging

The importance of mitochondrial damage for cellular senescence became evident only recently, when new knowledge was established on the cause of their own particular vulnerability to oxidative stress (97–99). Being the site of the oxidative reactions needed for ATP generation, mitochondria more than other cellular compartments are exposed to ROS that may damage their macromolecules, for example, DNA and inner-membrane structures, including the electron transport complexes. In addition, the bacterial-type DNA in mitochondria is not protected by histones, as is the case in the nucleus, further increasing mtDNA susceptibility to oxidative stress (49).

Another important reason for the high vulnerability of mitochondria is due to the properties of mtDNA that specify parts of the components of the respiratory chain. The electron transport chain consists of five protein complexes. Complexes I and IV and, partially, complex III (i.e., NADH-ubiquinone oxidoreductase, cytochrome oxidase, and ubiquinone-cytochrome-c oxidoreductase complexes, respectively) are coded for by mtDNA. The mtDNA is completely expressed, unlike the nuclear DNA, in which only about 7% of the genes are active (100). Therefore the estimated rate of mutations in mtDNA with effects on protein anabolism is 10 to 12 times as high as that of chromosomes (101). Furthermore, the mtDNA repair system is much less efficient than that of nuclear DNA. As a consequence, mutations that determine ensuing changes in the encoded respiratory chain proteins accumulate in mitochondria at a comparatively high rate. This explains why aging primarily affects complexes I and IV of the electron transport chain, coded by mtDNA, but not other proteins which are mainly coded by nuclear genes (102–106). Mitochondria with mutated mtDNA are believed to produce an increased amount of ROS, further damaging the mitochondrial genome, membranes, and enzymes, creating a spiral of increasing damage (48–52) if they are not removed by autophagocytosis.

Affected mitochondria are supposed to be autophagocytosed and digested, whereas the normal ones replicate and maintain energy production at a stable level (107,108). However, this may not be the truth in all cases. Evidence suggests that inborn mitochondrial mutations often accumulate with age, especially in postmitotic cells, causing a number of severe diseases, such as chronic progressive external ophthalmoplegia, Kearns–Sayre syndrome, Alzheimer's disease, Parkinson's disease, late diabetes mellitus (type II), myopathy, dystonia, hypertrophic cardiomyopathy, and others (52,109–113). One possible reason for this is that macroautophagocytosis—the mechanism by which mitochondria are removed by digestion within the acidic vacuolar compartment—seems to be a rather nonselective process (114), while the replication of defective mitochondria is not affected because the genes responsible for mitochondrial reproduction are located in the nucleus and not in the mitochondria themselves (101,115). Another reason may be that autophagocytosis may function less well in aged cells. Aging cells are thus supposed to accumulate increasing quantities of damaged mitochondria that inevitably disturb the process of ATP generation and increase cellular oxidative stress, maybe even dramatically and in an exponential manner at the very end of the life span.

C. Mechanisms of Lipofuscinogenesis

The material forming lipofuscin in postmitotic cells enters the lysosomal compartment by way of macro- or microautophagocytosis (in some cases by heterophagocytosis, like in retinal pigment epithelial cells). It is then subjected to the action of a battery of hydrolases (54,67,69,70,116,117). Most substances that become auto-/heterophagocytosed are completely degraded, and the building blocks are reutilized by the cell after being transported through the lysosomal membranes out in the cytosol. Those that become peroxidized and polymerized will, however, accumulate and constitute lipofuscin since iron-catalyzed oxidative reactions convert the material into a nondegradable product—lipofuscin (72).

That lipofuscin/ceroid is neither degraded nor exocytosed has been shown recently using two different cellular models. In the work of Elleder et al. (118), different cells in culture were exposed to ceroid granules, isolated from animals with Batten's disease. The ceroid was phagocytosed and remained unchanged within the lysosomal apparatus for 3 weeks. In our own study (119), accumulation of lipofuscin/ceroid in cultured rat cardiac myocytes was induced by addition of the cysteine-protease inhibitor leupeptin for some time. The cells were then returned to normal cell culture conditions, and the resultant increased amounts of lipofuscin/ceroid were found to remain for at least 2 weeks. For a comparison, normal intralysosomal degradation, or exocytosis, is known to take no more than one or a few hours (114,116,120).

Accumulated evidence suggests that oxidative stress plays a leading role in the conversion of autophagocytosed material into the undegradable lipofuscin/ceroid. Increased formation of lipofuscin/ceroid was observed in fibroblasts, glial cells, cardiac myocytes, and retinal pigment epithelial cells when cultured at high (40%) ambient oxygen (121–125). Consistent with these results, enhanced amounts of lysosomal redox active iron—endocytosed from medium enriched with an unsoluble iron complex of low-molecular weight—also increased the lipofuscin/ceroid accumulation in cultured human glial cells (123) and neonatal rat cardiac myocytes (121). In rat brain neurons, an increase of lipofuscin/ceroid was obtained by intralumbar injections of iron (126). In contrast, the iron chelator desferrioxamine, various antioxidants (e.g., vitamin E, selenium, glutathione, etc.), and caloric restriction (known to slow down oxidative processes) retarded lipofuscin/ceroid accumulation (121,123,127–132), while vitamin E deficiency has been found to be associated with an increase of lipofuscin/ceroid (128,133–136).

Brunk et al. (137) proposed a scenario on the involvement of oxidative stress in lipofuscin formation. It was assumed that ROS (mainly H_2O_2, which is produced in relation to mitochondria by dismutation of $O_2^{\bullet-}$), unless eliminated by catalase or glutathione peroxidase, easily diffuse into secondary lysosomes which contain various autophagocytosed macromolecules and redox active low-molecular weight iron; the latter released from a variety of metalloproteins while under intralysosomal degradation. The interaction between H_2O_2 and iron would result in Fenton-type chemistry, possible with the formation of hydroxyl radicals, with ensuing cross-linking of surrounding macromolecules and resultant lipofuscin formation (Fig. 1). Consistent with this assumption, peroxidation products showing lipofuscin-like fluorescence were obtained in test tube experiments, when cysteine (that reduces ferric to ferrous iron and thus activates Fenton reactions) was added to a rat liver lysosomal-mitochondrial fraction (138).

Apparently, autophagocytosis of mitochondria plays an important role in lipofuscinogenesis (54,137,139). Autophagocytosed mitochondria may provide some already peroxidized, undegradable macromolecules (as a consequence of earlier oxidative attacks while still in the cytosol), $O_2^{\bullet-}$ and iron (e.g., from heme groups in cytochromes). Peroxidation of macromolecules would continue (or start) within the lysosomal compartment to a degree that would be estimated by the available amounts of H_2O_2 and intralysosomal iron in redox active form. That oxidatively damaged mitochondria may contribute to lipofuscinogenesis is clear

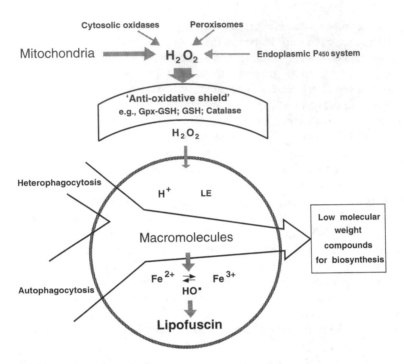

Figure 1 Schematic representation of the mechanisms behind lipofuscin formation within the lysosomal vacuome of postmitotic cells. A large variety of biomolecules are degraded by lysosomal hydrolytic enzymes (LE) within the acidic vacuolar compartment of the cell. Some of the macromolecules are iron-containing metalloproteins which would release iron in low molecular weight form when degraded. Secondary lysosomes have a low pH and are rich in reducing substances such as cysteine. Some ferric iron would thus be reduced into ferrous form. Hydrogen peroxide, which is diffusing into the compartment, may consequently undergo homolytic cleavage with formation of hydroxyl radicals which, in turn, would induce peroxidative reactions and formation of lipofuscin from macromolecules undergoing degradation.

from recent studies on cultured human fibroblasts (140,141). In these works, isolated liver mitochondria were exposed to photooxidation by UV light, homogenized to granules of approximately 1 μm, and added to the culture medium. The cells phagocytosed the granules and the appearance of abundant electron-dense lysosomal inclusions with typical lipofuscin-like autofluorescence was the result. In other studies (142,143), an abnormal mitochondrial protein—ATP synthase subunit *c*—was found to be responsible for the accumulation of ceroid pigment in Batten's disease. This aberrant protein appears to be undegradable, and consequently it accumulates within lysosomes and finally is converted to ceroid over time, presumably by peroxidation according to the above-described mechanisms.

Since autophagocytosis and oxidative stress are normally present throughout life, the process of lipofuscin accumulation is most likely independent of age. Alternatively, it could be assumed that the accumulation of lipofuscin is caused by age-related alterations of lysosomal functions, such as enhanced autophagocytosis, decreased intralysosomal degradation, or decreased exocytosis. However, evidence suggests that autophagocytosis decreases with age (144,145), while exocytosis of lipofuscin/ceroid of any importance, as noted above, does not seem to occur. Some studies testify to a decrease of intralysosomal degradation and, in particular, proteolytic activity with age (144,146–148). Also, there are indications that the activity of some lysosomal cysteine proteases (cathepsins B and L) decreases with age in cultured human fibroblasts (149–150) and in some rat and human neurons (151,152). A possible role of decreased lysosomal proteolysis in lipofuscin accumulation was hypothesized by Ivy et al. (153–155) on the basis of their findings that the lysosomal cysteine protease inhibitors leupeptin and E-64 cause an accumulation of lipofuscin-like osmiophilic inclusions in brain, retina, liver, and kidneys of experimental animals. However, in the work of Porta et al. (156), no correlation was found between cathepsin B activity (decreases only from the age of 14 months) and lipofuscin accumulation (linearly increases throughout life) in rat brain and heart. These results show that a decrease of proteolysis, if it occurs at all, may contribute to lipofuscinogenesis only late in life. Also, it should be pointed out that an association between decreased proteolysis and an accumulation of lipofuscin does not necessarily mean that the former directly causes the latter. We could equally well suppose that enhanced lipofuscin accumulation at old age rather would be the cause of decreased lysosomal degradative capacity. If so, autophagocytosis would be delayed, and an increase of cellular oxidative stress is to be expected because of prolonged survival of damaged mitochondria producing increased amounts of $O_2^{\cdot-}$ (Fig. 2).

IV. DISTRIBUTION OF LYSOSOMAL ENZYMES WITHIN A LIPOFUSCIN-LOADED ACIDIC VACUOLAR APPARATUS AND ITS EFFECT ON AUTOPHAGOCYTOSIS

We envisage that lipofuscin accumulation within the acidic vacuolar apparatus of aged postmitotic cells finally may reach a level where it affects the flow of lysosomal hydrolytic enzymes to autophagic vacuoles. The rationale behind this hypothesis is that primary lysosomes, newly produced from the Golgi complex, may fuse with any other vacuole within the acidic apparatus (114,157,158). In a case where the number of lipofuscin-containing secondary lysosomes is very

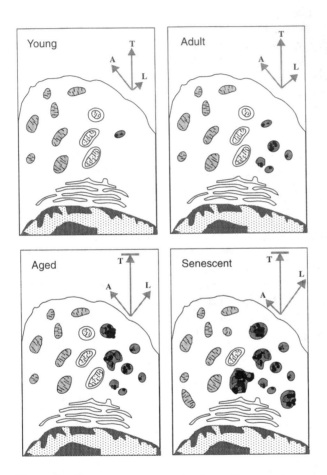

Figure 2 The arrows represent the flow of lysosomal enzymes, packed within primary lysosomes, from the Golgi apparatus to secondary lysosomes, either autophagolysosomes (A) or lipofuscin-loaded lysosomes (L). In senescent cells a major part of the newly produced lysosomal enzymes is believed to be delivered to lipofuscin-loaded lysosomes and thus lost for useful purposes, since lipofuscin cannot be degraded. Autophagocytosis would thus be hampered when the total production (T) of lysosomal enzymes cannot be increased any further in order to compensate for the diminishing proportion of enzymes delivered to autophagolysosomes. The result would be that old mitochondria, producing enhanced amounts of $O_2^{\bullet-}$ and H_2O_2 have to remain for a longer than normal period within the cell, increasing the level of oxidative stress.

large, most primary lysosomes would, for purely statistical reasons, fuse with those organelles than with the less numerous autophagosomes. With reduced amounts of lysosomal enzymes directed to autophagocytotic activities, and more ending up in lipofuscin-containing lysosomes, the aged postmitotic cells would not be able to sustain efficient turnover of worn out or damaged organelles, including mitochondria. The aged postmitotic cell may thus contain large amounts of lysosomal enzymes, although wrongly located within lipofuscin-containing secondary lysosomes where their degrading capacity is wasted on futile trials to degrade the undegradable lipofuscin. Such a scenario would result in insufficient autophagocytotic degradation of mitochondria, with ensuing enhanced production of ROS from aged and damaged mitochondria, that would further increase cellular oxidative stress. This, in turn, would lead to increased mitochondrial damage, even more enhanced oxidative stress, and pronounced intralysosomal iron-catalyzed oxidative reactions because of increased diffusion of H_2O_2 into the acidic vacuolar apparatus. In the case of substantially enhanced intralysosomal Fenton-type reactions, peroxidative destabilization of the lysosomal surrounding membranes with consequent leakage of degradative lytic enzymes into the cytosol would follow (Fig. 3). The result would be cellular degeneration and eventually apoptotic cell death (159–162).

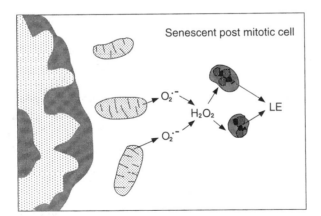

Figure 3 In senescent postmitotic cells, failing autophagocytotic activity would result in prolonged survival of old mitochondria, and thus enhanced oxidative stress. Increased cytosolic concentration of H_2O_2 would result in enhanced intralysosomal Fenton-type chemistry (compare with Fig. 1) with the formation of hydroxyl radicals in quantities high enough to rupture lipofuscin-containing, and thus iron-rich, secondary lysosomes with resultant leakage of lytic enzymes (LE) to the cytosol. Such leakage is known to induce apoptosis, perhaps by activation of the caspase system.

V. INCREASED OXIDATIVE STRESS AND APOPTOTIC CELL DEATH IN AGED POSTMITOTIC CELLS

Present knowledge about mitochondrial and lysosomal involvement in cellular aging offers an exciting scenario that may explain senescent changes. Since O_2^{-} and H_2O_2 are constantly formed intracellularly, oxidative modification of macromolecules is an inevitable result of normal oxygen metabolism. Therefore, even in young and healthy cells, some alterations will always take place and, as evidence suggests, they are not totally repaired. These alterations may disturb the functions of macromolecules and cellular organelles, as well as their degradability, leading to the accumulation of waste products (i.e., lipofuscin) within lysosomes. Lipofuscin is not substantially exocytosed from the cells. As a consequence, damaged, functionally defective, subcellular structures and lipofuscin will accumulate progressively in postmitotic cells.

Although macromolecular damage may occur in all cellular compartments, it is particularly pronounced in mitochondria, which are the main source of ROS and possess an autonomous genetic system with increased probability of mutations and limited repair capacity. Mitochondrial damage disturbs the function of the electron transport chain with an ensuing decrease of energy production. Affected mitochondria also produce more ROS which causes additional damage. However, they may replicate well, because the genes responsible for mitochondrial reproduction are located in the nucleus. As a result, the number of damaged mitochondria, the degree of oxidative stress, and ATP deficiency will increase with age.

Containing altered macromolecules and autophagocytosed in large numbers, mitochondria are clearly the main source of undegradable material, that is, lipofuscin, within secondary lysosomes. Probably the generation of ROS by mitochondria persists temporarily inside lysosomes producing more undegradable substances. The increasing lipofuscin deposits may prevent normal function of cellular organelles and their interaction. Obviously the accumulation of lipofuscin increases the cell volume, thus disturbing diffusion of nutrients to the central parts of the cells. Since lipofuscin-loaded lysosomes are functionally insufficient and tend to withdraw lysosomal enzymes from useful purposes, increased amounts of lysosomal enzymes would be produced. However, this compensation mechanism is not of an unlimited size, and the efficiency of intralysosomal autophagocytotic degradation, and therefore the renewal of subcellular structures, including mitochondria, will decrease in very old postmitotic cells. This, again, will contribute to energy deficiency, increased oxidative stress, and accelerated formation of lipofuscin at the end of the life span. Over time, cells with numerous defective mitochondria, other damaged structures, and a lipofuscin-loaded vacuolar apparatus not capable of keeping up with the necessary degradation will be unable to maintain their functions.

Enhanced cellular oxidative stress not only would result in an acceleration of lipofuscin formation, but also, if pronounced enough, in lysosomal membrane destabilization due to oxygen radical attack, with leakage of lysosomal enzymes into the cytosol, ensuing deterioration of cellular components, and even cell death. Depending on the degree of lysosomal oxidative damage, the result could be either apoptotic or necrotic cell death (159–162). Contributing to the destabilization of lipofuscin-loaded secondary lysosomes under oxidative stress may be the iron content of lipofuscin (163). Thus, according to our hypothesis, within postmitotic cells, age-related mitochondrial damage on the one hand and lipofuscin accumulation on the other together constitute a sequence of interrelated events—along a mitochondrial-lysosomal axis—leading to cellular and finally the organism death.

REFERENCES

1. Lewin MB, Timiras PS. Lipid changes with aging in cardiac mitochondrial membranes. Mech Ageing Dev 1984; 24:343–351.
2. Benedetti A, Ferretti G, Curatola G, Jezequel AM, Orlandi F. Age and sex related changes of plasma membrane fluidity in isolated rat hepatocytes. Biochem Biophys Res Commun 1988; 156:840–845.
3. Yu BP, Chen JJ, Kang CM, et al. Mitochondrial aging and lipoperoxidative products. Ann NY Acad Sci 1996; 786:44–56.
4. Frolkis VV, Frolkis RA, Mkhitarian LS, et al. Contractile function and Ca^{2+} transport system of myocardium in ageing. Gerontology 1988; 34:64–74.
5. Cepeda C, Lee N, Buchwald NA, Radisavljevic Z, Levine MS. Age-induced changes in electrophysiological responses of neostriatal neurons recorded in vitro. Neuroscience 1992; 51:411–423.
6. Potier B, Lamour Y, Dutar P. Age-related alterations in the properties of hippocampal pyramidal neurons among rat strains. Neurobiol Aging 1993; 14:17–25.
7. Cartee GD. Myocardial GLUT-4 glucose transporter protein levels of rats decline with advancing age. J Gerontol 1993; 48:B168–B170.
8. Mednikova YS, Kopytova FV. Some physiological characteristics of motor cortex neurons of aged rabbits. Neuroscience 1994; 63:611–615.
9. Chaturvedi MM, Kanungo MS. Analysis of conformation and function of the chromatin of the brain of young and old rats. Mol Biol Rep 1985; 10:215–219.
10. Bhaskar MS, Rao KS. Altered conformation and increased strand breaks in neuronal and astroglial DNA of aging rat brain. Biochem Mol Biol Int 1994; 33:377–384.
11. Gaubatz JW, Tan BH. Aging affects the levels of DNA damage in postmitotic cells. Ann NY Acad Sci 1994; 719:97–107.
12. Murrell WG, Masters CJ, Willis RJ, Crane DI. Chromatin structure and the expression of cardiac genes. Mech Ageing Dev 1995; 82:1–17.
13. Parhad IM, Scott JN, Cellars LA, et al. Axonal atrophy in aging is associated

with a decline in neurofilament gene expression. J Neurosci Res 1995; 41:355–366.

14. Bohr VA, Anson RM. DNA damage, mutation and fine structure DNA repair in aging. Mutat Res 1995; 338:25–34.

15. Barnett YA, King CM. An investigation of antioxidant status, DNA repair capacity and mutation as a function of age in humans. Mutat Res 1995; 338:115–128.

16. Gilchrest BA, Bohr VA. Aging processes, DNA damage, and repair. FASEB J 1997; 11:322–330.

17. Strehler BL. Time, Cells and Aging, 2nd ed. New York Academic Press, 1977.

18. Tashiro T, Komiya Y. Maturation and aging of the axonal cytoskeleton: biochemical analysis of transported tubulin. J Neurosci Res 1991; 30:192–200.

19. Vickers JC, Delacourte A, Morrison JH. Progressive transformation of the cytoskeleton associated with normal aging and Alzheimer's disease. Brain Res 1992; 594: 273–278.

20. Fifkova E, Morales M. Aging and the neurocytoskeleton. Exp Gerontol 1992; 27: 125–136.

21. Strehler BL, Chang MP, Johnson LK. Loss of hybridizable ribosomal DNA from human post-mitotic tissues during aging: I. Age-dependent loss in human myocardium. Mech Ageing Dev 1979; 11:371–378.

22. Nosal G. Neuronal involution during ageing. Ultrastructural study in the rat cerebellum. Mech Ageing Dev 1979; 10:295–314.

23. Ekstrom R, Liu DS, Richardson A. Changes in brain protein synthesis during the life span of male Fischer rats. Gerontology 1980; 26:121–128.

24. Spoerri PE, Glees P, Spoerri O. Neuronal regression during ageing: an ultrastructural study of the human cortex. J Hirnforsch 1981; 22:441–446.

25. Blazejowski CA, Webster GC. Decreased rates of protein synthesis by cell-free preparations from different organs of aging mice. Mech Ageing Dev 1983; 21:345–356.

26. Pluskal MG, Moreyra M, Burini RC, Young VR. Protein synthesis studies in skeletal muscle of aging rats. I. Alterations in nitrogen composition and protein synthesis using a crude polyribosome and pH 5 enzyme system. J Gerontol 1984; 39:385–391.

27. Romero MT, Silverman AJ, Wise PM, Witkin JW. Ultrastructural changes in gonadotropin-releasing hormone neurons as a function of age and ovariectomy in rats. Neuroscience 1994; 58:217–225.

28. Frolkis VV, Kvitnitskaya-Ryzhova TY, Martynenko OA. Aging of neurons in the mollusc *Lymnaea stagnalis* small parietal ganglion: a morpho-functional comparison in the same neuron. Exp Gerontol 1995; 30:533–544.

29. Johnson JE Jr. Fine structure of IMR-90 cells in culture as examined by scanning and transmission electron microscopy. Mech Ageing Dev 1979; 10:405–443.

30. La Velle A, Buschmann MT. Nuclear envelope invaginations in hamster facial motor neurons during development and aging. Brain Res 1983; 312:171–175.

31. de la Roza C, Cano J, Satorre J, Reinoso-Suarez F. A morphologic analysis of

neurons and neuropil in the dorsal lateral geniculate nucleus of aged rats. Mech Ageing Dev 1986; 34:233–248.

32. Timiras PS. Physiological Basis of Aging and Geriatrics. New York: Macmillan, 1988.

33. Vandewoude MF, Buyssens N. Effect of ageing and malnutrition on rat myocardium. I. The myocyte. Virchows Arch A 1992; 421:179–188.

34. Sachs HG, Colgan JA, Lazarus ML. Ultrastructure of the aging myocardium: a morphometric approach. Am J Anat 1977; 150:63–71.

35. Beregi E, Regius O, Huttl T, Gobl Z. Age-related changes in the skeletal muscle cells. Z Gerontol 1988; 21:83–86.

36. Solmi R, Pallotti F, Rugolo M, et al. Lack of major mitochondrial bioenergetic changes in cultured skin fibroblasts from aged individuals. Biochem Mol Biol Int 1994; 33:477–484.

37. Tate EL, Herbener GH. A morphometric study of the density of mitochondrial cristae in heart and liver of aging mice. J Gerontol 1976; 31:129–134.

38. Vanneste J, van den Bosch de Aguilar P. Mitochondrial alterations in the spinal ganglion neurons in ageing rats. Acta Neuropathol 1981; 54:83–87.

39. Tanaka M, Kovalenko SA, Gong JS, et al. Accumulation of deletions and point mutations in mitochondrial genome in degenerative diseases. Ann NY Acad Sci 1996; 786:102–111.

40. Wei YH, Pang CY, You BJ, Lee HC. Tandem duplications and large-scale deletions of mitochondrial DNA are early molecular events of human aging process. Ann N Y Acad Sci 1996; 786:82–101.

41. Filburn CR, Edris W, Tamatani M, et al. Mitochondrial electron transport chain activities and DNA deletions in regions of the rat brain. Mech Ageing Dev 1996; 87:35–46.

42. Muscari C, Giaccari A, Stefanelli C, et al. Presence of a DNA-4236 bp deletion and 8-hydroxy-deoxyguanosine in mouse cardiac mitochondrial DNA during aging. Aging 1996; 8:429–433.

43. Zhang C, Bills M, Quigley A, et al. Varied prevalence of age-associated mitochondrial DNA deletions in different species and tissues: a comparison between human and rat. Biochem Biophys Res Commun 1997; 230:630–635.

44. Paradies G, Ruggiero FM, Petrosillo G, Quagliariello E. Age-dependent decline in the cytochrome c oxidase activity in rat heart mitochondria: role of cardiolipin. FEBS Lett 1997; 406:136–138.

45. Yamada K, Sugiyama S, Kosaka K, Hayakawa M, Ozawa T. Early appearance of age-associated deterioration in mitochondrial function of diaphragm and heart in rats treated with doxorubicin. Exp Gerontol 1995; 30:581–593.

46. Muller-Hocker J, Schafer S, Link TA, Possekel S, Hammer C. Defects of the respiratory chain in various tissues of old monkeys: a cytochemical-immunocytochemical study. Mech Ageing Dev 1996; 86:197–213.

47. Rooyackers OE, Adey DB, Ades PA, Nair KS. Effect of age on in vivo rates of mitochondrial protein synthesis in human skeletal muscle. Proc Natl Acad Sci USA 1996; 93:15364–15369.

48. Bandy B, Davison AJ. Mitochondrial mutations may increase oxidative stress:

implications for carcinogenesis and aging? Free Radic Biol Med 1990; 8:523–539.

49. Richter C. Oxidative damage to mitochondrial DNA and its relationship to ageing. Int J Biochem Cell Biol 1995; 27:647–653.

50. Papa S. Mitochondrial oxidative phosphorylation changes in the life span. Molecular aspects and physiopathological implications. Biochim Biophys Acta 1996; 1276: 87–105.

51. Beal MF. Mitochondria, free radicals, and neurodegeneration. Curr Opin Neurobiol 1996; 6:661–666.

52. Wallace DC. Mitochondrial DNA in aging and disease. Sci Am 1997; 277:22–29.

53. Miyagishi T, Takahata N, Iizuka R. Electron microscopic studies on the lipo-pigments in the cerebral cortex nerve cells of senile and vitamin E deficient rats. Acta Neuropathol 1967; 9:7–17.

54. Brunk U, Ericsson JLE. Electron microscopical studies of rat brain neurons. Localization of acid phosphatase and mode of formation of lipofuscin bodies. J Ultrastruct Res 1972; 38:1–15.

55. Ikeda H, Tauchi H, Sato T. Fine structural analysis of lipofuscin in various tissues of rats of different ages. Mech Ageing Dev 1985; 233:77–93.

56. Jayne EP. Histochemical studies of age pigments in the human heart. J Gerontol 1950; 5:319–325.

57. Strehler BL, Mark DD, Mildvan AS, Gee MV. Rate of magnitude of age pigment accumulation in the human myocardium. J Gerontol 1959; 14:430–439.

58. Reichel E, Holander J, Clark HJ, Strehler BL. Lipofuscin pigment accumulation as a function of age and distribution in rodent brain. J Gerontol 1968; 23:71–78.

59. Brizzee KR, Ordy JM, Kaack B. Early appearance and regional differences in intraneuronal and extraneuronal lipofuscin accumulation with age in the brain of a nonhuman primate (*Macaca mulatta*). J Gerontol 1974; 29:366–381.

60. Porta EA, Sablan HM, Joun NS, Chee G. Effects of the type of dietary fat at two levels of vitamin E in Wistar male rats during development and aging. IV. Biochemical and morphometric parameters of the heart. Mech Ageing Dev 1982; 18:159–199.

61. Munnel JF, Getty R. Rate of accumulation of cardiac lipofuscin in the aging canine. J Gerontol 1968; 23:154–158.

62. Nakano M, Mizuno T, Gotoh S. Accumulation of cardiac lipofuscin in crab-eating monkeys (*Macaca fasicularis*): the same rate of lipofuscin accumulation in several species of primates. Mech Ageing Dev 1993; 66:243–248.

63. Nakano M, Oenzil F, Mizuno T, Gotoh S. Age-related changes in the lipofuscin accumulation of brain and heart. Gerontology 1995; 41:69–79.

64. Hannover A. Mikroskopiske undersögelser af nervesystemet. Kgl. Danske Vidensk. Kabernes Selskobs Naturv Math Afh (Copenhagen) 1842; 10:1–112.

65. Koneff H. Beiträge zur Kenntniss der Nervenzellen den peripheren Ganglien. Mitt Naturforsch Gesellsch Bern 1886; 44:13–14.

66. Essner E, Novikoff AV. Human hepatocellular pigments and lysosomes. J Ultrastruct Res 1960; 3:374–391.

67. de Duve C, Wattiaux R. Functions of lysosomes. Annu Rev Physiol 1966; 28:435–492.

68. Brunk U, Ericsson JL, Ponten J, Westermark B. Residual bodies and "aging" in cultured human glia cells. Effect of entrance into phase 3 and prolonged periods of confluence. Exp Cell Res 1973; 79:1–14.

69. Dean RT, Barrett AJ. Lysosomes. Essays Biochem 1976; 12:1–40.

70. Novikoff AB, Shin WY. Endoplasmic reticulum and autophagy in rat hepatocytes. Proc Natl Acad Sci USA 1978; 75:5039–5042.

71. Porta EA. Advances in age pigment research. Arch Gerontol Geriatr 1991; 212:303–320.

72. Terman A, Brunk UT. Lipofuscin: mechanisms of formation and increase with age. APMIS 1998; 106:265–276.

73. Ghadially FN. Ultrastructural Pathology of the Cell and Matrix, 2nd ed. London: Butterworths, 1975.

74. Hamberg H, Brunk U, Ericsson J, Jung B. Cytoplasmic effects of x-irradiation on cultured cells in a nondividing stage. II. Alterations in lysosomes, plasma membrane, Golgi apparatus and related structures. Acta Pathol Microbiol Scand A 1977; 85:625–639.

75. Armstrong D, Koppang N. Ceroid-lipofuscinosis, a model for aging. In: Sohal RS, ed. Age Pigments. Amsterdam: Elsevier, 1981:355–382.

76. Carpenter KL, van der Veen C, Taylor SE, et al. Macrophages, lipid oxidation, ceroid accumulation and alpha-tocopherol depletion in human atherosclerotic lesions. Gerontology 1995; 41:53–67.

77. de Gritz BG, Rahko T. Diet-induced residual formation in pigs. Gerontology 1995; 41:305–317.

78. Goebel HH, Klein H, Santavuori P, Sainio K. Ultrastructural studies of the retina in infantile neuronal ceroid-lipofuscinosis. Retina 1988; 8:59–66.

79. Haltia M, Tyynela J, Baumann M, Henseler M, Sandhoff K. Immunological studies on sphingolipid activator proteins in the neuronal ceroid-lipofuscinoses. Gerontology 1995; 41:239–248.

80. Kitani K, Senda M, Toyama H, et al. Decline in glucose metabolism in the brain in neuronal ceroid lipofuscinosis (NCL) in English setter—evidence by positron emission tomography (PET). Gerontology 1995; 41:249–257.

81. Ohbayashi C, Kanomata N, Imai Y, Ito H, Shimasaki H. Hermansky-Pudlak syndrome: a case report with analysis of auto-fluorescent ceroid-like pigments. Gerontology 1995; 41:297–303.

82. Oliver C. Lipofuscin and ceroid accumulation in experimental animals. In: Sohal RS, ed. Age Pigments. Amsterdam: Elsevier, 1981:335–353.

83. Porta EA, Mower HF, Moroye M, Lee C, Palumbo NE. Differential features between lipofuscin (age pigment) and various experimentally produced 'ceroid pigments'. In: Zs.-Nagy I, ed. Lipofuscin-1987: State of the Art Amsterdam:Excerpta Medica, 1988:341–374.

84. Wolfe LS, Gauthier S, Haltia M, Palo J. Dolichol and dolichyl phosphate in the neuronal ceroid-lipofuscinoses and other diseases. Am J Med Genet 1988; 5:233–242.

85. Tappel AL. Lipid peroxidation damage to cell components. Fed Proc 1973; 32: 1870–1874.
86. Kikugawa K, Nakahara T, Sakurai K. Fluorescence of 1,4-dihydropyridine derivatives relevant to age pigments. Chem Pharm Bull 1987; 35:4656–4660.
87. Chen P, Wiesler D, Chmelik J, Novotny M. Substituted 2-hydroxy-1,2-dihydropyrrol-3-ones: fluorescent markers pertaining to oxidative stress and aging. Chem Res Toxicol 1996; 9:970–979.
88. Eldred GE, Lasky MR. Retinal age pigments generated by self-assembling lysosomotropic detergents. Nature 1993; 361:724–726.
89. Szweda LI. Age-related increase in liver retinyl palmitate. Relationship to lipofuscin. J Biol Chem 1994; 269:12–15.
90. Yin DZ, Brunk UT. Microfluorometric and fluorometric lipofuscin spectral discrepancies: a concentration-dependent metachromatic effect? Mech Ageing Dev 1991; 59:95–109.
91. Harman D. Aging: a theory based on free radical and radiation chemistry. J Gerontol 1956; 211:298–300.
92. Yu BP. Cellular defenses against damage from reactive oxygen species. Physiol Rev 1994; 74:139–163.
93. Harman D. Free-radical theory of aging. Increasing the functional life span. Ann NY Acad Sci 1994; 717:1–15.
94. Sohal RS, Weindruch R. Oxidative stress, caloric restriction, and aging. Science 1996; 273:59–63.
95. Halliwell B, Gutteridge JMC. Free Radicals in Biology and Medicine. Oxford: Clarendon Press, 1989.
96. Yu BP, Yang R. Critical evaluation of the free radical theory of aging. A proposal for the oxidative stress hypothesis. Ann NY Acad Sci 1996; 786:1–11.
97. Linnane AW, Marzuki S, Ozawa T, Tanaka M. Mitochondrial DNA mutations as an important contributor to ageing and degenerative diseases. Lancet 1989; 8639: 642–645.
98. Richter C. Reactive oxygen and DNA damage in mitochondria. Mutat Res 1992; 275:249–255.
99. Ozawa T. Mechanism of somatic mitochondrial DNA mutations associated with age and diseases. Biochim Biophys Acta 1995; 1271:177–189.
100. Watson JD, Hopkins NH, Roberts JW, Steitz JA. Molecular Biology of the Gene, 4th ed. Menlo Park, CA: Benjamin/Cummings, 1987.
101. Ozawa T. Genetic and functional changes in mitochondria associated with aging. Physiol Rev 1997; 77:425–464.
102. Torii K, Sugiyama S, Takagi K, Satake T, Ozawa T. Age-related decrease in respiratory muscle mitochondrial function in rats. Am J Respir Cell Mol Biol 1992; 6: 88–92.
103. Trounce I, Byrne E, Marzuki S. Decline in skeletal muscle mitochondrial respiratory chain function: possible factor in ageing. Lancet 1989; 8639:637–639.
104. Sugiyama S, Takasawa M, Hayakawa M, Esumi H, Ozawa T. Detrimental effects of 2-amino-1-methyl-6-phenylimidazo[4,5-b]pyridine, a mutagenic agent, on mitochondrial respiration among various rat tissues. Biochem Mol Biol Int 1993; 30: 797–805.

105. Takasawa M, Hayakawa M, Sugiyama S, et al. Age-associated damage in mitochondrial function in rat hearts. Exp Gerontol 1993; 28:269–280.

106. Bowling AC, Schulz JB, Brown RH Jr, Beal MF. Superoxide dismutase activity, oxidative damage, and mitochondrial energy metabolism in familial and sporadic amyotrophic lateral sclerosis. J Neurochem 1993; 61:2322–2325.

107. Skulachev VP. Why are mitochondria involved in apoptosis? Permeability transition pores and apoptosis as selective mechanisms to eliminate superoxide-producing mitochondria and cell. FEBS Lett 1996; 397:7–10.

108. de Grey AD. A proposed refinement of the mitochondrial free radical theory of aging. Bioessays 1997; 19:161–166.

109. Holt IJ, Harding AE, Morgan-Hughes JA. Deletions of muscle mitochondrial DNA in patients with mitochondrial myopathies. Nature 1988; 331:717–719.

110. Wallace DC, Singh G, Lott MT, et al. Mitochondrial DNA mutation associated with Leber's hereditary optic neuropathy. Science 1988; 242:1427–1430.

111. Wallace DC, Zheng XX, Lott MT, et al. Familial mitochondrial encephalomyopathy (MERRF): genetic, pathophysiological, and biochemical characterization of a mitochondrial DNA disease. Cell 1988; 55:601–610.

112. Yoneda M, Chomyn A, Martinuzzi A, Hurko O, Attardi G. Marked replicative advantage of human mtDNA carrying a point mutation that causes the MELAS encephalomyopathy. Proc Natl Acad Sci USA 1992; 89:11164–11168.

113. Wallace DC, Shoffner JM, Trounce I, et al. Mitochondrial DNA mutations in human degenerative diseases and aging. Biochim Biophys Acta 1995; 1271:141–151.

114. Seglen PO, Bohley P. Autophagy and other vacuolar protein degradation mechanisms. Experientia 1992; 48:158–172.

115. Kowald A, Kirkwood TB. A network theory of ageing: the interactions of defective mitochondria, aberrant proteins, free radicals and scavengers in the ageing process. Mutat Res 1996; 316:209–236.

116. Holtzman E. Lysosomes. New York: Plenum Press, 1989.

117. Wihlmark U, Wrigstad A, Roberg K, Nilsson SEG, Brunk UT. Lipofuscin accumulation in cultured retinal pigment epithelial cells causes enhanced sensitivity to blue light irradiation. Free Radic Biol Med 1997; 22:1229–1234.

118. Elleder M, Drahota Z, Lisa V, et al. Tissue culture loading test with storage granules from animal models of neuronal ceroid-lipofuscinosis (Batten disease): testing their lysosomal degradability by normal and Batten cells. Am J Med Genet 1995; 57:213–221.

119. Terman A, Brunk UT. On the degradability and exocytosis of ceroid/lipofuscin in cultured rat cardiac myocytes. Mech Ageing Dev 1998; 100:145–156.

120. Kovacs J, Laszlo L, Kovacs AL. Regression of autophagic vacuoles in pancreatic acinar, seminal vesicle epithelial, and liver parenchymal cells: a comparative morphometric study of the effect of vinblastine and leupeptin followed by cycloheximide treatment. Exp Cell Res 1988; 174:244–251.

121. Marzabadi MR, Sohal RS, Brunk UT. Effect of ferric iron and desferrioxamine on lipofuscin accumulation in cultured rat heart myocytes. Mech Ageing Dev 1988; 46:145–157.

122. Sohal RS, Marzabadi MR, Galaris D, Brunk UT. Effect of ambient oxygen concen-

tration on lipofuscin accumulation in cultured rat heart myocytes—a novel in vitro model of lipofuscinogenesis. Free Radic Biol Med 1989; 6:23–30.

123. Thaw HH, Collins VP, Brunk UT. Influence of oxygen tension, pro-oxidants and antioxidants on the formation of lipid peroxidation products (lipofuscin) in individual cultivated human glial cells. Mech Ageing Dev 1984; 24:211–223.

124. von Zglinicki T, Saretzki G, Döcke W, Lotze C. Mild hyperoxia shortens telomeres and inhibits proliferation of fibroblasts: a model for senescence? Exp Cell Res 1995; 220:186–193.

125. Wihlmark U, Wrigstad A, Roberg K, Brunk UT, Nilsson SE. Lipofuscin formation in cultured retinal pigment epithelial cells exposed to photoreceptor outer segment material under different oxygen concentrations. APMIS 1996; 104:265–271.

126. Zs.-Nagy I, Steiber J, Jeney F. Induction of age pigment accumulation in the brain cells of young male rats through iron-injection into the cerebrospinal fluid. Gerontology 1995; 41:145–158.

127. Constantinides P, Harkey M, McLaury D. Prevention of lipofuscin development in neurons by anti-oxidants. Virchows Arch A 1986; 409:583–593.

128. Nandy K, Mostofsky DI, Idrobo F, Blatt L, Nandy S. Experimental manipulations of lipofuscin formation in aging mammals. In: Zs.-Nagy I, ed. Lipofuscin-1987: State of the Art. Amsterdam: Excerpta Medica, 1988:289–304.

129. Marzabadi MR, Sohal RS, Brunk UT. Effect of alpha-tocopherol and some metal chelators on lipofuscin accumulation in cultured neonatal rat cardiac myocytes. Anal Cell Pathol 1990; 2:333–346.

130. Marzabadi MR, Sohal RS, Brunk UT. Mechanisms of lipofuscinogenesis: effect of the inhibition of lysosomal proteinases and lipases under varying concentrations of ambient oxygen in cultured rat neonatal myocardial cells. APMIS 1991; 99:416–426.

131. Gao G, Öllinger K, Brunk UT. Influence of intracellular glutathione concentration of lipofuscin accumulation in cultured neonatal rat cardiac myocytes. Free Radic Biol Med 1994; 16:187–194.

132. Ma XY, Su YB, Zhang FR, Li JF. Effects of vitamin E on the blastogenic response of splenocytes and lipofuscin contents in the hearts and brains of aged mice. J Environ Pathol Toxicol Oncol 1996; 15:51–53.

133. Martin AJP, Moore T. Some effects of prolonged vitamin E deficiency in rats. J Hygiene 1939; 39:643–650.

134. Moore T, Wang YL. Formation of fluorescent pigment in vitamin E deficiency. Br J Nutr 1947; 1:53–64.

135. Mason KE. The tocopherols, effects of deficiency: pharmacology. In: Serbell WH, Harris RS, eds. Vitamins New York: Academic Press, 1954:514–565.

136. Bertoni-Freddari C, Meier-Ruge W, Ulrich J. Aging-like alterations at the neuronal membrane of young rats fed a vitamin E deficient diet. In: Zs.-Nagy I, ed. Lipofuscin-1987: State of the Art. Amsterdam: Excerpta Medica, 1988:439–444.

137. Brunk UT, Jones CB, Sohal RS. A novel hypothesis of lipofuscinogenesis and cellular aging based on interactions between oxidative stress and autophagocytosis. Mutat Res 1992; 275:395–403.

138. Yin D, Yuan XM, Brunk UT. Test-tube simulated lipofuscinogenesis. Effect of

oxidative stress on autophagocytotic degradation. Mech Ageing Dev 1995; 81:37–50.

139. Collins VP, Arborgh B, Brunk U, Schellens JP. Phagocytosis and degradation of rat liver mitochondria by cultivated human glial cells. Lab Invest 1980; 42:209–216.

140. von Zglinicki T, Nilsson E, Döcke WD, Brunk UT. Lipofuscin accumulation and ageing of fibroblasts. Gerontology 1995; 41:95–108.

141. Nilsson E, Yin D. Preparation of artificial ceroid/lipofuscin by UV-oxidation of subcellular organelles. Mech Ageing Dev 1997; 99:61–78.

142. Kominami E, Ezaki J, Wolfe LS. New insight into lysosomal protein storage disease: delayed catabolism of ATP synthase subunit c in Batten disease. Neurochem Res 1995; 20:1305–1309.

143. Ezaki J, Wolfe LS, Higuti T, Ishidoh K, Kominami E. Specific delay of degradation of mitochondrial ATP synthase subunit c in late infantile neuronal ceroid lipofuscinosis (Batten disease). J Neurochem 1995; 64:733–741.

144. Terman A. The effect of age on formation and elimination of autophagic vacuoles in mouse hepatocytes. Gerontology 1995; 41:319–326.

145. Vittorini S, Paradiso C, Masini M, et al. Age-related decline of macroautophagy and liver protein breakdown in the Sprague-Dawley rat: protective effect of caloric restriction. Aging Clin Exp Res 1995; 7:476–477.

146. Reznick AZ, Gershon D. The effect of age on the protein degradation system in the nematode *Turbatrix aceti*. Mech Ageing Dev 1979; 11:403–415.

147. Ward W, Richardson A. Effect of age on liver protein synthesis and degradation. Hepatology 1991; 14:935–948.

148. Dice JF. Altered intracellular protein degradation in aging: a possible cause of proliferative arrest. Exp Gerontol 1989; 24:451–459.

149. Jahani M, Gracy RW. Cathepsin-B activity in human-skin fibroblasts from young, old and premature aging syndromes. Fed Proc 1985; 44:876–886.

150. di Paolo BR, Pignolo RJ, Cristofalo VJ. Overexpression of the two-chain form of cathepsin B in senescent WI-38 cells. Exp Cell Res 1992; 201:500–505.

151. Nakanishi H, Tominaga K, Amano T, et al. Age-related changes in activities and localizations of cathepsins D, E, B, and L in the rat brain tissues. Exp Neurol 1994; 126:119–128.

152. Amano T, Nakanishi H, Kondo T, et al. Age-related changes in cellular localization and enzymatic activities of cathepsins B, L and D in the rat trigeminal ganglion neuron. Mech Ageing Dev 1995; 83:133–141.

153. Ivy GO, Schottler F, Wenzel J, Baudry M, Lynch G. Inhibitors of lysosomal enzymes: accumulation of lipofuscin-like dense bodies in the brain. Science 1984; 226:985–987.

154. Ivy GO, Kanai S, Ohta M, et al. Lipofuscin-like substances accumulate rapidly in brain, retina and internal organs with cysteine protease inhibition. Adv Exp Med Biol 1989; 266.31–47.

155. Ivy GO, Kanai S, Ohta M, et al. Leupeptin causes an accumulation of lipofuscin-like substances in liver cells of young rats. Mech Ageing Dev 1991; 57:213–231.

156. Porta E, Llesuy S, Monserrat AJ, Benavides S, Travacio M. Changes in cathepsin B and lipofuscin during development and aging in rat brain and heart. Gerontology 1995; 41:81–93.

157. Schellens JP, Vreeling-Sindelarova H, Plomp PJ, Meijer AJ. Cytochemical and morphometric analysis of autophagy in energy depleted rat hepatocytes. Cell Biol Int Rep 1990; 14:805–814.

158. Fengsrud M, Roos N, Berg T, et al. Ultrastructural and immunocytochemical characterization of autophagic vacuoles in isolated hepatocytes: effects of vinblastine and asparagine on vacuole distributions. Exp Cell Res 1995; 221:504–519.

159. Brunk UT, Dalen H, Roberg K, Hellquist HB. Photooxidative disruption of lysosomal membranes causes apoptosis of cultured human fibroblasts. Free Radic Biol Med 1997; 23:616–626.

160. Hellquist HB, Svensson I, Brunk UT. Oxidant-induced apoptosis: a consequence of lethal lysosomal leak? Redox Rep 1997; 3:65–70.

161. Yuan XM, Li W, Olsson AG, Brunk UT. The toxicity to macrophages of oxidized low-density lipoprotein is mediated through lysosomal damage. Atherosclerosis 1997; 133:153–161.

162. Li W, Yuan XM, Olsson AG, Brunk UT. Uptake of oxidized LDL by macrophages results in partial lysosomal enzyme inactivation and relocalization. Atheroscler Thromb Vasc Biol 1998; 18:177–184.

163. Brun A, Brunk U. Heavy metal localization and age related accumulation in the rat nervous system. A histochemical and atomic absorption spectrophotometric study. Histochemie 1973; 34:333–342.

13

Quantitative Analysis of Mutations of Mitochondrial DNA During Human Aging

Jörg Napiwotzki, Annette Reith, Andreas Becker, Sandra Leist, and Bernhard Kadenbach
Philipps-Universität, Marburg, Germany

I. INTRODUCTION

Aging represents a multifactorial process involving progressive decline of energetic capabilities of the organism. Eukaryotic cells include two systems for the synthesis of chemical energy in the form of ATP: (1) the glycolytic system which is independent of oxygen and leads in its absence to the production of lactic acid, and (2) the oxidative phosphorylation system (OXPHOS), located in the inner mitochondrial membrane, and essentially dependent on oxygen. More than 90% of cellular ATP is synthesized in mitochondria by OXPHOS. During evolution nature has preserved a nuclear independent genome in mitochondria, the genetic remnant of an aerobic bacterium which, according to the endosymbiotic hypothesis (1) fused with an obligate anaerobic cell resulting in the eucaryotic cell. During evolution most of the genetic content of the aerobic cell has been transfered to the nucleus of the eukaryotic cell. The mitochondrial DNA (mtDNA) contains only 13 structural genes, all of which code for protein subunits of proton pumps, the energy converting enzyme complexes of OXPHOS. Surprisingly, eukaryotic cells can grow and multiply in the absence of mtDNA (2,3). Thus OXPHOS is not required for basic cellular life, but is essential for highly-energy-consuming functions of specialized cells in, for example, the heart, skeletal muscle, and brain.

  In this review we summarize data supporting our teleological view on the biological function of the nuclear independent mitochondrial genome in eukaryotic cells: to ensure smooth aging and death of the individual. This view on aging suggests stochastic somatic mutations, preferentially in mtDNA, as the basic molecular event. These are propagated in individual cells, leading to OXPHOS-deficient cells, accompanied by a slow decrease of the energetic capabilities of tissues with increasing age. Some aspects of this proposal have been published previously (4,5).

II. CORRELATION BETWEEN AGING AND OXPHOS ENZYME ACTIVITIES

Aging is associated with a progressive decline of tissue mass and performance (6–8). The loss of mass is observed in skeletal muscle, brain, and other tissues, in particular in those having a high demand for OXPHOS. This phenomenon coincides with the decay of mitochondria in these tissues (9,10). The decay is characterized by the decrease of oxidative capacity and mitochondrial content (11), and by the increase of mutations in mtDNA (12). An increasing number of respiratory deficient cells (fibers), distributed in a mosaic pattern, has been found histochemically with increasing age in the heart and skeletal muscle (13) and in particular in extraocular muscle (14).

Several studies have shown an age-linked decrease in the activity of respiratory chain enzyme complexes (15–21) and F_0F_1-ATP-synthase (22,23). The decline of respiratory chain activity in skeletal muscle is mainly associated with decreased cytochrome-c oxidase. Decreased ubiquinol-dehydrogenase (complex III) was observed in females, but not in males (20). In intact hepatocytes, an age-dependent decrease of the mitochondrial membrane potential was shown by flow cytometry, together with increased mitochondrial peroxide generation (24).

III. THE ROLE OF REACTIVE OXYGEN SPECIES

Reactive oxygen species (ROS) are generated from cellular sources like neutrophils and macrophages as compounds of a defense system (25) and as by-products of the respiratory chain in mitochondria, where single electron transfer from radicals of flavin or ubiquinone to dioxygen results in formation of O_2^- radicals (26), in particular at high membrane potentials (27). Hydrogen peroxide and OH radicals are generated by further reactions of the O_2^- radicals (28,29). Additional ROS are generated by nonenzymatic production of OH radicals during exposure to ionizing radiation (30,31). Although the O_2^- radical and H_2O_2 are metabolized by various and efficient mitochondrial radical scavenge systems such as superoxide dismutase (SOD), catalase, glutathione, and glutathione reductase, oxidative damage to mitochondria seems to be abundant (32).

Many studies have clearly demonstrated the sensitivity of mitochondrial proteins, lipids, and DNA to oxidative stress (33–36). DNA damage by ROS results in single- and double-strand breaks, abasic sites, and base damages (37,38). MtDNA contains an advanced level of 8-oxoguanine compared with nuclear DNA (39), but both genomes accumulate base damage with increasing age (10). However, after exposure of human cells to H_2O_2, mtDNA damage was shown to be more extensive and to persist longer than nuclear damage (40).

IV. THE MITOCHONDRIAL GENOME

MtDNA is a circular supertwisted double-stranded DNA of only 16,569 base pairs coding for two ribosomal RNAs, 22 tRNAs, and 13 proteins. These proteins are all subunits of the four proton pumping complexes of OXPHOS: seven subunits of NADH dehydrogenase (complex I), one of ubiquinol-dehydrogenase (complex III), three of cytochrome-c oxidase (complex IV), and three of ATP-synthase (complex V). In Table 1 are summarized the most important differences between the nuclear and the mitochondrial genome. The unique D-loop replication mechanism may support spontaneous mutational events in the absence of mutagens like ROS (see Chapter 5). Mutations of mtDNA, occurring 10 to 20 times more frequently than mutations of nuclear DNA (9,41), represent the essential basis of aging according to the "mitochondrial hypothesis of aging" (Chap. 8). Of particular importance for the role of mtDNA in human aging is its continuous turnover in cells until death, taking place also in postmitotic cells like brain, skeletal muscle, and heart, allowing propagation of mutated mtDNA particularly

Table 1 Comparison of the Human Mitochondrial and Nuclear Genome

	mtDNA	Nuclear DNA
Structure	Circular	Linear
Complexity (nucleotide pairs)	16,569	~3,900,000,000
Genome copies per cell	1000–10,000	2
Number of structural genes	13	~40,000
Replication mechanism	D-loop, single strand	Semidiscontinuous
Type of inheritance	Maternal	Mendelian
Turnover in postmitotic cells	Continuous	Blocked
Mutation rate	10–20 times more frequent than nuclear DNA	—

For detailed references, see Ref. 9.

in postmitotic cells. The restricted information content only for subunits of mito-chondrial proton pumps, which are not essential for basic cellular life (2,3), en-sures "smooth aging." According to the mitochondrial hypothesis of aging, the maximum energetic capacity of tissues decreases with increasing life rather harm-lessly, because cells can survive without OXPHOS, and skeletal muscles can produce high power in aged muscles for short times by using glycolytically syn-thesized ATP. The maternal inheritance of mtDNA (9) could have its biological significance in the process of rejuvenescence from generation to generation. Healthy newborns do not carry mutations in their mtDNA (42,43). In contrast, in unfertilized oocytes of women more than 30 years of age, the common deletion has been identified (44,45). From the approximately million oocytes formed in the female embryo by midgestation, only about 400 become fertil oocytes during the whole life of a women. It may thus be suggested that during the process of oocyte development, those oocytes with no mutations in their mtDNA are selected.

V. MUTATIONS OF mtDNA

The 10 to 20 times higher mutational rate of human mtDNA in comparison to nuclear DNA could be based on several reasons:

> mtDNA is located at the matrix surface of the inner mitochondrial mem-brane, where ROS are generated by the respiratory chain (46).
> Mitochondria have a less efficient DNA repair system in comparision to the nucleus. In contrast to previous assumptions (47), several reports confirmed the existence of DNA repair enzymes in mammalian mito-chondria (48–51). But the DNA repair system of mitochondria is less efficient than the nuclear one (40).
> In contrast to the nucleus, no DNA-protecting proteins like histones occur in mitochondria.
> A higher error rate of mtDNA polymerase-γ, compared to nuclear DNA polymerase, was measured (52,53).
> The mtDNA contains an unusually high amount of direct repeats which could support spontaneous deletions of mtDNA via a recombinational event (54–56).
> The single-stranded D-loop replication mechanism of mtDNA could favor the intermediate formation of clover-leaf structures at the tRNA genes, which could increase the error rate of DNA polymerase-γ at hot spots, resulting in point mutations of tRNA genes (57).
> The mtDNA undergoes a continuous turnover both in mitotic and postmi-totic cells (58) and this increases the probability of mutations (39,59).

In all human cells, including postmitotic cells like in brain, heart, and skeletal muscle, mitochondria grow continuously and are completely degraded by lysosomal phagocytosis, thus keeping its total amount constant. This continuous turnover of mitochondria and mtDNA allows amplification of spontaneous or maternally inherited mutations of mtDNA. The molecular mechanism of preferred amplification of mutated versus wild-type mtDNA is not known. However, de Grey (60) suggested a plausible mechanism, assuming enhanced lysosomal phagocytosis and degradation of those mitochondria containing wild-type mtDNA as compared to mitochondria with defective mtDNA and thus defective OXPHOS:

Intact mitochondria produce more ROS and contain more damaged surface molecules than mitochondria with defective or blocked respiratory chain. The defective mitochondrial surface molecules could represent signals for accelerated lysosomal phygocytosis.

The consequences of mutations in mtDNA are complex and have strong effects on the energy metabolism of affected cells and tissues. Deleterious rearrangements or point mutations in structural genes will only impair the affected enzyme complex, while those in tRNA genes would impair the translation of all 13 mitochondrial encoded proteins, because each mitochondrial tRNA is essential for protein synthesis.

VI. HUMAN AGING IS ASSOCIATED WITH STOCHASTIC MUTATIONS OF mtDNA

Mitochondrial diseases, a group of rare diseases based on defective mitochondria, and aging have in common the decline of energetic efficiency of tissues with increasing age. Both are based on defective mitochondrial functions, either due to mutations of nuclear genes, coding for mitochondrial proteins, or deletions or deleterious mutions of mtDNA (9,61,62), with a consequent decrease of OXPHOS activity. More than a dozen different mtDNA deletions have been identified in various tissues of elderly humans. Some deletions were found in a particular tissue, but others were detected in more than one tissue (63–65). The best investigated deletion of mtDNA is the so-called common deletion, a 4977 bp ΔmtDNA species (nt 8482–13,459), often found in cases of sporadic deletion (54), and frequently identified as the molecular basis of Kearns–Sayre syndrome (KSS) and chronic progressive external opthalmoplegia (CPEO) (66–68). Another reported deletion of mtDNA comprises 7.4 kb (nt 8637–16,084) and was found in patients with cardiomyopathy and in aged hearts (69). In both cases the deleted regions are flanked by direct repeats in the wild-type mtDNA. (54).

More than 50 different point mutations of mtDNA have been associated with various kinds of mitochondrial diseases (5). Some of these point mutations, for example, those characteristic for MERRF, MELAS, NARP, and CIPO, have also been identified in aged human tissues (5,42,43,70,71). In mitochondrial diseases generally one type of mtDNA mutation occurs in all tissues. In contrast, multiple point mutations and deletions have been shown to occur in the same tissue during aging (59,43,72). This is confirmed by Melov et al. (12,73), who showed in postmitotic tissues of human and mice an age-dependent accumulation of a heterogeneous array of mtDNA rearrangements. Nevertheless, in a single cell, usually only one type of deletion or point mutation is accumulating during aging, as shown by in situ hybridization (74). This observation suggests propagation of a single mutational event until all wild-type mtDNA molecules of that cell are replaced by mutated mtDNA, resulting in a respiratory-deficient cell. A slow amplification of mutated mtDNA with increasing age was also shown in a patient with KSS, where histochemically the cytochrome-c oxidase–deficient area in muscle fibers increased with age (75). Therefore, we propose that during aging a stochastic and mosaic pattern of respiratory-deficient cells accumulate, in particular in respiratory-active tissues, based on multiple different mutations in different individual cells. Each OXPHOS-defective cell contains in general only one type of mtDNA mutation.

This view is different from the "vicious circle" proposition (8,76,77), where continuous oxidative damage of mtDNA is assumed, with consequent decline in respiratory chain function, which in turn enhances the production of more ROS to cause further damage. This vicious circle would result in multiple mutations of mtDNA in every cell, which is not the case (74).

In mitochondrial diseases a threshold of 70–90% mutated mtDNA is required before clinical symptoms become evident (78,79). In contrast, in human aging the amount of a specific mutation is low, generally not exceeding 1% of total mtDNA. Since various different mutations of mtDNA accumulate in a tissue with increasing age, the sum of all mutations of mtDNA would result in a decreased maximum energetic capacity, but without the appearance of defined clinical symptoms.

VII. QUANTITATION OF DELETED mtDNA IN TISSUES
FROM ELDERLY HUMANS

In various investigations the amount of mutated mtDNA in tissues of aged humans was quantitated by PCR methods (9,64,65,80). Comparison of the results of different groups is difficult, because the change of a single parameter of the PCR, such as the buffer composition, type of polymerase, primer pair, cycle number, elongation time, annealing temperature, or the detection method of PCR products, could influence the amount of determined mtDNA (81,82). For the common deletion Chen et al. (83) and Marin-Garcia et al. (69) determined an amount of about 0.005% from total mtDNA in tissues of older humans, while Ozawa (84) and Wallace et al. (59) found up to 7% and 10%, respectively (for a review see reference 85).

In order to investigate the reason for the large variations, we determined the percentage of common deletion in muscle biopsies from 15 humans of different ages by using two different methods. The data of Figure 1A were obtained by PCR cycle titrations of deleted and wild-type mtDNA, whereas the data of Figure 1B were obtained by a method combining serial dilution PCR of reference samples to determine the efficiency, with cycle titrations of human samples as described by Becker et al. (87). The PCR conditions and the samples were the same in both investigations, except that in Figure 1A an average efficiency of 1.8 was assumed for calculation of deleted and wild-type mtDNA, whereas in Figure 1B individual efficiencies were determined for calculation of initial amounts of wild-type and deleted mtDNA, using an external standard (87). As seen in the graphs, the amounts of deleted mtDNA determined by cycle titration (Fig. 1A) are about 1000-fold higher than those determined by the combined method (Fig. 1B). Nevertheless the relative proportions of percent deleted mtDNA between different individuals, obtained by both methods, are similar. In both cases an accumulation of deleted mtDNA with increasing age can be concluded, as found by other groups. We assume that the results obtained by the more direct method (Fig. 1A) are closer to reality than those obtained by using a reference DNA sample (e.g., plasmid DNA) to determine the efficiency of PCR. It is well known that the efficiency of PCR is strongly dependent on the quality of template DNA. Thus extrapolation to zero cycles, using efficiencies determined with reference DNA, could produce an error of several orders of magnitude.

VIII. THE MITOCHONDRIAL HYPOTHESIS OF AGING

The mitochondrial hypothesis of aging (5,9,89) suggests stochastic somatic mutations of mtDNA, in particular in postmitotic tissues, as the primary basis of aging. A single deleteriously mutated mtDNA molecule in a mitochondrion, containing 5–10 mtDNA molecules, will have no effect on the biosynthesis of subunits of OXPHOS enzyme complexes because defective mRNAs or tRNAs are complemented by corresponding intact RNAs. Only when all 5 to 10 molecules of wild-type mtDNA are replaced by mutated mtDNA, mainly due to propagation via turnover of mitochondria as described above, OXPHOS in that mitochondrion becomes defective. A similar complementation of defective mitochondria by

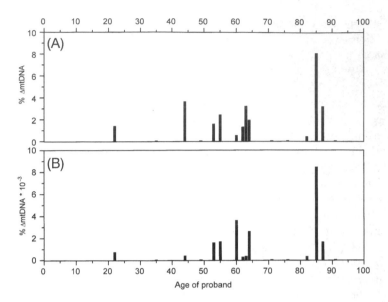

Figure 1 Comparison of the percentage of ΔmtDNA in skeletal muscle samples from individuals of different ages, as determined by two different methods. The five very low balks represent samples with values below 0.01 of scale. (**A**) The denaturation, annealing, and elongation phases were repeated for 14, 16, 18, 20, 22, and 24 cycles in the case of wild-type DNA (5 ng DNA), for 18, 20, 22, 24, 26, and 30 cycles in the case of deleted DNA (50 ng DNA). The amount of PCR product was detected by using an NIH image program and the ratio of common deletions to total mtDNA (wild type + common deletion) was calculated by extrapolation to zero cycle by using an average efficiency of 1.8/cycle for both PCRs. (**B**) To determine the wild-type DNA the denaturation, annealing, and elongation phase were performed for 13, 14, and 15 cycles and for deleted DNA for 21, 22, and 24 cycles. The initial amount of wild-type and deleted mtDNA was determined using an NIH image program and placenta mtDNA as standard for wild-type DNA, as described by Becker et al. (87), and plasmid DNA (pCII) as standard for the common deletion, as described by Kim et al. (88). Methods: Total DNA from different biopsies and autopsies (10–50 mg) was isolated as described by Wallace et al. (86). PCR was performed in a Perkin Elmer cycler TC1 in 50 μl sample volume containing 50 ng total DNA (common deletion in A and B, and wild type in B) or 5 ng total DNA (wild type in A) covered by paraffin by using the following primer pairs: wild-type DNA: P1 (nt 8865–8881) GATTATAGGCTTTCGCTC, P2 (9361–9343) AGTGTGTTGGTTAG-TAGGC; deleted DNA: P3 (8260–8280) TACCCTATAGCACCCCCTCTA, P4 (13647–13631) GAAGCGAGGTTGACCTG. The PCR program consisted of following parts: hot start (94°C, 4 min), denaturation (94°C, 1 min), annealing (57°C, 45 s), and elongation (72°C, 45 s).

functional ones occurs in cells containing between several hundred and several thousand mitochondria.

Under normal conditions only a small percentage of the cell's maximum oxidative capacity is used. Recently the control of cell respiration by the ATP: ADP ratio, rather than by ADP or the mitochondrial membrane potential, was shown (90). ATP binds to subunit IV of cytochrome-c oxidase (complex IV) and leads to allosteric inhibition of respiration. Thus even a large percentage of respiratory-defective mitochondria will have no influence on the energy supply of the cell, unless the energy demand exceeds the total OXPHOS capacity. Similarly, in whole tissues like the heart, respiratory-deficient cells can be complemented by respiratory-competent cells. In young individuals the maximum OXPHOS capacity of tissues exceeds the normal energy demand. In old individuals the maximum OXPHOS capacity is decreased due to an increasing number of respiratory-deficient cells (12–14). Therefore, under high energy load of the heart, ATP synthesis in respiratory-deficient cells must be complemented by increased glycolysis. Respiration-independent synthesis of ATP by glycolysis results in the production of lactic acid (5), which impairs muscle contraction (91). It was proposed that the death of old individuals could be the consequence of such rare situations (4).

REFERENCES

1. Margulis L. Symbiosis in Cell Evolution. Microbial Evolution in the Archean and Proterozoic Eons, 2nd ed. New York: W. H. Freeman, 1992.
2. King MP, Attardi G. Human cells lacking mtDNA: repopulation with exogenous mitochondria by complementation. Science 1989; 246:500–503.
3. Herzberg NH, Middlekoop E, Adorf M, et al. Mitochondria in cultured human cells depleted of mitochondrial DNA. Eur J Cell Biol 1993; 61:400–408.
4. Kadenbach B, Müller-Höcker J. Mutations of mitochondrial DNA and human death. Naturwiss 1990; 77:221–225.
5. Kadenbach B, Münscher C, Frank V, Müller-Höcker J, Napiwotzki J. Human aging is associated with stochastic somatic mutations of mitochondrial DNA. Mutat Res 1995; 338:161–172.
6. Davies CT, Thomas DO, White MJ. Mechanical properties of young and elderly human muscle. Acta Med Scand Suppl 1986; 711:219–226.
7. Hurley BF. Age, gender, and muscular strength. J Gerontol 1995; 50:41–44.
8. Papa S. Mitochondria oxidative phosphorylation changes in the life span. Molecular aspects and physiopathological implications. Biochim Biophys Acta 1996; 1276: 87–105.
9. Wallace DC, Diseases of the mitochondrial DNA. Annu Rev Biochem 1992; 61: 1175–1212.

10. Shigenaga MK, Hagen TM, Ames BN. Oxidative damage and mitochondrial decay in aging. Proc Natl Acad Sci USA 1994; 91:10771–10778.
11. Poggi P, Marchetti C, Scelsi R. Automatic morphometric analysis of skeletal muscle fibers in the aging man. Anat Rec 1987; 217:30–34.
12. Melov S, Shoffner JM, Kaufman A, Wallace DC. Marked increase in the number and variety of mitochondrial DNA rearrangements in aging human skeletal muscle. Nucleic Acids Res 1995; 23:4122–4126.
13. Müller-Höcker J. Cytochrome c oxidase deficient fibres in the limb muscle and diaphragm of man without muscular disease: an age related alteration. J Neurol Sci 1990; 100:14–21.
14. Müller-Höcker J, Schneiderbanger K, Stefani FH, Kadenbach B. Progressive loss of cytochrome c oxidase in the human extraocular muscles in ageing—a cytochemical-immunohistochemical study. Mutat Res 1992; 275:115–124.
15. Trounce I, Byrne E, Marzuki S. Decline in skeletal muscle mitochondrial respiratory chain function: possible factor in ageing. Lancet 1989; I:637–639.
16. Yen T-C, Chen Y-S, King K-L, Yeh S-H, Wei Y-H. Liver mitochondrial respiratory functions decline with age. Biochem Biophys Res Commun 1989; 165:994–1003.
17. Byrne E, Dennett X, Trounce I. Oxidation energy failure in post-mitotic cells: a major factor in senescence. Rev Neurol 1991; 147:532–535.
18. Cooper JM, Mann VM, Schapira AVH. Analyses of mitochondrial respiratory chain function and mitochondrial DNA deletion in human skeletal muscle: effect of ageing. J Neurol Sci 1992; 113:91–98.
19. Boffoli D, Scacco SC, Vergari R, et al. Decline with age of the respiratory chain activity in human skeletal muscle. Biochim Biophys Acta 1994; 1226:73–82.
20. Boffoli D, Scacco SC, Vergari R, et al. Ageing is associated in females with a decline in the content and activity of the b-cl complex in skeletal muscle mitochondria. Biochim Biophys Acta 1996; 1315:66–72.
21. Hsieh RH, Hou JH, Wei YH. Age-dependent respiratory function decline and DNA deletions in human muscle mitochondria. Biochem Mol Biol Int 1994; 32:1009–1022.
22. Guerrieri F, Capozza G, Kalous M, et al. Age-dependent changes in the mitochondrial F_0F_1 ATP-synthase. Arch Gerontol Geriatr 1992; 14:299–308.
23. Guerrieri F, Capozza G, Fratello A, Zanotti F, Papa S. Functional and molecular changes in F_0F_1 ATP-synthase of cardiac muscle during aging. Cardioscience 1993; 4:93–98.
24. Sastre J, Federico VP, Pla R, et al. Aging of the liver: age-associated mitochondrial damage in intact hepatocytes. Hepatology 1996; 24:1199–1205.
25. Janssen YM, Van Houten B, Borm PJ, and Mossman BT. Cell and tissue responses to oxidative damage. Lab Invest 1993; 69:261–274.
26. Chance B, Sies H, Boveris A. Hydroperoxide metabolism in mammalian organs. Physiol Rev 1979; 59:527–605.
27. Korshunov SS, Skulachev VP, Starkov AA. High protonic potential actuates a mechanism of production of reactive oxygen species in mitochondria. FEBS Lett 1997; 416:15–18.
28. Boveris A, Chance B. The mitochondrial generation of hydrogen peroxide. General properties and effect of hyperbaric oxygen. Biochem J 1973; 34:707–716.

29. Boveris A, Cadenas E, Stoppani AOM. Role of ubiquinone in the mitochondrial generation of hydrogen peroxide. Biochem J 1976; 156:435–444.
30. Teoule R, Bert C, Bonicel A. Thymine fragment damage retained in the DNA polynucleotide chain after gamma irradiation in aerated solutions. II. Radiat Res 1977; 72:190–200.
31. Clark JM, Pattabiraman N, Jarvis W, Beardsley GP. Modeling and molecular mechanical studies of the *cis*-thymine glycol radiation damage lesion in DNA. Biochemistry 1987; 26:5404–5409.
32. Richter C. Reactive oxygen and DNA damage in mitochondria. Mutat Res 1992; 275:249–255.
33. Musci G, Bonaccorsi di Patti MC, Fagiolo U, Calabrese L. Age-related changes in human ceruloplasmin. Evidence for oxidative modifications. J Biol Chem 1993; 268:13388–13395.
34. Rouach H, Clement M, Orfanelli MT, et al. Hepatic lipid peroxidation and mitochondrial susceptibility to peroxidative attacks during ethanol inhalation and withdrawal, Biochim Biophys Acta 1983; 753:439–444.
35. Klimek J, Schaap AP, Kimura T. The relationship between NADPH-dependent lipid peroxidation and degradation of cytochrome P-450 in adrenal cortex mitochondria. Biochem Biophys Res Commun 1983; 110:559–566.
36. Marcillat O, Zhang Y, Davies KJA. Oxidative and non-oxidative mechanisms in the inactivation of cardiac mitochondrial electron transport chain components by doxorubicin. Biochem J 1989; 259:181–189.
37. Demple B, and Harrison L. Repair of oxidative damage to DNA: enzymology and biology. Annu Rev Biochem 1994; 63:915–948.
38. Friedberg EC, Walker GC, Siede W. DNA Repair and Mutagenesis. Washington, DC: American Society of Microbiology 1995.
39. Richter C, Park J-W, Ames BN. Normal oxidative damage to mitochondrial and nuclear DNA is extensive. Proc Natl Acad Sci USA 1988; 85:6465–6467.
40. Yakes FM, Van Houten B. Mitochondrial DNA damage is more extensive and persists longer than nuclear DNA damage in humans cells following oxidative stress. Proc Natl Acad Sci USA 1997; 94:514–519.
41. Merriwether DA, Clark AG, Bellinger SW, et al. The structure of human mitochondrial DNA variation. J Mol Evol 1991; 33:543–555.
42. Münscher C, Rieger T, Müller-Höcker J, Kadenbach B. The point mutation of mitochondrial DNA characteristic for MERRF disease is found also in healthy people of different ages. FEBS Lett 1993; 317:27–30.
43. Münscher C, Müller-Höcker J, Kadenbach B. Human aging is associated with various point mutations in tRNA genes of mitochondrial DNA. Biol Chem Hoppe Seyler 1993; 374:1099–1104.
44. Suganuma N, Kitagawa T, Nawa A, Tomoda Y. Human ovarian aging and mitochondrial DNA deletion. Horm Res 1993; 39(suppl 1):16–21.
45. Chen X, Prosser R, Simonetti S, et al. Rearranged mitochondrial genomes are present in human oocytes. Am J Hum Genet 1995; 57:239–247.
46. Fleming JE, Miquel J, Cottrell SF, Yengoyan LS, Economos AC. Is cell aging caused by respiration-dependent injury to the mitochondrial genome? Gerontology 1982; 28:44–53.

47. Clayton DA. Replication of mitochondrial DNA. Cell 1982; 28:693–705.

48. Domena JD, Mosbaugh DW. Purification of nuclear and mitochondrial uracil-DNA glycosylase from rat liver; identification of two distinct subcellular forms. Biochemistry 1985; 24:7320–7328.

49. Tomkinson AE, Bonk RT, Linn S. Mitochondrial endonuclease activities specific for apurinic/apyrimidinic sites in DNA from mouse cells. J Biol Chem 1988; 263: 12532–12537.

50. Tomkinson AE, Bonk RT, Kim J, Bartfeld N, Linn S. Mammalian mitochondrial endonuclease activities specific for ultraviolet-irradiated DNA. Nucleic Acids Res 1990; 18:929–935.

51. Pettepher CC, Ledoux SP, Bohr VA, Wilson GL. Repair of alkali-labile sites within the mitochondrial DNA of RINr 38 cells after exposure to the nitrosourea streptozotocin. J Biol Chem 1991; 266:3113–3117.

52. Kunkel TA, Loeb LA. Fidelity of mammalian DNA polymerases. Science 1981; 213:765–768.

53. Pinz KG, Shibutani S, Bogenhagen DF. Action of mitochondrial DNA polymerase γ at sites of base loss or oxidative damage. J Biol Chem 1995; 279:9202–9206.

54. Schon EA, Rizzuto R, Moreas CT, et al. A direct repeat is a hotspot for large-scale deletion of human mitochondrial DNA. Science 1989; 244:346–349.

55. Shoffner JM, Lott MT, Voljavec AS, et al. Spontaneous Kearns-Sayre chronic external ophthalmoplegia plus syndrome associated with a mitochondrial DNA deletion: a slip replication model and metabolic therapy. Proc Natl Acad Sci USA 1989; 86:7952–7956.

56. Johns DR, Rutledge SL, Stine OC, Hurko O. Directly repeated sequences associated with pathogenic mitochondrial DNA deletions. Proc Natl Acad Sci USA 1989; 86: 8059–8062.

57. Lauber J, Marsac C, Kadenbach B, Siebel P. Mutations in mitochondrial tRNA genes: a frequent cause of neuromuscular disease. Nucleic Acids Res 1991; 19: 1393–1397.

58. Menzies RA, Gold PH. The turnover of mitochondria in a variety of tissues of young adult and aged rats. J Biol Chem 1971; 246:2425–2429.

59. Wallace DC, Shoffner JM, Trounce I, et al. Mitochondrial DNA mutations in human degenerative diseases and aging. Biochim Biophys Acta 1995; 1271:141–151.

60. de Grey AD. A proposed refinement of the mitochondrial free radical theory of aging. Bioessays 1997; 19:161–166.

61. Holt IJ, Harding AE, Morgan-Hughes JA. Deletions of muscle mitochondrial DNA in patients with mitochondrial myopathies. Nature 1988; 331:717–719.

62. Wei YH. Mitochondrial DNA alterations as ageing-associated molecular events. Mutat Res 1992; 275:145–155.

63. Cortopassi GA, Arnheim N. Detection of a specific mitochondrial DNA deletion in tissues of older humans. Nucleic Acids Res 1990; 18:6927–6933.

64. Cortopassi GA, Shibata D, Soong N-W, Arnheim N. A pattern of accumulation of a somatic deletion of mitochondrial DNA in aging human tissues. Proc Natl Acad Sci USA 1992; 89:7370–7374.

65. Corral-Debrinski M, Horton T, Lott MT, et al. Mitochondrial DNA deletions in

human brain: regional variability and increase with advanced age. Nat Genet 1992; 2:324–329.

66. Moraes CT, DiMauro S, Zeviani M. Mitochondrial DNA deletions in progressive external ophthalmoplegia and Kearns-Sayre syndrome. N Engl J Med 1989; 320: 1293–1299.

67. Nelson I, Degoul F, Obermaier-Kusser B, et al. Mapping of heteroplasmic mitochondrial deletions in Kearns-Sayre syndrome. Nucleic Acids Res 1989; 17:8117–8124.

68. Müller-Höcker J, Seibel P, Schneiderbanger K, et al. In situ hybridization of mitochondrial DNA in the heart of a patient with Kearns-Sayre syndrome and dilatative cardiomyopathy. Hum Pathol 1992; 23:1431–1437.

69. Marin-Garcia J, Goldenthal MJ, Ananthakrishnan R, et al. Specific mitochondrial DNA deletions in idiopathic dilated cardiomyopathy. Cariovasc Res 1996; 31:306–313.

70. Pallotti F, Chen X, Bonilla E, Schon EA. Evidence that specific mtDNA point mutations may not accumulate in skeletal muscle during normal human aging. Am J Hum Genet 1996; 59:591–602.

71. Zhang C, Linnane AW, Nagley P. Occurrence of a particular base substitution (3243 A to G) in mitochondrial DNA of tissues of ageing humans. Biochem Biophys Res Commun 1993; 195:1104–1110.

72. Zhang C, Baumer A, Maxwell R. Multiple mitochondrial DNA deletions in an elderly human individual. FEBS Lett 1992; 297:34–38.

73. Melov S, Hinerfeld D, Esposito L, Wallace DC. Multi-organ characterization of mitochondrial genomic rearrangements in ad libitum and caloric restricted mice show striking somatic mitochondrial DNA rearrangements with age. Nucleic Acids Res 1997; 25:974–982.

74. Müller-Höcker J, Seibel P, Schneiderbanger K, Kadenbach B. Different in situ hybridization patterns of mitochondrial DNA in cytochrome c oxidase-deficient extraocular muscle fibers in the elderly. Virchows Archiv 1993; 422:7–15.

75. Bresolin N, Moggio M, Bet L, et al. Progressive cytochrome c oxidase deficiency in a case of Kearns-Sayre syndrome: morphological, immunological and biochemical studies in muscle biopsies and autopsy tissues. Ann Neurol 1987; 21:564–572.

76. Ozawa T. Mechanism of somatic mitochondrial DNA mutations associated with age and diseases. Biochim Biophys Acta 1995; 1271:177–189.

77. Shoffner JM, Wallace DC. Oxidative phosphorylation diseases. In: Scicer CR, Beaudet AL, Sly WS, Valle D, eds. The Metabolic and Molecular Bases of Inherited Disease, 7th ed. New York: McGraw-Hill, 1995:1535–1610.

78. Seibel P, Degoul F, Bonne G, et al. Genetic, biochemical and pathophysiological characterization of a familial mitochondrial encephalomyopathy (MERRF). J Neurol Sci 1991; 105:217–224.

79. Hayashi J-I, Ohta S, Kikuchi A, et al. Introduction of disease related mitochondrial DNA deletions into HeLa cells lacking mitochondrial DNA results in mitochondrial dysfunction. Proc Natl Acad Sci USA 1991; 88:10614–10618.

80. Simonetti S, Chen X, DiMauro S, Schon EA. Accumulation of deletions in human mitochondrial DNA during normal aging: analysis by quantitative PCR, Biochim Biophys Acta 1992; 1180:113–122.

81. Hamblet NS, Castora FJ. Mitochondrial DNA deletion analysis: a comparison of
     PCR quantitative methods. Biochem Biophys Res Commun 1995; 207:839–847.
82. Zhang C, Peters LE, Linnane AW, Nagley P. Comparison of different quantitative
     PCR procedures in the analysis of the 4977-bp deletion in human mitochondrial
     DNA. Biochem Biophys Res Commun 1996; 223:450–455.
83. Chen X, Bonilla E, Sciacco M, Schon EA. Paucity of deleted mitochondrial DNAs
     in brain regions of Huntington's disease patients. Biochim Biophys Acta 1995;
     1271:229–233.
84. Ozawa T. In: Papa S, Tager JM, eds. Biochemistry of Cell Membranes. Basel:
     Birkhäuser, 1995:339–361.
85. Gadaleta MN, Kadenbach B, Lezza AMS, et al. Age-linked changes in the genotype
     and phenotype of mitochondria. In: Papa S, Guerrieri F, Tager JM, Frontiers of
     Cellular Bioenergetics: Molecular Biology, Biochemistry and Physiopathology.
     eds. New York: Plenum, 1998.
86. Wallace DC, Zheng X, Lott MT, et al. Familial mitochondrial encephalomyopathy
     (MERRF): genetic, pathophysiological and biochemical characterization of a mito-
     chondrial DNA disease. Cell 1988; 55:601–610.
87. Becker A, Reith A, Napiwotzki J, Kadenbach B. A quantitative method of determin-
     ing initial amounts of DNA by polymerase chain reaction cycle titration using digi-
     tal imaging and a novel stain. Anal Biochem 1996; 237:204–207.
88. Kim SH, Chi JG, Reith A, Kadenbach B. Quantitative analysis of mitochondrial
     DNA deletion in paraffin embedded muscle tissues from patients with KSS and
     CPEO. Biochim Biophys Acta 1997; 1360:193–195.
89. Wallace DC, Bohr VA, Cortopassi G, et al. Group report: the role of bioenergetics
     and mitochondrial DNA mutations in aging and age-related diseases. In: Esser K,
     Martin GM, eds. Molecular Aspects of Aging. New York: John Wiley, 1995:199–
     225.
90. Arnold S, Kadenbach B. Cell respiration is controlled by ATP, an allosteric inhibi-
     tor of cytochrome c oxidase. Eur J Biochem 1997; 249:350–354.
91. Gulati J, Babu A. Effect of acidosis on Ca^{2+} sensitivity of skinned cardiac muscle
     with troponin C exchange. Implications for myocardial ischemia. FEBS Lett 1989;
     245:279–282.

14

Oxidative Damage and Fragility of Mitochondrial DNA

Takayuki Ozawa
University of Nagoya, Nagoya, Japan

I. INTRODUCTION

Aging involves many factors and processes, and hence there are many proposals for the mechanism of aging. These mechanisms can be placed in two categories: (1) aging is programmed in cellular genomes, and (2) aging is not genetically programmed; naturally occurring processes are responsible for aging.

The first mechanism has found support in publications such as Hayflick's limit of cell division (1) and Allsopp et al.'s (2). However, the mechanism seems to be not well suited to the stable tissue, such as muscle and nerve, where no cell division occurs after birth. Support for the second mechanism can be found in the "free radical theory of aging" by Harman (3), who proposes that nonspecific free radical damage to macromolecules is the cause of aging. The theory seems to be feasible for the postmitotic stable tissues. However, despite much supporting evidence, this theory has been not quite persuasive, because oxygen-damaged macromolecules, such as lipids and proteins, usually do not accumulate steadily with age. Namely, an appropriate biomarker has eluded researchers.

Linnane et al. (4) pointed out in the "mitochondrial theory of aging" that the somatic accumulation of mitochondrial DNA (mtDNA) mutations during human life is a major cause of human aging and geriatric processes in the stable tissues, such as the gradual loss of cellular bioenergetic capacity, muscle weakness of senescence, declining mental capacity, and age-related progressive decline of ventricular performance. The theory found substantial support in publications that document many kinds of mtDNA deletions in human stable tissues associated with age and degenerative diseases. However, the mechanism for somatic accumulation of deletions remains unsolved.

Comprehensive analyses of mtDNA in human tissue reveals that hydroxyl radicals leaked from mitochondrial electron transfer chains (ETCs) attacks mtDNA forming base modifications, strand breaks, and deletions (5). The oxidative damage to wild-type (ω) mtDNA leads it to disintegration into hundreds of types of deleted (Δ) mtDNA fragments (6). Hence the formation of reactive oxygen species (ROS) largely depends on the redox state of mtETC. The processes are proposed as the "redox mechanism of mitochondrial aging" (7). The oxidatively modified base of mtDNA in the stable tissues is demonstrated to be a superb biomarker of the mechanism. The documented extreme fragility of mtDNA to oxidative stress could explain naturally occurring cell death (apoptosis) without vascular involvement.

Mitochondrial control of nuclear apoptosis, through attenuation of mitochondrial membrane potential and release of apoptotic protease activating factors into cytosol, is documented by recent publications (7–9). Apoptosis could be induced by the exogenous factors, such as interaction of leukocytes' cytotoxic ligands (Fas-L) with cytoplasmic membrane receptor (Fas). On the other hand, apoptosis of constitutive cells by endogenous factors, such as energy deficit due to oxidative mtDNA disintegration, could be a major cause of geriatric processes in the stable tissues.

II. OXYGEN TOXICITY

A. Oxygen Free Radicals

As early as the time of the discovery of oxygen, Priestley (10) noticed that "oxygen might burn the candle of life too quickly, and soon exhaust the animal powers within."

In early studies on the mitochondrial respiratory chain, Wieland (11) favored a theory of "hydrogen activation" of substrates by dehydrogenase that activate hydrogen atoms of substrate molecules with the result that they become labile and can be transferred to a suitable hydrogen acceptor such as oxygen to form H_2O_2. However, at that time, Wieland was never able to identify H_2O_2, a highly toxic substance, in animal tissues. Opposing Wieland's theory, Warburg (12) presented the idea of "oxygen activation" by respiratory ferment, where four electrons are transferred to the oxygen molecule, which is then completely reduced and combines with hydrogen ions to form water. Warburg's "atmungsferment" was identified with Keilin's cytochrome oxidase (13), and was recognized as the terminal catalyst of the respiratory chain (14). This landmark discovery eclipsed for several decades the investigation of intermediates in oxygen reduction.

From studies on free radicals (15), Michaelis proposed in 1946 (16) that the free radical intermediates O_2^- and H_2O_2 are two obligatory univalent steps in the reduction of oxygen to water, and thus are a universal attribute of aerobic life. Years after the proposal, not peroxisomal or microsomal enzymes, but mtETC enzymes with high affinity to oxygen are pointed to as the H_2O_2 generator under intracellular oxygen partial pressure of less than 50 μM O_2 (17). The intermediates of oxygen reduction in cytochrome oxidase were exposed by Chance et al. (18) with the optical studies of oxy- and peroxy-cytochrome oxidase, indicating that the intermediates of oxygen reduction remain within the active site of intact cytochrome oxidase until the final reaction stage of water is achieved. From the general properties of mitochondrial H_2O_2 generation and the effect of hyperbaric oxygen, it is argued (19,20) that besides the flavin reaction in reduced mtETC, formation of H_2O_2 may be due to interaction with an energy-dependent component of mtETC at the cytochrome-b level. These findings indicate that complexes III and IV of mtETC play a crucial role not only for cellular energy production, but also for protection against cellular intoxication derived from ROS. Hence attenuation of the active site encoded by mtDNA could result in a serious outcome in cellular viability.

From the mechanism of oxygen reduction, the leak of electron from mtETC and the generation of ROS is expected not only by the genetic defect, but also by physiological attenuation of the cellular redox state. Hyperoxia (21) increases H_2O_2 release by lung mitochondria because of too much oxygen supply over enzymic capability to dispose of ROS. A similar ROS increase is observed at the time of reperfusion into ischemic tissues causing injuries, clinically termed "reperfusion injury." Skulachev (22) pointed out that mammalian uncoupled respiration or plant noncoupled respiration is an effective device to prevent oxidative damage and cell death by maintaining a safely low level of oxygen and its one-electron reductants. Boveris and Chance (19) demonstrated that H_2O_2 production by animal mitochondria, negligible in active respiration (state 3) or in the presence of uncouplers, becomes quite measurable in resting respiration (state 4). State 4 increases reduced electron carriers such as flavins, NADH oxidoreductase, CoQ, cytochrome b, and nonheme iron proteins (23–25) which are the main target for one-electron oxidation by oxygen. State 4 mitochondria from rat liver or from pigeon heart generate about 0.3–0.6 nmol H_2O_2/min/mg protein. This H_2O_2 generation represents approximately 2% of the total oxygen utilization under these conditions (19), which could raise tissue H_2O_2 concentration approximately 1 μM in balance with ROS scavenging enzymic activities. Thus, during the life of the individual, a vast sum of the redox energy is consumed for the generation of ROS. In addition, it was found (26) that cytochrome c is a 20 times more efficient catalyst than ferrous ion to promote the formation of hydroxyl radical (OH) by the Fenton reaction. Hence, when mtETC becomes defective and leaky, it turns out to be an efficient plant of OH production. The OH production is demonstrated by interaction of O_2^- with nitric oxide (NO) (27), of which synthase (NOS) is extensively inducted by Fas-L, such as tumor necrosis factor (TNF)

(28,29) and interleukin-1β (IL-1β) (30,31), leading the target cell to apoptotic death under physiological conditions. In this respect, OH⁻ could be regarded as not a mere by-product of the mitochondrial respiration, but as an active one with an important bioenergetic role to induce the physiological cell death and to eliminate unwanted/transformed cells.

B. Oxidative Damage to Mitochondrial Genome

The mtDNA locates inside of the mitochondrial inner membrane where ROS continuously leak from mtETC (32), hence being directly susceptible to attack by ROS, despite the cellular defenses against damage from ROS (33). During evolution from yeast to mammals, mtDNA has downsized to one-fifth, losing its intron in which mutations are inert. Hence human mtDNA becomes extremely more fragile to acquire the oxidative damage and large deletions than a single-cell organism. It seems to be reasonable that the genes coding the cellular energy plant have to be manageably fragile in order to be the primary target of the apoptotic process (34).

The underlying mechanism for the large deletion is double-strand separation by oxidatively modified bases, resulting in the generation of long stretches of single-strand DNA. Breaks form endonuclease-sensitive sites (35) or polymerase chain reaction (PCR)–detectable strand breaks (36) with a consequence of oxygen radical attack. In nuclear DNA, where an efficient repair system operates, the oxidized nucleosides such as a hydroxyl radical adduct of deoxyguanosine [8-hydroxy-2′-deoxyguanosine (8-OH-dG)] are rapidly excised and excreted into urine (37). However, in mitochondria where the repair system is inefficient, they accumulate with age, especially in the postmitotic stable tissue, leading mtDNA to the random point mutations, the double-strand separation, single-strand break, the large deletion, and to fragmentation (Figs. 1 and 2). A study (38) on the mutanogenesis of 8-OH-dG in a mammalian cell clearly demonstrates that a syn-

---

Figure 1   Size distribution of ΔmtDNA in the cardiomyocyte mtDNA of a patient with mtCM. To survey ΔmtDNA, the TD system (6) is applied to the mtDNA specimen in the cardiomyocytes of a patient with mtCM. Her heart at age 7 (pointed by arrow 4 in Fig. 2) is excised at the time of heart transplantation. Detected 212 types of ΔmtDNA are classified into four groups: OriL⁻/OriH⁻, OriL⁺/OriH⁺, OriL⁻, and OriH⁻ according to the preservation of the Ori of L-stand (L) or H-strand (H). White bars indicate ΔmtDNA, and dark shading bars indicate deleted regions. They are arranged according to their sizes. Remarkable mirror image is noted between ΔmtDNA in OriL⁻/OriH⁻ group and in OriL⁺/OriH⁺, and between OriL⁻ group and OriH⁻. Genomes in mtDNA are schematically illustrated at the bottom of the panel.

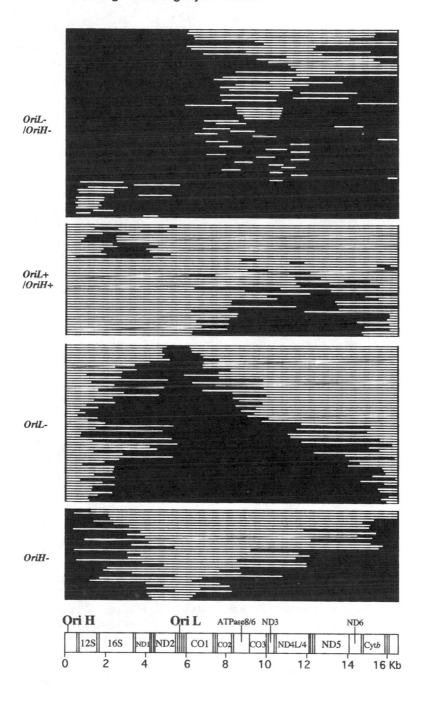

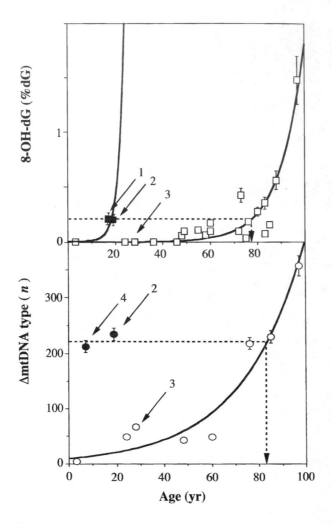

Figure 2 Age-associated correlative increase in the total number of ΔmtDNA type and oxygen damage. Upper panel: mtDNAs are extracted from autopsied cardiac muscles of 21 human subjects, 8 males and 13 females aged 3 to 97 years, without cardiological symptoms are obtained at random. Samples of mtDNA are enzymatically hydrolyzed into nucleosides, and analyzed by the microHPLC/MS system (39). The amount of 8-OH-dG (percent of dG) increases exponentially with the ages of subjects ($r = 0.84$). Overlaid plots are the 8-OH-dG percentage of mtCM patients who died by heart failure at age 17 (female, a closed square pointed by arrow 1) and at age 19 (male, a closed square pointed by arrow 2) (43), respectively. The 8-OH-dG percentage of a negative control harboring no severe point mutations (female, an open square pointed by arrow 3) is a marginal level. An arrow with dashed line indicates that the 8-OH-dG percentage of mtCM patients is

thetic protooncogene containing 8-OH-dG induces random point mutations at the modified site and adjacent positions on the gene replication. It is reported (36) that oxygen stress with 200 μM H_2O_2 to cultured fibroblasts results in arrested cell growth associated with PCR-detectable lesions in DNA. The oxidative mtDNA damage within 15 min of oxygen stress is repaired with similar kinetics as nuclear DNA, however, the mtDNA damage after 60 min oxygen stress is not repaired, but persists in contrast to nuclear DNA.

C. Genomic Damage and Aging

With precise quantitative determination of 8-OH-dG in mitochondrial genome by using a microHPLC/mass spectrometer (39), it is found (5,40) that 8-OH-dG in mtDNA of postmitotic cells is a superb biomarker of oxygen free radical damage. The amount of 8-OH-dG in mtDNA in human diaphragm muscle progressively increases to reach levels of 0.5% of dG at age 85, a 25-fold increase with age (40) associated with an increase in the incidence of mtDNA deletion (41), in heart muscle to reach levels of 1.5% of dG at age 97, a 75-fold increase (5). A similar accumulation of 8-OH-dG is reported in human brain to 0.87% at age 90, a 15-fold increase (42). Using the same specimen used for PCR detection of ∆mtDNA, it was found (5) that exponential increase in oxidative mtDNA damage closely correlates with that in ∆mtDNA with 7.4 kbp deletion. Hence it is argued (5) that the oxidative damage leads to mtDNA deletions that result in dysfunctional mitochondria.

A recently devised system for the total detection of mtDNA deletions (TD system) (6) reveals that human mtDNA is extremely fragile to OH attack, disintegrating wild-type (ω) mtDNA into hundreds of minicircles associated with age and mitochondrial diseases (6,43–46), as shown in Figure 1. The TD system documented that a progressive age-related increase in the total number (n) of ∆mtDNA types correlates with accumulation of 8-OH-dG in heart mtDNA re-

---

equivalent to that of the normal subjects of 78-year-old. Lower panel: The number of ∆mtDNA types (n) is determined by the total detection system (6) on 8 mtDNA specimens (from 3 males and 5 females) of enough quantity out of 21 samples after the 8-OH-dG analyses, shown in the upper panel. ∆mtDNA type n increases exponentially with the ages of subjects, resulting decrease of wild-type mtDNA down to 11% of the total mtDNA with a strong negative correlation with age ($r = 0.89$). Overlaid plots are ∆mtDNA type n in the excised heart of a female mtCM patient (44) who received heart transplantation at age 7 (a closed circle pointed by arrow 4), and that of a male mtCM patient died at age 19 (a closed circle pointed by arrow 2, the same as shown in the upper panel). Mean ∆mtDNA type n of the patients is equivalent to the normal subject, an 82-year-old.

flecting a long-term accumulation of the oxidative damage (Fig. 2, Table 1). There is a remarkable mirror image of the size distribution of ΔmtDNA (45,46) (Fig. 1) and a strong linear correlation ($r = 0.97$) between minicircles that lack either one of replication origins (Ori) or both, and ΔmtDNA preserving both origins. The fact suggests random occurrence of ΔmtDNA without a preferential strand-break site. Hence it seems reasonable to presume random double-strand separation by accumulated 8-OH-dG (5), single-strand breaks by OH attacks (35,36), and rejoining of mtDNA as a preferable mechanism for its fragmentation into hundreds of ΔmtDNAs, which further accelerates the oxygen damage. Experimentally the damage to mtDNA is correlated with the decline in the mtETC activity in laboratory animals (47–50).

Extensive oxygenation of human skeletal muscle indicates mitochondrial dysfunction causing suppressed oxygen utilization, hence relative tissue hyperoxia is demonstrated noninvasively among senescent individuals and the patients with mitochondrial cardiomyopathy (mtCM) and/or myopathy harboring hazardous point mutations (51). Similar reduced oxidative metabolism is reported in the cortex of Alzheimer's-type dementia (52). Defective mtETC encoded by the mutated mtDNA enhances the OH formation resulting in progressive oxidative damage, fragmentation of wild-type mtDNA, defected mtETC, and relative tissue hyperoxia. Such a vicious cycle of the OH damage and mutations in mtDNA seems to result in those changes to be synergistic and exponential, as illustrated in Figure 2.

The oxidative damage to mtDNA leads cells to a progressive decline of bioenergetic activities, and to the naturally occurring chronic death in the aging process. In the case of mtCM, inherited/acquired serious mutations in mtDNA could result in abnormal subunits of the mitochondrial energy transducing system and/or in pleiotropic defects of the system leading cells to the abnormal production of ROS and somatic mutations, and to subacute cell death. The genetic analyses on autopsied myocardia of pediatric patients have documented the abnormally increased oxidative mtDNA damage and ΔmtDNA leading to mtDNA fragmentation into hundred of minicircles (43,53) (Table 1, Fig. 2). These deleterious mutations could cause maladaptive growth and death of cardiac myocytes, and heart failure.

III. MECHANISMS OF AGING

A. The Free Radical Theory of Aging

In context with free radicals theoretically predicted by Michaelis, Harman (3) in 1956 proposed the "free radical theory of aging" that free radicals cause nonspecific damage to macromolecules, such as DNA, lipids, and proteins, being a principal cause of aging. The theory has attracted much attention with the develop-

Table 1 Types of mtDNA Among Subjects/Cells

Subject	Sex	Age	Disease	ΔmtDNA type (n)	Subtype of ΔmtDNA				ωmtDNA (%)	8-OH-dG per 10^4 dG
					OriL$^+$/H$^+$	OriL$^-$	OriH$^-$	OriL$^-$/H$^-$		
A.K.	F	3	VSD	5	4	1	0	0	>99	<1
S.T.	M	24	Accident	49	16	15	8	10	85	<1
N.N.	F	28	Pul. Emb.	67	23	14	8	22	71	<1
Y.I.	F	48	Thymoma	43	13	8	4	118	73	5.9
Y.T.	M	60	Gastric ca.	49	13	23	1	12	71	17.5
Y.Y.	F	76	SAH	218	66	68	37	47	47	18.6
K.A.	M	85	Colon ca.	230	61	64	33	72	58	15.3
H.M.	F	97	Gastric ca.	358	78	88	63	129	11	148
M.K.	F	7	DCM	212	37	58	38	79	47*	38.8*
T.K.	M	19	mtCM	235	48	59	31	97	16	20.1
ρ$^+$			normoxia	49	14	15	5	15	80*	<1*
ρ$^+$			95% O$_2$	187	35	55	28	69	53*	29.9*

All samples of mtDNA are extracted from autopsied heart muscle, except cultured cell line. Abbreviations: ΔmtDNA, mtDNA, with deletions; ωmtDNA, wild-type mtDNA; Ori, replication origin; F, female; M, male; VSD, ventricular septal defect; Pul. Emb., pulmonary embolism; ca., cancer; SAH, subarachnoidal hemorrhage; DCM, dilated cardiomyopathy (152); mtCM, mitochondrial cardiomyopathy (43); ρ$^+$, a cultured human cell line, 701.2.8 c (74); *, calculated from the regression formula (45).

ment in free radical biology [reviewed by Packer (54).] However, the theory has been not quite persuasive, because biologically active oxygen-damaged macromolecules usually turn over with certain metabolic rate, such as in the case of lipid peroxides (55) or oxidatively damaged proteins (56), except mere biologically inactive degenerative products such as lipofuscin or amyloid. Oxidized nucleosides, such as 8-OH-dG in nuclear DNA are rapidly repaired and excreted into urine (37). Thus the level of 8-OH-dG in rat liver DNA stays as low as 1 per 10^5 bases despite "extensive" oxidative damage (57). The level shows a small (two- to threefold) increase with age in rat liver (58). Few biomarkers to support the theory have been established. Hence the link between the oxidative damage of macromolecule and cell death has been missing.

B. The Mitochondrial Theory of Aging

Based on the gross genetic structure of human mtDNA reported by Anderson et al. in 1981 (59), in 1985–1988 our group elucidated using northern blot analyses (60–62) that the mitochondrial gene–encoded subunits, but not the nuclear gene–encoded subunits, are selectively defected in mtETC complexes of skeletal muscle from the patients with mitochondrial myopathy, which is named after morphological and biochemical abnormalities of muscular mitochondria (63). The specific defect is demonstrated to arise from inherited genetic abnormality by Southern blot analyses (64) that documented large deletions in mtDNA of a patient with mitochondrial myopathy and her mother; namely, the ρ^- mutation as in the case of yeast. The mutation associated with several degenerative diseases is also reported by others (65,66).

Expanding these biochemical and molecular biological studies on human tissues as well as yeast, it was proposed in 1989 (4) that the accumulation of mtDNA mutations during life is a underlying cause of human aging and degenerative diseases. This "mitochondrial theory of aging" stems from the following: the high frequency of gene mutation in mtDNA, analogous to the extremely high mutational rate of yeast; the small size of the mitochondrial genome and its known information content; the lack of protective histone; and inefficient repair system for mtDNA, unlike nuclear DNA; and the somatic segregation of individual mtDNA during eukaryotic cell division. At that time PCR technology (67) was brought into practical use, making it possible to amplify mtDNA of a small quantity. With PCR survey, an extensive array of age-dependent accumulation of deleted (Δ) mtDNA was documented among many human tissues, especially in postmitotic stable tissues such as nerve and muscle (5,68–72). However, the mechanism to generate large deletions has not been elucidated (73).

C. The Redox Mechanism of Mitochondrial Aging

Accumulated evidence to date exhorts to unify both ideas of the free radical theory of aging and the mitochondrial theory of aging to be the "redox mechanism of mitochondrial aging'' proposed by the author (7): ROS production is largely attenuated by mitochondrial redox state under physiological/pathological conditions. Oxidative mtDNA damage results in a cumulative increase in somatic mtDNA mutations synergistically leading to mtDNA fragmentation, bioenergetic deficit, cell death, and aging. Germ-line point mutations specific for the patients with mitochondrial diseases extensively accelerate these changes, resulting in premature aging (Fig. 2).

The fragmentation of mtDNA and apoptotic cell death under oxygen stress could be mimicked in the cultured human fibroblast cells (74): the exposure of a cultured fibroblast cell line (ρ^+) under oxygen stress, 95% oxygen, for 3 days leads the great majority of the ρ^+ cells to an apoptotic death with mtDNA fragmentation into 187 types of ΔmtDNA, whereas its derivative cells (ρ^0) lacking mtDNA are relatively immune; more than 80% of the ρ^0 cells survived. The results indicate mtDNA is the principal target of this cell death. Similarly, mtDNA strand break is observed after 60 min exposure of fibroblast to 200 μM H_2O_2, leading to arrested cell growth (35).

IV. MITOCHONDRIAL DNA AND MUTATIONS

A. Genetic Structure

In 1949, Ephrussi et al. (75) discovered a respiratory-deficient strain of yeast forming a small colony, called cytoplasmic ''petite'' mutants. Their studies showed that mutations resulting in the petite phenotype are inherited in a non-Mendelian fashion, and they therefore postulated the lesions to be in an extrachromosomal or cytoplasmic element, designated as the rho (ρ) factor. Hence cytoplasmic petite mutants are referred to as ρ^- mutants. Although this was not known at the time of the discovery, the ρ factor was subsequently shown to be identical to mtDNA and ρ^- mutants to arise from large deletions in mtDNA (76).

In 1963, electron micrographic study by Nass and Nass (77) indicated that the mitochondria of chick embryo cells contained threadlike structures that could be digested by DNase but not by RNase. At the same time, Schatz et al. (78) first detected DNA and quantitated it in purified yeast mitochondria by biochemical procedures. The first circular genetic map of yeast mitochondrial genome using petite mutants was published in 1975 by Molloy et al. (79). In 1981, a group of Cambridge researchers (59) reported the gross genetic structure of human mtDNA of closed circular duplex: the position of the 16,596 bp molecule is numbered as nucleotide position (np). The genes located on mtDNA encode some important subunits of mtETC and ATP synthetase: seven of complex I, three of complex IV, two of complex V, and one of complex III. Each 13 structural genes and 2 ribosomal RNA genes are economically tight-packed, being punctuated not by intron as in the case of nuclear genes, but by 22 transfer RNA genes.

B. Mutations

Many kinds of mtDNA mutations, such as point mutations and large deletions, have been reported to be the cause of degenerative diseases, named as diseases of the mtDNA (80). However, the cause-effect relationship between the reported mutational genotype and clinical phenotype remained unclear in these early studies, mainly because the mutation survey was carried out within limited regions of mtDNA. Grivell (73) stated that, curiously, there is no obvious correlation between the severity of the clinical symptoms or biochemical abnormality and either the location of the reported deletion or the number of deleted genes. In retrospect, the curiosity was caused by the survey of deletion within a limited region of mtDNA using a particular mtDNA probe for the Southern blot analyses or a single PCR primer pair. The problem has been settled by the TD system using 180 PCR primer pairs covering all around the wild-type mtDNA duplex (6,45).

1. Point Mutation

It has been disclosed that mitochondria of patients with mitochondrial myopathy harbor point mutations either maternally inherited (81–83) or somatically acquired (84) (also cf. Ref. 34). With the direct base sequencing of the entire mtDNA (85), accumulated data clarified the dominant feature of the mtDNA diseases that their clinical signs and symptoms are triggered by maternally inherited or somatically acquired multiple point mutations locating over wild-type mtDNA (53). It has become clear that the mutational genotypes based on the severity of point mutations and on their combination correspond to the clinical phenotypes, ranging from the asymptomatic to incapacitating symptoms (for more details, cf. Ref. 34). The combination of point mutations synergistically accelerate the accumulation of the somatic mtDNA lesions, the oxidative damage, and large deletions (5).

In addition to the maternally inherited germ-line mutations, nucleotide substitution is documented to occur in a single generation of Holstein cows, probably due to a genetic "founder effect" during oogenesis (86)—that is, amplification of one or a few mtDNA molecules as template will yield one predominant genotype in the mature oocyte that contains 100 to 1000 times more mtDNA than is found in somatic cells (87). In humans, a somatically acquired point mutation at np3243 A-to-G transition was reported in the cells of an individual (88). The somatically acquired point mutations were also detected in the cloned skeletal muscle mtDNA from an MELAS patient (10 clones/60 clones), being significantly higher than in those from a normal skeletal muscle (0/60) as well as in a normal placenta (2/60) (84).

2. Deletions

There are two types of mtDNA deletions either inherited or somatically acquired: the Southern blot–detectable deletion and the PCR-detectable one.

Soon after the practical use of PCR, multiple populations of ΔmtDNA are detected in human myocardium using a single PCR primer pair (89). Hence the author noted (90) that the PCR-detectable multiple forms of ΔmtDNA pleioplasmically coexist with wild-type mtDNA in a tissue. They are in contrast to the Southern blot-detectable one that is often detected among patients with early-onset mitochondrial myopathies (65), with chronic progressive ophthalmoplegia (64), or with Pearson's syndrome (66); a single ΔmtDNA of a large quantity, 20–85% of the total mtDNA detectable by the Southern blot, heteroplasmically coexists with wild-type mtDNA, presumably originated from a clonal expansion of an initial deletion event occurring early in oogenesis.

The "mitochondria theory of aging" (4) predicts the somatic accumulation of mtDNA mutations during human life as a major cause of human aging and geriatric process. Accumulation of PCR-detectable ΔmtDNA has been documented among many human tissues, especially in postmitotic stable tissues, such as nerve and muscle (5;68–72). Postmitotic cells in the stable tissues tend to accumulate somatic mtDNA deletions during an individual's life span, being different from mitotic cell with short life span, such as intestinal microvilli cells that are continuously replaced with newly divided cells. Quantitative data on a single ΔmtDNA detected by PCR among various aged individuals indicate that there are four orders of magnitude fewer ΔmtDNA in infancy as compared to old age (91,92), and that a newborn harbors only very low amounts of the 5 kbp deletion that is commonly observed in different tissues of adults (69), being undetected in the corresponding fetal tissues (70). Hence PCR-detectable multiform of ΔmtDNA seem to arise afresh with each generation, and to accumulate with age.

3. Disintegration into Fragments

The "redox mechanism of mitochondrial aging" (53) predicts that the somatic mutations harbor the following characteristics:

The absolute level of accumulated mutations is accountable for age-related decline of mitochondrial function and bioenergetic deficit.

The mutations accumulate age-dependently correlating closely with oxidative damage and cell death.

The mutations could be reproduced experimentally in human cultured cells.

A cell harbors hundreds of mitochondria and mtDNA copies, and the fractional concentration of each ΔmtDNA detected by the conventional PCR using a single primer pair is usually 0.01–0.3% of the total mtDNA (93–95). Based on the low absolute level of a single ΔmtDNA, a question arises whether an observed ΔmtDNA is the cause or the effect of the aging (94,95). However, the number and the size of ΔmtDNA visualized by PCR depend on the particular primer pair used, such that the more distantly separated primers enable detection of the larger deletions (96). Hence, a PCR-detectable ΔmtDNA is suggested to be the "tip of the iceberg" of the spectrum of somatic mutations (97). To settle the problem and to draw the whole figure of somatic deletions, the TD system that enables us to detect all possible ΔmtDNA over 0.5 kbp is recently devised (6), and applied to mtDNA specimens from the human tissues of various ages (45). Surprisingly, the whole "iceberg" in cardiomyocytes of age 97 is visualized as 358 types of ΔmtDNAs including 280 types of "minicircles" that lack either one of Ori or both (cf. Fig. 1), associated with decrease in wild-type mtDNA down to 11%.

The system documents the extreme fragility of the wild-type mtDNA with oxidative damage that leads the wild-type mtDNA molecule to disintegrate into hundreds of ΔmtDNA fragments associated with normal aging (45), with Parkinson's disease (46), and even with young mtCM patients harboring severe point mutations at age 7 and 19 to be premature aging equivalent to the normal subjects of age over 80 (43–45) (Figs. 1 and 2, Table 1). Similar to fragmentation, rearrangement which is shown to be the preceding step for the deletion (98), and depletion of wild-type mtDNA below the detection limit is demonstrated by applying long PCR to mtDNA in skeletal muscles of aged subjects (99).

The fragmentation of mtDNA and apoptotic cell death could be mimicked within 3 days in the cultured human fibroblast cells under 95% oxygen stress (74). Arrested cell growth associated with mtDNA strand breaks are observed in fibroblasts under 200 μM H_2O_2 within 60 min (36).

V. OXIDATIVE GENETIC LESION AND APOPTOSIS

Accumulated evidence suggests that there is a close relation between the redox mechanism of mitochondrial aging and apoptosis. Recent findings (8,9,100) allow outlining of the apoptosis cascade (Fig. 3). The active mtETC creates an electrochemical proton gradient ($\Delta\mu_H^+$) that can be in the form of mitochondrial transmembrane potential ($\Delta\Psi_m$) or ΔpH that drive ATP synthesis (101). The oxidative mtDNA damage and/or exogenous ligands lead cells to the depletion of $\Delta\mu_H^+$, hence to $\Delta\Psi_m$ collapse (22). Subsequently, $\Delta\Psi_m$-dependent permeability transition (PT) pore [also termed the "mitochondrial megachannel" (MMC) (102,103)] opens releasing intra-mitochondrial apoptotic protease activating fac-

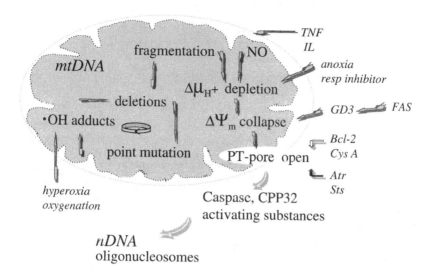

Figure 3 Cascade of cellular apoptosis. Bioenergetics and the cascade of cellular apoptosis are schematically illustrated based on the following reports: In mtDNA, point mutations, either inherited (7) or somatically acquired (84), accerelate oxidative damage and deletions leading wild-type mtDNA to fragmentation (43,44). Extensive tissue oxygenation, focal hyperoxia, associates with mtDNA mutations and with age (51). Hyperoxia induces an apoptotic cell death associated with fragmentation of mtDNA (74). Oxidative damage and deletions of mtDNA increase synergetically in human tissues with age (5,40). Extensive mtDNA fragmentation and oxidative damage associate with age (45). Age-associated accumulation of oxidative damage and deletions lead tissues to the loss of mitochondrial respiratory chain activities (48,49), hence to $\Delta\mu_H^+$ depletion. NO dose-dependently inhibits cytochrome oxidase activity (153), and its synthase (NOS) is extensively inducted by cytolytic factors such as TNF (28,29) and IL-1 (30,31) leading the target cell to apoptosis. Anoxia or respiratory inhibitor causes an acute apoptosis (146,150). A drop in $\Delta\Psi_m$ is one of the first events in apoptosis (154). Stimulated Fas released a ganglioside that collapses $\Delta\Psi_m$ (137). The PT induction leading PT pore to open is prevented by Bcl-2 (8) or by *Cys A* (136), a specific ligand of ANT. The PT pore opening is accelerated by an ANT ligand *Atr* or an apoptosis-inducing reagent *Sts* (8). The PT pore opening leads mitochondria to elute mitochondrial solutes such as dATP and cytochrome *c* that activate an inactive form of a apoptotic protease CPP32 to its active form (9). Activated proteinase cleaves various substrate including nuclear lamin (155) exposing nDNA to nuclease digestion (127) down into nucleosomes. The hatched line represents, outer or inner mitochondrial membrane. The shaded area represents intermembrane space, and the darker shading represent matrix.

tors into cytosol (9), then to nuclear DNA digestion into oligonucleosomes, and finally to cell death [reviewed by Ozawa (34,104)]:

oxidative damage → mtDNA fragmentation

$$\to \Delta\Psi_m \text{ collapse} \to \text{apoptosis}$$

The cascade links oxidative mtDNA damage to geriatric processes in tissues.

A. Apoptosis

The term apoptosis was named from morphological observations (105), and is functionally termed as programmed cell death or naturally occurring cell death under physiological conditions. Apoptosis is considered to play an important role in the elimination of unnecessary cells in human morphogenesis and harmful cells such as radical-producing or -transforming cells. Recently, with the advance of techniques to document DNA strand break on microscopic specimens, cell apoptosis has been reported to be prominent in many types of cells, as a possible pathophysiological cause of age- or disease-related tissue degeneration and atrophy; for example, as cardiac myocytes in the failing heart (106); in chronic heart failure (107); in dilated cardiomyopathy (108); sinus node cells in the patients with complete heart block and fatal arrhythmias (109); and neuronal cells in neurodegenerative diseases (110). However, the proposed mechanism of apoptosis had remained unclear, even controversial, until recently, because of the existence of many apoptosis-inducible factors and survival factors that affect at different points among the apoptosis cascade reaction, and moreover because of the extreme fragility of mtDNA.

The most readily measurable morphological features of apoptotic cell death are nuclear; namely, chromatin condensation and endonuclease-mediated nuclear DNA fragmentation producing a ''ladder'' of oligonucleosomal-sized nuclear DNA fragments visible by gel electrophoresis (111), which were considered to be the hallmark of apoptosis. To the contrary, no obvious morphological changes of mitochondria or of other organelle were observed until the end stage of apoptosis. Hence it has been tempting to consider that nuclear DNA plays a primary and causative role in apoptotic cell death. However, no clear evidence existed to support it. Conversely, the fact that the target cell nucleus is not required for cell-mediated granzyme- or Fas-based cytotoxicity (112) raised the possibility that apoptotic nuclear damage may be an epiphenomenon with respect to cell death. A series of reports (112–114) indicated that anucleate cytoplasts can undergo apoptosis and that the antiapoptotic protein Bcl-2 (115) and other extracellular survival signals can protect them, indicating that the nuclear signaling is not required for apoptosis or for Bcl-2/survival factor protection.

B. Apoptosis Program

The cell suicide program is best illustrated by genetic studies in the nematode *Caenorhabditis elegans* (116). Two genes involved in the control of apoptosis in *C. elegans* have been well characterized. One gene, *ced-9*, encodes a protein that prevents cells from undergoing apoptosis, while another gene, *ced-3*, encodes a protease whose activity is required to initiate apoptosis. The *bcl-2* family of genes are the mammalian counterparts of *ced-9* (117). The *ced-3* protein is a cysteine protease related to the interleukin-1β-converting enzyme (ICE) family of proteases (also called caspases) in mammalian cells (118–122). The closest mammalian homologue of *ced-3* protein is CPP32 in terms of sequence identity and substrate specificity (123). Like *ced-3* protein, CPP32 normally exists in the cytosolic fraction as its inactive form of 32 kDa. In cells undergoing apoptosis, the protease is activated proteolytically to be a 17/11 kDa or 20/11 kDa active form (120,124), named as apopain (121). Apopain cleaves several substrates including death substrates, poly (ADP-ribose) polymerase (125) and nuclear lamin (126), exposing nucleus to Ca^{2+}-endonuclease (127) and/or DNase I (128) digestion. However, the intracellular factors and/or bioenergetics that initially activate CPP32 have remained unknown.

Recently a cell-free system was established to duplicate the features of the apoptotic program, including activation of CPP32 and nuclear DNA fragmentation (9): the system consisted of nuclei added in cytosol from normally growing cells and intact mitochondria, carefully prepared by the method of Hayakawa et al. (49), of which PT pore is well sustained by $\Delta\Psi_m$. By addition of an apoptosis-inducing reagent, staurosporine (*Sts*) or atractyloside (*Atr*), which disrupts the intact PT pore, nuclear apoptosis could be initiated by the release of apoptotic protease activating factors from mitochondria into cytosol (Fig. 3). A phosphocellulose column separates the protease activating factors into two kinds, viz., deoxyATP and cytochrome *c*. Depletion of cytochrome *c* with anticytochrome *c* antibody loses the dATP-dependent activation of CPP32 and the nuclear DNA fragmentation. Conversely, without dATP, cytochrome *c* could not activate CPP32. Nuclear apoptosis with the presence of the intact mitochondria is induced by the external addition of dATP and cytochrome *c* to the system. Hence both substances are required and sufficient for the activation of CPP32 to apopain.

C. Redox Factors for Apoptosis

Intact mitochondrial inner membrane is quasi-impermeable for small solute allowing the creation of $\Delta\mu_H^+$ in the form of $\Delta\Psi_m$, thus it can keep the apoptotic protease activation factors inside of the membrane, and the collapse of the $\Delta\Psi_m$-dependent PT pore and mitochondrial swelling releases them into cytosol. Associated with PT, small molecules such as glutathione are rapidly effluxed before large amplitude swelling (129), and the apoptosis-inducing protein with mito-chondrial swelling and disruption of the mitochondrial outer membrane (130).

The precise biochemical mechanism of cytochrome c function in the activation of CPP32 remains to be determined. However, cytochrome c is a 20 times more efficient catalyst than ferrous ion to promote the formation of OH⁻ by the Fenton reaction both in NADPH-driven respiration of leukocyte plasma membrane and in nonenzymatic H_2O_2 solution (26). Reported molar ratio of cytosolic inorganic plus organic Fe (131) and mitochondrial cytochrome c (132) in rat liver cells is approximately 20:1, however, efficiency to produce OH⁻ is 1:20. Hence the efflux of cytochrome c into cytosol would be highly cytotoxic due to the OH⁻ formation from H_2O_2. Sensible cells translocate apo-cytochrome c, which is translated in the cytoplasmic ribosomes, into the mitochondrial intermembrane space through a unique pathway (133), and it is loosely attached to the surface of the inner membrane (134). Finally heme is installed by cytochrome c heme lyase, a peripheral protein of the inner membrane, and the holo-protein folds into its native structure. Hence catalytically active cytochrome c locates exclusively in mitochondria, as has been shown by using radiolabeled cytochrome c (135).

The release of apoptotic protease activating factors from mitochondria into cytosol is caused by the bioenergetic crisis and the collapse of $\Delta\Psi_m$. Cyclosporin (*Cys*) A, a specific ligand of adenine nucleotide translocator (ANT) that is a component of the PT pore, prevents PT and apoptosis (22,136). Mitochondrial control of nuclear apoptosis is demonstrated from the fact that PT constitutes a critical early event of the apoptotic process (8): in a cell-free system combining purified mitochondria and nuclei, mitochondria undergoing PT in response to an ANT ligand, atractyloside (*Atr*), is prevented by other specific ligand, bonkretic acid, or by hyperexpression of Bcl-2. $\Delta\Psi_m$ is also disrupted by ganglioside derived from stimulated CD95 (called also Fas or APO-1) by anti-Fas immunoglobulin (137). However, besides these exogenous ANT ligands or respiratory inhibitors, the endogenous factors that cause the bioenergetic crisis and the collapse of $\Delta\Psi_m$ in naturally occurring chronic cell death had remained unknown until the extreme fragility of mtDNA was recently disclosed (Figs. 2 and 3) (6,43–45,74).

Bcl-2 and a splice variant of Bcl-x (138) have been shown to heterodimerize with other members of the Bcl-2 protein family, including Bax. These oncoproteins are proposed to manipulate oxygen free radical damage and cell death (139). Consistent with the proposal, it was shown that the antioxidant suppressed dopamine-induced apoptosis in mouse thymocytes (140). Several studies have shown that Bcl-2 can protect cellular membranes from oxidative damage (141,142), and the generation of ROS (143). The NO-induced apoptosis is effectively protected by overexpression or transfection of antiapoptosis gene *bcl-2* (144,145). On the other hand, studies on the Bcl-2 protection to cell apoptosis under hypoxia concluded that ROS are not essential for apoptosis, and that Bcl-2 protects against apoptosis in ways that do not depend on the inhibition of ROS production or activity (146,147). These facts could be interpreted as both the hypoxia-induced apoptosis and the naturally occurring cell death with age are triggered by the bioenergetic crisis of cells, the decrease in $\Delta\Psi_m$, and PT pore opening.

D. Apoptosis Cascade

From the reports, the apoptosis cascade based on molecular genetics and bioenergetics is illustrated in Figure 3. At the point of $\Delta\Psi_m$ collapse in the upstream of the cascade, one of the main tributaries of apoptosis with aging, hyperoxia, and mtDNA fragmentation, merges with other tributaries of apoptotic/necrotic acute cell death with hypoxia (146,147), the respiratory inhibitors (129,148,149), the depletion of reductants (129,150,151), the oxidative-phosphorylation uncouplers (8), the divalent cations (8,129), or NO (144,145). Here in the bioenergy-dependent apoptotic cascade, apparently contradictory apoptosis-inducible factors, for example, hyperoxia and hypoxia, could be localized intelligibly.

VI. CONCLUSIONS AND PERSPECTIVES

The comprehensive analyses of the entire mtDNA, including the detection of the inherited/acquired point mutations, of the somatic oxidative damage, and of the total deletions, could reveal the mutational genotype unique to an individual. The analyses disclosed that the types and combination of point mutations decide the severity of somatic oxidative damage and mutations. Hence there is the definite correspondence between the point mutational genotype and the phenotype of the patients. The practicable survey of point mutations will be useful for genetic diagnosis predicting the patients' life span and for the management of patients, such as cardiac transplantation and/or gene therapy.

The survey of the somatic oxidative damage and the total deletions in mtDNA disclosed that mitochondrial genes coding cellular energy plant are unexpectedly fragile to hydroxyl radical damage, hence to the oxygen stress. Cellular wild-type mtDNA easily fragments into over one hundred types of ΔmtDNA resulting in defective mitochondrial energy transducing system and in cellular bioenergetic crisis. This could be the missing endogenous link in the cascade of naturally occurring cell death under physiological condition without vascular involvement. Namely, the fragmentation of wild-type mtDNA due to the oxidative stress leads to cellular bioenergetic crisis, to the collapse of $\Delta\Psi_m$, to the release of the apoptotic protease activating factors into cytosol, to uncontrolled cell death, to tissue degeneration and atrophy, and to aging and degenerative diseases. Further elucidation of precise mechanism of this apoptotic cascade will enable us to protect unwanted cell death, hence to arrest of degenerative diseases and to accelerate harmful/transformed cell death.

REFERENCES

1. Hayflick L. The limited in vitro lifetime of human diploid cell strains. Exp Cell Res 1965; 37:614–636.
2. Allsopp RC, Vaziri H, Patterson C, et al. Telomere length predicts replicative capacity of human fibroblasts. Proc Natl Acad Sci USA 1992; 89:10114–10118.
3. Harman D. Aging: a theory based on free radical and radiation chemistry. J Gerontol 1956; 11:298–300.
4. Linnane AW, Marzuki S, Ozawa T, Tanaka M. Mitochondrial DNA mutations as an important contributor to ageing and degenerative diseases. Lancet 1989; i:642–645.
5. Hayakawa M, Hattori K, Sugiyama S, Ozawa T. Age-associated oxygen damage and mutations in mitochondrial DNA in human hearts. Biochem Biophys Res Commun 1992; 189:979–985.
6. Hayakawa M, Katsumata K, Yoneda M, et al. Mitochondrial DNA minicircles, lacking replication origins, exist in the cardiac muscle of a young normal subject. Biochem Biophys Res Commun 1995; 215:952–960.
7. Ozawa T. Mechanism of somatic mitochondrial DNA mutations associated with age and diseases [review]. Biochim Biophys Acta 1995; 1271:177–189.
8. Zamzami N, Susin SA, Marchetti P, et al. Mitochondrial control of nuclear apoptosis. J Exp Med 1996; 183:1533–1544.
9. Liu X, Kim CN, Yang J, Jemmerson R, Wang X. Induction of apoptotic program in cell-free extracts: requirement for dATP and cytochrome c. Cell 1996; 86:147–157.
10. Priestley J. Experiments and observations on different kinds of air. London: J. Johnson at St Paul's Churchyard, 1775:101–102.
11. Wieland H. Über den mechanismus der oxydationsvorgänge. Ergebn Physiol 1922; 20:477–518.
12. Warburg O. Über eisen, den sauerstoff übertragenden bestandteil des atmungsferments. Biochem Z 1924; 152:479–494.
13. Keilin D. On cytochrome, a respiratory pigment, common to animals, yeast, and higher plants. Proc R Soc Lond 1925; 98:312–339.
14. Warburg O, Negelein E. Über das absorptionsspektrum des atmungsferments. Biochem Z 1929; 214:64–100.
15. Michaelis L. Free radicals as intermediate steps of oxidation-reduction. Cold Spring Harb Symp Quant Biol 1939; 7:33–49.
16. Michaelis L. Fundamentals of oxidation and reduction. In: Green DE, ed. Currents in Biochemical Research. New York: Interscience, 1946; 207–227.
17. Boveris A, Oshino N, Chance B. The cellular production of hydrogen peroxide. Biochem J 1972; 128:617–630.

18. Chance B, Saronio C, Leigh JSJ. Functional intermediates in the reaction of membrane bound cytochrome oxidase with oxygen. J Biol Chem 1975; 250:9226–9232.

19. Boveris A, Chance B. The mitochondrial generation of hydrogen peroxide : general properties and effect of hyperbaric oxygen. Biochem J 1973; 134:707–716.

20. Boveris A, Cadenas E. Mitochondrial production of superoxide anions and its relationship to the antimycin insensitive respiration. FEBS Lett 1975; 54:311–314.

21. Turrens JF, Freeman BA, Crapo JD. Hyperoxia increases H_2O_2 release by lung mitochondria and microsomes. Arch Biochem Biophys 1982; 217:411–421.

22. Skulachev VP. Role of uncoupled and non-coupled oxidations in maintenance of safely low level of oxygen and its one-electron reductants [review]. Q Rev Biophys 1996; 29:169–202.

23. Ksenzenko M, Konstantinov AA, Khomutov GB, Tikhonov AN, Ruuge EK. Effect of electron transfer inhibitors on superoxide generation in the cytochrome bc1 site of the mitochondrial respiratory chain. FEBS Lett 1983; 155:19–24.

24. Massey V. Activation of molecular oxygen by flavins and flavoproteins [review]. J Biol Chem 1994; 269:22459–22462.

25. Cross AR, Jones OT. Enzymic mechanisms of superoxide production [review]. Biochim Biophys Acta 1991; 1057:281–298.

26. Hayakawa M, Ogawa T, Sugiyama S, Ozawa T. Hydroxyl radical and leukotoxin biosynthesis in neutrophil plasma membrane. Biochem Biophys Res Commun 1989; 161:1077–1085.

27. Hogg N, Darley-Usmar VM, Wilson MT, Moncada S. Production of hydroxyl radicals from the simultaneous generation of superoxide and nitric oxide. Biochem J 1992; 281:419–424.

28. Geng YJ, Wu Q, Muszynski M, Hansson GK, Libby P. Apoptosis of vascular smooth muscle cells induced by in vitro stimulation with interferon-gamma, tumor necrosis factor-alpha, and interleukin-1 beta. Arterioscler Thromb Vasc Biol 1996; 16:19–27.

29. Dong HD, Kimoto Y, Takai S, Taguchi T. Apoptosis as a mechanism of lectin-dependent monocyte-mediated cytotoxicity. Immunol Invest 1996; 25:65–78.

30. Ankarcrona M, Dypbukt JM, Brune B, Nicotera P. Interleukin-1 beta-induced nitric oxide production activates apoptosis in pancreatic RINm5F cells. Exp Cell Res 1994; 213:172–177.

31. Dunger A, Augstein P, Schmidt S, Fischer U. Identification of interleukin 1-induced apoptosis in rat islets using in situ specific labelling of fragmented DNA. J Autoimmun 1996; 9:309–313.

32. Chance B, Sies H, Bovferies A. Hydroperoxide metabolism in mammalian organs. Physiol Rev 1979; 59:527–605.

33. Yu BP. Cellular defenses against damage from reactive oxygen species. Physiol Rev 1994; 74:139–162.

34. Ozawa T. Genetic and functional changes in mitochondria associated with aging. Physiol Rev 1997; 77:425–464.

35. Deng RY, Fridovich I. Formation of endonuclease III sensitive sites as a consequence of oxygen radical attack on DNA. Free Radic Biol Med 1989; 6:123–129.

36. Yakes FM, Van Houten B. Mitochondrial DNA damage is more extensive and

persists longer than nuclear DNA damage in human cells following oxidative stress. Proc Natl Acad Sci USA 1997; 94:514–519.

37. Fraga CG, Shigenaga MK, Park JW, Degan P, Ames BN. Oxidative damage to DNA during aging: 8-hydroxy-2'-deoxyguanosine in rat organ DNA and urine. Proc Natl Acad Sci USA 1990; 87:4533–4537.

38. Kamiya H, Miura K, Ishikawa H, et al. c-Ha-*ras* containing 8-hydroxyguanine at codon 12 induces point mutations at the modified and adjacent positions. Cancer Res 1992; 52:3483–3485.

39. Hayakawa M, Ogawa T, Sugiyama S, Tanaka M, Ozawa T. Massive conversion of guanosine to 8-hydroxy-guanosine in mouse liver mitochondrial DNA by administration of azidothymidine. Biochem Biophys Res Commun 1991; 176:87–93.

40. Hayakawa M, Torii K, Sugiyama S, Tanaka M, Ozawa T. Age-associated accumulation of 8-hydroxydeoxyguanosine in mitochondrial DNA of human diaphragm. Biochem Biophys Res Commun 1991; 179:1023–1029.

41. Torii K, Sugiyama S, Tanaka M, et al. Aging-associated deletions of human diaphragmatic mitochondrial DNA. Am J Respir Cell Mol Biol 1992; 6:543–549.

42. Mecocci P, MacGarvey U, Kaufman AE, et al. Oxidative damage to mitochondrial DNA shows marked age-dependent increases in human brain. Ann Neurol 1993; 34:609–616.

43. Katsumata K, Hayakawa M, Tanaka M, Sugiyama S, Ozawa T. Fragmentation of human heart mitochondrial DNA associated with premature aging. Biochem Biophys Res Commun 1994; 202:102–110.

44. Ozawa T, Katsumata K, Hayakawa M, et al. Genotype and phenotype of severe mitochondrial cardiomyopathy: a recipient of heart transplantation and the genetic control. Biochem Biophys Res Commun 1995; 207:613–620.

45. Hayakawa M, Katsumata K, Yoneda M, et al. Age-related extensive fragmentation of mitochondrial DNA into minicircles. Biochem Biophys Res Commun 1996; 226: 369–377.

46. Ozawa T, Hayakawa M, Katsumata K, et al. Fragile mitochondrial DNA: the missing link in the apoptotic neuronal cell death in Parkinson's disease. Biochem Biophys Res Commun 1997; 235:158–161.

47. Torii K, Sugiyama S, Takagi K, Satake T, Ozawa T. Age-related decrease in respiratory muscle mitochondrial function in rats. Am J Respir Cell Mol Biol 1992; 6: 88–92.

48. Sugiyama S, Takasawa M, Hayakawa M, Ozawa T. Changes in skeletal muscle, heart and liver mitochondrial electron transport activities in rats and dogs of various ages. Biochem Mol Biol Int 1993; 30:937–944.

49. Hayakawa M, Sugiyama S, Hattori K, Takasawa M, Ozawa T. Age-associated damage in mitochondrial DNA in human hearts. Mol Cell Biochem 1993; 119:95–103.

50. Takasawa M, Hayakawa M, Sugiyama S, et al. Age-associated damage in mitochondrial function in rat hearts. Exp Gerontol 1993; 28:269–280.

51. Ozawa T, Sahashi K, Nakase Y, Chance B. Extensive tissue oxygenation associated with mitochondrial DNA mutations. Biochem Biophys Res Commun 1995; 213: 432–438.

52. Hoyer S. Senile dementia and Alzheimer's disease. Brain blood flow and metabolism. Prog Neuropsychopharmacol Biol Psychiatry 1986; 10:447–478.

53. Ozawa T. Mitochondrial DNA mutations associated with aging and degenerative diseases [review]. Exp Gerontol 1995; 30:269–290.

54. Packer L. Oxidative stress, antioxidant, aging and disease. In: Oxidative Stress and Aging. Cutler RG, Packer L, Bertram J, Mori A, eds. Basel: Birkhauser Verlag, 1995; 1–14.

55. Bindoli A. Lipid peroxidation in mitochondria. Free Radic Biol Med 1988; 5:247–261.

56. Davies KJ. Intracellular proteolytic systems may function as secondary antioxidant defenses: an hypothesis. J Free Radic Biol Med 1986; 2:155–173.

57. Richter C, Park J-W, Ames BN. Normal oxidative damage to mitochondrial and nuclear DNA is extensive. Proc Natl Acad Sci USA 1988; 85:6465–6467.

58. Ames BN, Shigenaga MK, Hagen TM. Mitochondrial decay with age. Biochim Biophys Acta 1995; 1271:165–170.

59. Anderson S, Bankier AT, Barrell BG, et al. Sequence and organization of the human mitochondrial genome. Nature 1981; 290:457–465.

60. Tanaka M, Nishikimi M, Suzuki H, et al. Multiple cytochrome deficiency and deteriorated mitochondrial polypeptide composition in fatal infantile mitochondrial myopathy and renal dysfunction. Biochem Biophys Res Commun 1986; 137:911–916.

61. Tanaka M, Nishikimi M, Ozawa T, Miyabayashi S, Tada K. Lack of subunit II of cytochrome c oxidase in a patient with mitochondrial myopathy. Ann NY Acad Sci 1987; 488:503–504.

62. Ichiki T, Tanaka M, Nishikimi M, et al. Deficiency of subunits of complex I and mitochondrial encephalomyopathy. Ann Neurol 1988; 23:287–294.

63. DiMauro S, Bonilla E, Zeviani M, Nakagawa M, DeVivo DC. Mitochondrial myopathies. Ann Neurol 1985; 17:521–538.

64. Ozawa T, Yoneda M, Tanaka M, et al. Maternal inheritance of deleted mitochondrial DNA in a family with mitochondrial myopathy. Biochem Biophys Res Commun 1988; 154:1240–1247.

65. Holt IJ, Harding AE, Morgan-Hughes JA. Deletions of muscle mitochondrial DNA in patients with mitochondrial myopathies. Nature 1988; 331:717–719.

66. Rotig A, Colonna M, Binnefont JP, et al. Mitochondrial DNA deletion in Peason's marrow/pancreas syndrome. Lancet 1989; i:902–903.

67. Saiki RK, Gelfand DH, Stoffel S, et al. Primer-directed enzymatic amplification of DNA with a thermostable DNA polymerase. Science 1988; 239:487–494.

68. Ikebe S, Tanaka M, Ohno K, et al. Increase of deleted mitochondrial DNA in the striatum in Parkinson's disease and senescence. Biochem Biophys Res Commun 1990; 170:1044–1048.

69. Linnane AW, Baumer A, Maxwell RJ, et al. Mitochondrial gene mutation: the ageing process and degenerative diseases. Biochem Int 1990; 22:1067–1076.

70. Cortopassi GA, Arnheim N. Detection of a specific mitochondrial DNA deletion in tissues of older humans. Nucleic Acids Res 1990; 18:6927–6933.

71. Hattori K, Tanaka M, Sugiyama S, et al. Age-dependent increase in deleted mitochondrial DNA in the human heart: possible contributing factor to "presbycardia." Am Heart J 1991; 121:1735–1742.

72. Corral-Debrinski M, Shoffner JM, Lott MT, Wallace DC. Association of mitochon-

drial DNA damage with aging and coronary atherosclerotic heart disease. Mutat Res 1992; 275: 169–180.

73. Grivell LA. Mitochondrial DNA: small, beautiful and essential. Nature 1989; 341: 569–571.

74. Yoneda M, Katsumata K, Hayakawa M, Tanaka M, Ozawa T. Oxygen stress induces an apoptotic cell death associated with fragmentation of mitochondrial genome. Biochem Biophys Res Commun 1995; 209: 723–729.

75. Ephrussi B, Hottinguer H, Chimenes AM. Action de l'acriflavine sur les levures. I. La mutation "petite colonie." Ann Inst Pasteur 1949; 76:351–367.

76. Tzagoloff A. Mitochondrial Genetics. In: Mitochondria. New York: Plenum Press, 1982:267–322.

77. Nass S, Nass MMK. Intramitochondrial fibers with DNA characteristics II. Enzymatic and other hydrolytic treatments. J Cell Biol 1963; 19:613–629.

78. Schatz G, Halsbrunner E, Tuppy H. Deoxyribonucleic acid associated with yeast mitochondria. Biochem Biophys Res Commun 1964; 15:127–132.

79. Molloy PL, Linnane AW, Lukins HB. Biogenesis of mitochondria: analysis of deletion of mitochondrial antibiotic resistance markers in petite mutants of *Saccharomyces cerevisiae*. J Bacteriol 1975; 122:7–18.

80. Wallace DC. Disease of the mitochondrial DNA [review]. Annu Rev Biochem 1992; 61:1175–1212.

81. Wallace DC, Singh G, Lott MT, et al. Mitochondrial DNA mutation associated with Leber's hereditary optic neuropathy. Science 1988; 242:1427–1430.

82. Wallace DC, Zheng X, Lott MT, et al. Familial mitochondrial encephalomyopathy (MERRF): genetic, pathophysiological, and biochemical characterization of a mitochondrial DNA disease. Cell 1988; 66:601–610.

83. Yoneda M, Tsuji S, Yamauchi T, Mitochondrial DNA mutation in family with Leber's hereditary optic neuropathy [letter]. Lancet 1989; i:1076–1077.

84. Kovalenko SA, Tanaka M, Yoneda M, Iakovlev AF, Ozawa T. Accumulation of somatic nucleotide substitutions in mitochondrial DNA associated with the 3243 A-to-G tRNA(leu)(UUR) mutation in encephalomyopathy and cardiomyopathy. Biochem Biophys Res Commun 1996; 222:201–207.

85. Tanaka M, Hayakawa M, Ozawa T. Automated sequencing of mitochondrial DNA. In: Attardi GM, Chomyn A, eds. Methods in Enzymology. vol. 264. Mitochondrial genetics and biogenesis. Orlando, FL: Academic Press, 1996:407–421.

86. Hauswirth WW, Laipis P. Mitochondrial DNA polymorphism in a maternal lineage of Holstein cows. Proc Natl Acad Sci USA 1982; 79:4686–4690.

87. Piko L, Matsumoto L. Number of mitochondria and some properties of mitochondrial DNA in the mouse egg. Dev Biol 1976; 19:1–10.

88. Zhang C, Linnane A, Nagley P. Occurrence of a particular base substitution (3243 A to G) in mitochondrial DNA of tissues of ageing humans. Biochem Biophys Res Commun 1993; 195: 1104–1110.

89. Sato W, Tanaka M, Ohno K, et al. Multiple populations of deleted mitochondrial DNA detected by a novel gene amplification method. Biochem Biophys Res Commun 1989; 162:664–672.

90. Ozawa T, Tanaka M, Sato, W., et al. Types and mechanism of mitochondrial DNA mutations in mitochondrial myopathy and related diseases. In: Gorrod JW, Albano

O, Ferrari E, Papa S, eds. Molecular Basis of Neurological Disorders and Their Treatment. London: Chapman & Hall, 1991:171–190.

91. Sugiyama S, Hattori K, Hayakawa M, Ozawa T. Quantitative analysis of age-associated accumulation of mitochondrial DNA with deletion in human hearts. Biochem Biophys Res Commun 1991; 180:894–899.

92. Simonetti S, Chen X, DiMauro S, Schon EA. Accumulation of deletions in human mitochondrial DNA during normal aging: analysis by quantitative PCR. Biochim Biophys Acta 1992; 1180:113–122.

93. Ozawa T, Tanaka M, Ikebe S, et al. Quantitative determination of deleted mitochondrial DNA relative to normal DNA in parkinsonian striatum by a kinetic PCR analysis. Biochem Biophys Res Commun 1990; 172:483–489.

94. Cooper JM, Mann VM, Schapira AH. Analyses of mitochondrial respiratory chain function and mitochondrial DNA deletion in human skeletal muscle: effect of ageing. J Neurol Sci 1992; 113:91–98.

95. Remes AM, Hassinen IE, Ikaheimo MJ, et al. Mitochondrial DNA deletions in dilated cardiomyopathy: a clinical study employing endomyocardial sampling. J Am Coll Cardiol 1994; 23:935–942.

96. Zhang C, Baumer A, Maxwell RJ, Linnane AW, Nagley P. Multiple mitochondrial DNA deletions in an elderly human individual. FEBS Lett 1992; 297:34–38.

97. Soong NW, Hinton DR, Cortopassi G, Arnheim N. Mosaicism for a specific somatic mitochondrial DNA mutation in adult human brain. Nat Genet 1992; 2:318–323.

98. Poulton J, Deadman ME, Bindoff L, et al. Families of mtDNA rearrangements can be detected in patients with mtDNA deletions: duplications may be a transient intermediate form. Hum Mol Genet 1993; 2:23–30.

99. Kovalenko SA, Kopsidas JM, Kelso JM, Linnane AW. Deltoid human muscle mtDNA is extensively rearranged in old age subjects. Biochem Biophys Res Commun 1997; 232:147–152.

100. Shimizu S, Eguchi Y, Kamiike W, Matsuda H, Tsujimoto Y. Bcl-2 expression prevents activation of the ICE protease cascade. Oncogene 1996; 12:2251–2257.

101. Mitchell P. David Keilin's respiratory chain concept and its chemiosmotic consequences. Les Prix Nobel en 1978. Stockholm: The Nobel Foundation, 1979.

102. Bernardi P, Vassanelli S, Veronese P, et al. Modulation of the mitochondrial permeability transition pore: effect of protons and divalent cations. J Biol Chem 1992; 267:2934–2939.

103. Szabo I, Bernardi P, Zoratti M. Modulation of the mitochondrial megachannel by divalent cations and protons. J Biol Chem 1992; 267:2940–2946.

104. Ozawa T. Oxidative damage and fragmentation of mitochondrial DNA in cellular apoptosis. Biosci Rep 1997; 17:237–250.

105. Kerr JFRY, Wyllie AH, Currie AR. Apoptosis: a basic biological phenomenon with wide-ranging implications in tissue kinetics. Br J Cancer 1972; 26:239–257.

106. Katz AM. The cardiomyopathy of overload: an unnatural growth response [review]. Eur Heart J 1995; 16:110–114.

107. Sharov VG, Sabbah HN, Shimoyama H, et al. Evidence of cardiocyte apoptosis in myocardium of dogs with chronic heart failure. Am J Pathol 1996; 148:141–149.

108. Yao M, Keogh A, Spratt P, dos Remedios CG, Kiessling PC. Elevated DNase I

levels in human idiopathic cardiomyopathy: an indicator of apoptosis? J Mol Cell Cardiol 1996; 28:95–101.

109. James TN, St. Martin E, Willis PW, Lohr TO. Apoptosis as a possible cause of gradual development of complete heart block and fatal arrhythmias associated with absence of the AV node, sinus node, and internodal pathways. Circulation 1996; 93:1424–1438.

110. Simonian NA, Coyle JT. Oxidative stress in neurodegenerative diseases. [review]. Annu Rev Pharmacol Toxicol 1996; 36:83–106.

111. Wyllie AH. Glucocorticoid-induced thymocyte apoptosis is associated with endogenous endonuclease activation. Nature 1980; 284:555–556.

112. Nakajima H, Golstein P, Henkart PA. The target cell nucleus is not required for cell-mediated granzyme- or Fas-based cytotoxicity. J Exp Med 1995; 181:1905–1909.

113. Jacobson MD, Burne JF, Raff MC. Programmed cell death and Bcl-2 protection in the absence of a nucleus. EMBO J 1994; 13:1899–1910.

114. Schulze-Osthoff K, Walczak H, Droge W, Krammer PH. Cell nucleus and DNA pragmentation are not required for apoptosis. J Cell Biol 1994; 127:15–20.

115. Tsujimoto Y, Croce CM. Analysis of the structure, transcripts, and protein products of Bcl-2, the gene involved in human follicular lymphoma. Proc Natl Acad Sci USA 1986; 83:5214–5218.

116. Hengartner MO, Horvitz RH. The ins and outs of programmed cell death during C. elegans development. Phil Trans R Soc Lond B 1994; 345:243–246.

117. Hengartner MO, Horvitz RH. C. elegans cell survival gene ced-9 encodes a functional homolog of the mammalian protooncogene Bcl-2. Cell 1994; 76:665–676.

118. Yuan J-Y, Shaham S, Ledoux S, Ellis MH, Horvitz RH. The C. elegans cell death gene cod-3 encodes a protein similar to mammalian interleukin-1β converting enzyme. Cell 1993; 75:641–652.

119. Lazebnik YA, Kaufmann SH, Desnoyers S, Poirier GG, Eamshaw WC. Cleavage of poly(ADP-ribose) polymerase by a proteinase with properties like ICE. Nature 1994; 371:346–347.

120. Nicholson DW, Ali A, Thornberry NA, et al. Identification and inhibition of the ICE/CED-3 protease necessary for mammalian apoptosis [see comments]. Nature 1995; 376:37–43.

121. Schlegel J, Peters I, Onrenlus S, et al. CPP32/apopain is a key interleukin 1β converting enzyme-like protease involved in Fas-mediated apoptosis. J Biol Chem 1996; 271:1841–1844.

122. Wang X, Zelenski NG, Yang J, et al. Cleavage of sterol regulatory element binding proteins (SREBPs) by CPP32 during apoptosis. EMBO J 1996; 15:1012–1020.

123. Xue D, Horvitz HR. Inhibition of the Caenorhabditis elegans cell-death protease CED-3 by a CED-3 cleavage site in baculovirus p35 protein. Nature 1995; 377: 248–251.

124. Wang X, Pai J-T, Wiedenfeld EA, et al. Purification of an interleukin-1β converting enzyme-related cysteine protease that cleaves sterol regulatory element-binding proteins between the leucine zipper and transmembrane domains. J Biol Chem 1995; 270:18044–18050.

125. Tewari M, Quan L, O'Rourke K, et al. Yama/CPP32β, a mammalian homolog of ced-3, is a CrmA-inhibitable protease that cleaves the death substrate poly(ADP-ribose) polymerase. Cell 1995; 81:801–809.

126. Lazebnik YA, Cole S, Cooke CA, Nelson WG, Earnshaw WC. Nuclear events of apoptosis in vitro in cell-free mitotic extracts: a model system for analysis of the active phase of apoptosis. J Cell Biol 1993; 123:7–22.

127. Gaido ML, Cidlowski JA. Identification, purification, and characterization of a calcium-dependent endonuclease (NUC18) from apoptotic rat thymocytes. NUC18 is not histone H2B. J Biol Chem 1991; 266:18580–18585.

128. Peitsch MC, Polzar B, Stephan H, et al. Characterization of the endogenous deoxyribonuclease involved in nuclear DNA degradation during apoptosis (programmed cell death). EMBO J 1993; 12:371–377.

129. Reed DJ, Savage MK. Influence of metabolic inhibitors on mitochondrial permeability transition and glutathione status. Biochim Biophys Acta 1995; 1271:43–50.

130. Skulachev VP. Why are mitochondria involved in apoptosis?: permeability transition pores and apoptosis as selective mechanisms to eliminate superoxide-producing mitochondria and cell. FEBS Lett 1996; 397:7–10.

131. Thiers RE, Vallee BL. Distribution of metals in subcellular fractions of rat. J Biol Chem 1957; 226:911–920.

132. Dainzani MU, Viti I. The content and distribution of cytochrome c in the fatty liver of rats. Biochem J 1955; 59:141–145.

133. Mayer A, Neupert W, Lill R. Translocation of apocytochrome c across the outer membrane of mitochondria. J Biol Chem 1995; 270:12390–12397.

134. Gonzales DH, Neupert W. Biogenesis of mitochondrial c-type cytochromes. J Bioenerg Biomembr 1990; 22:753–768.

135. Beinert H. The extent of artificial redistribution of cytochrome c in rat liver homogenate. J Biol Chem 1951; 190:287–292.

136. Zoratti M, Szabo I. The mitochondrial permeability transition. Biochim Biophys Acta 1995; 1241:139–176.

137. De Maria R, Lenti L, Malisan F, et al. Requirement for GD3 ganglioside in CD95- and ceramide-induced apoptosis. Science 1997; 277:1652–1655.

138. Boise LH, Gonzalez-Garcia M, Postema CE, et al. Bcl-x, a Bcl-2-related gene that functions as a dominant regulator of apoptotic cell death. Cell 1993; 74:597–608.

139. Korsmeyer SJ, Yin X-M, Oltvai ZN, Veis-Novack DJ, Linette GP. Reactive oxygen species and the regulation of cell death by the Bcl-2 gene family. Biochim Biophys Acta 1995; 1271:63–66.

140. Offen D, Ziv I, Gorodin S, et al. Dopamine-induced programmed cell death in mouse thymocytes. Biochim Biophys Acta 1995; 1268:171–177.

141. Kane D, Sarafian TA, Anton R, et al. Bcl-2 inhibition of neural death: decreased generation of reactive oxygen species. Science 1993; 262:1274–1277.

142. Hockenberg DM, Oltvai ZN, Yin X-M, Milliman CL, Korsmeyer SJ. Bcl-2 functions in an antioxidant pathway to prevent apoptosis. Cell 1993; 75:241–251.

143. Zamzami N, Marchetti P, Castedo M, et al. Sequential reduction of mitochondrial transmembrane potential and generation of reactive oxygen species in early programmed cell death. J Exp Med 1995; 182:367–377.

144. Albina JE, Martin BA, Henry W Jr, Louis CA, Reichner JS. B cell lymphoma-2

transfected P815 cells resist reactive nitrogen intermediate-mediated macrophage-dependent cytotoxicity. J Immunol 1996; 157:279–283.

145. Messmer UK, Reed UK, Brune B. Bcl-2 protects macrophages from nitric oxide-induced apoptosis. J Biol Chem 1996; 271:20192–20197.

146. Shimizu S, Eguchi Y, Kosaka H, et al. Prevention of hypoxia-induced cell death by Bcl-2 and Bcl-xL. Nature 1995; 374:811–813.

147. Jacobson MD, Raff MC. Programmed cell death and Bcl-2 protection in very low oxygen. Nature 1995; 374:814–816.

148. Shimizu S, Eguchi Y, Kamiike W, et al. Bcl-2 blocks loss of mitochondrial membrane potential while ICE inhibitors act at a different step during inhibition of death induced by respiratory chain inhibitors. Oncogene 1996; 13:21–29.

149. Shimizu S, Eguchi Y, Kamiike W, et al. Retardation of chemical hypoxia-induced necrotic cell death by Bcl-2 and ICE inhibitors: possible involvement of common mediators in apoptotic and necrotic signal transductions. Oncogene 1996; 12:2045–2050.

150. Jones DE. Mitochondrial dysfunction during anoxia and acute cell injury. Biochim Biophys Acta 1995; 1271:29–33.

151. Meister A. Mitochondrial changes associated with glutathione deficiency. Biochim Biophys Acta 1995; 1271:35–42.

152. Ozawa T. Mitochondrial cardiomyopathy [review]. Herz 1994; 19:105–118.

153. Sakai T, Ishizaki T, Nakai T, et al. Role of nitric oxide and superoxide anion in leukotoxin, 9, 10-epoxy-1,2-octadecenoate-induced mitochondrial dysfunction. Free Radic Biol Med 1996; 20:607–612.

154. Petit PX, Lecoeur H, Zorn E, et al. Alterations in mitochondrial structure and function are early events of dexamethasone-induced thymocyte apoptosis. J Cell Biol 1995; 130:157–167.

155. Lazebnik YA, Takahashi A, Moir RD, et al. Studies of the lamin proteinase reveal multiple parallel biochemical pathways during apoptotic execution. Proc Natl Acad Sci USA 1995; 92:9042–9046.

15

Mitochondria in Cellular Aging and Degeneration

Gino A. Cortopassi and Alice Wong
University of California, Davis, California

I. INTRODUCTION

Several lines of experimentation support the view that the genetic, biochemical, and bioenergetic functions of somatic mitochondria deteriorate during normal aging. Deletion mutations of the mitochondrial genome accumulate exponentially with age in nerve and muscle tissue of humans and several other species. In muscle, a tissue that undergoes age-related atrophy in humans, there is an exponential rise in the number of cytochrome-oxidase-deficient fibers which is first detectable in the fourth decile of age. Most biochemical studies of animal mitochondrial activity indicate a decline in electron transport activity with age, as well as decreased bioenergetic capacity with age, as measured by mitochondrial membrane potential. Mitochondrial mutations may be both the result of mitochondrial oxidative stress and cells bearing pure populations of pathogenic mitochondrial mutations are sensitized to oxidant stress. Oxidant stress to mitochondria is known to induce the mitochondrial permeability transition, which has recently been implicated in the release of cytochrome c and the initiation of apoptosis. Thus several lines of evidence support a contribution of mitochondrial dysfunction to the phenotypic changes associated with aging.

II. MULTIFACTORIAL DEFICITS OF FUNCTION

The multifactorial deficits of function that occur during human aging support the view that aging is the result of multiple underlying causes. Aging can be understood in an evolutionary context, in which natural selection strongly favors genetic variation that leads to survival to the age of reproduction (1), whereas positive or negative natural selection for traits that appear long after the age of reproduction is weak or neutral (2–7). Other support for the neutrality of genetic variation that underlies aging phenotypes comes from the consideration that for most of the history of modern humans [i.e., since about 140,000 years ago (8,9)] the mean human life span has been much shorter than it currently is, providing a relatively small pool of elderly individuals on which selection could act (Fig. 1). Since selection is likely to be effectively neutral on genetic polymorphisms that promote or attenuate survival long after the age of reproduction, the human gene pool is likely to contain substantial genetic variation that controls the rate of cellular maintenance (10). Genes that provide for cellular maintenance are likely to include those that protect molecules from damage, including genes for DNA repair and protection from oxidant stress. DNA repair of the nuclear and mitochondrial genomes, for example, is quite a costly process in terms of metabolic energy spent. Thus one might expect the "set point" of DNA repair processes to provide that genomes should maintain integrity until the age of reproduction, but that natural selection for maintenance of genomes long after the age

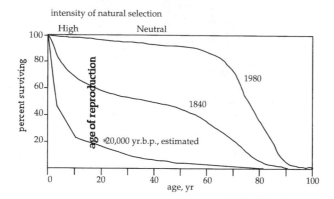

Figure 1 Ideogram of the historical age structure of the human population and the intensity of natural selection on genotypes which affect aging phenotypes. U.S. population census data from 1980 and 1840 are redrawn from Ref. 97. The age structure of preagricultural human hunter-gatherers is a matter of debate, but infant mortality (a major influence on mean life span) is likely to be at least twice as high as in 1840, and adult mortality is likely to be significantly increased as well.

of reproduction should be weak. Thus a corollary prediction of the evolutionary theory of aging is that the somatic mutations and other defects resulting from incomplete cellular maintenance should rise with age.

III. NUCLEAR SOMATIC MUTATIONS

There is substantial evidence that the frequency of nuclear somatic mutations rises with age in nononcogenes using phenotypic selection assays of mutagenesis (11). Using a PCR-based method of genotypic selection, it has been demonstrated that somatic mutations also rise in known oncogenes both with age and with exposure to carcinogens (12–14). In animal models, a twofold rise in nuclear somatic mutations with age has been reported (15).

IV. CONTRIBUTION OF MITOCHONDRIAL DEFICITS TO AGING

A contribution of mitochondrial deficits to age-related decline in cellular function of humans and other animals has been posited (16–23). Support for a decline in mitochondrial function with age has come from genetic assays of mitochondrial mutagenesis and biochemical assays of mitochondrial function.

V. AGE-RELATED RISE IN mtDNA DELETION MUTAGENESIS

The mitochondrial genomes of humans (20,24–28), monkeys (29), rats (30,31), mice (32–34), and nematodes (35) accumulate deletion and/or rearrangement mutations with age (Fig. 2). Some of the deletion mutations that accumulate in humans with age are identical to those that occur in human mitochondrial genetic disease. In humans, mtDNA mutations accumulate preferentially in postmitotic tissues that are oxidatively active (24); and aged brains cells of the substantia nigra accumulate the highest levels of mitochondrial mutation (36). It is possible that the higher level of deletion mutagenesis in these tissues is the result of increased mitochondrial oxidative stress, however, this has never been directly demonstrated. These deletion mutations appear to be identical to those observed in the mitochondrial disease Kearns–Sayre syndrome (37–40), and appear to rise exponentially with age (20,24,41–44).

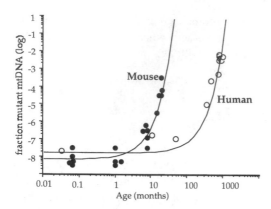

Figure 2 Exponential accumulation of mtDNA deletion mutations in mice and humans. Data redrawn from Ref. 48.

VI. ABSOLUTE FREQUENCY AND DISTRIBUTION OF AGE-RELATED mtDNA MUTATIONS

The highest mean concentration of any particular deletion mutation such as the 4977 deletion may be maximally about 1% in any particular tissue homogenate, and if the 4977 deletion were evenly distributed and the only mtDNA deletion to accumulate with age, it is not clear if this would have a major physiological effect. However, it appears quite unlikely that mtDNA mutations are evenly distributed; data are more consistent with an uneven distribution of mtDNA deletion mutations (45–47), as one might expect from a somatic mutational process. Further support for uneven distribution of age-related mtDNA mutagenesis comes from a comparison of apparent mtDNA mutation frequencies in DNA prepared from cells versus tissue homogenates using a quantitative assay (48) in which mutations were assayed from DNA prepared in two ways from the same-aged mouse hearts. From one heart half the DNA was prepared from homogenized tissue by the standard phenol/chloroform extraction method. From its sister half, cardiocyte cells were first disaggregated into pools of differing sizes, pools of 300 to 300,000 cardiocytes. The apparent deletion mutagenesis frequency is approximately 100-fold higher in the identical amount of DNA made from homogenates compared with cardiocytes, supporting the notion that mtDNA deletion mutations occur in groups or clusters (Fig. 3).

VII. POINT MUTATIONS OF mtDNA AND AGING

Other deletion and/or rearrangement mutations in addition to the 4977 deletion mutation have been reported to increase with age (reviewed in Ref. 49). Also

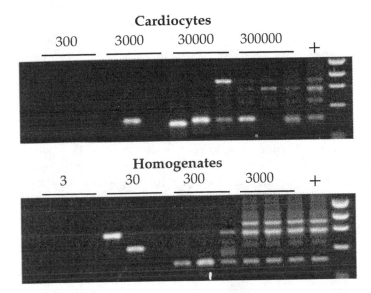

Figure 3 Differential apparent concentration of mtDNA deletions in homogenates versus cardiocytes is consistent with clustering of mtDNA mutations (E. Wang and G. A. Cortopassi, unpublished data). Upper panel: PCR of DNA extracted from cardiocytes and amplified by method described in Ref. 48; lower panel: PCR of DNA extracted from heart homogenates. The estimated number of cells are indicated on the top of each gel. +, positive control; M, size marker.

a new method for total mtDNA amplification has been developed that illustrates the rise of several mtDNA deletions with age (23); however, the assay is not quantitative and thus could overestimate or underestimate the total fraction of mutant mtDNA genomes. Point mutations identical to pathogenic mtDNA mutations have been reported to increase with age (50–53), however, using a different mutation assay for a different point mution this increase was not observed (54).

VIII. BIOCHEMICAL ANALYSES OF MITOCHONDRIAL FUNCTION WITH AGING

Three major types of analysis of the contribution of mitochondrial biochemical defects with age have been carried out, including single enzyme analysis in aging muscle fibers, analysis of the complete electron transport chain or individual complexes, and analysis of membrane potential with age.

A rise in cytochrome-oxidase-negative muscle fibers in humans over the age of 40 years has been reported by multiple groups (55–58); the data of Byrne and Dennett (56) are redrawn Figure 4. These data support an exponential rise in cytochrome-oxidase-negative fibers, which are not observable in individuals under age 40, and which exponentially increase in incidence after age 40. Although it is known that muscle atrophy and dropout of muscle fibers does occur with age, it is not clear if this is the result of loss of cytochrome oxidase activity.

The second type of analysis of mitochondrial function with respect to age is the analysis of enzymatic activity of the complete mitochondrial respiratory chain and/or its four major electron transport units, or the rate of ADP-stimulated molecular oxygen consumption (i.e., state 3 respiration). In this category the majority of studies report an age-related decline in electron transport chain activity in mice, rats, and humans (59–65). However, two groups have observed no decline in electron transport chain activity in rodents (66–68).

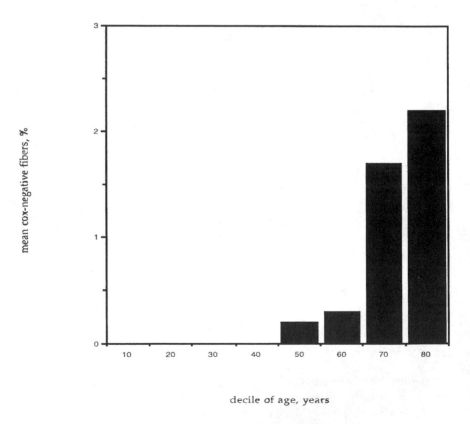

Figure 4   The incidence of cytochrome-c-oxidase-negative muscle fibers increases exponentially after the fourth decile of age. Data redrawn from Refs. 19 and 55.

The third type of analysis is to measure the mitochondrial membrane potential, the driving force for ATP synthesis, in young and old animals. Measurements of the mitochondrial membrane potential in young and old animals have supported the view that there is decreased mitochondrial membrane potential in mitochondria from aged species (69–71).

IX. OXIDANT STRESS AND MITOCHONDRIA

Because approximately 90% of cellular oxygen is consumed in mitochondria, and because approximately 3% of molecular oxygen is thought to escape complete reduction to water, mitochondria are considered to be the major intracellular contributor to superoxide (O_2^-) generation, and perhaps to oxidant stress in general. The potential involvement of oxidative stress in aging has been suggested and reviewed elsewhere (16,72; see Sohal, this volume). It has been demonstrated that there is an increased production of mitochondrial O_2^- and H_2O_2 from aged animals (72,73 and references therein). The results of knockout experiments in mice confirm the importance of protection from mitochondrially localized O_2^- in that knockouts of the mitochondrial superoxide dismutase (SOD) are lethal (74,75).

Because the 13 polypeptides encoded by the mtDNA are each involved in respiration, the most likely effect of age-related mtDNA mutagenesis is a decrease or dysfunction of the respiratory chain. Inhibition of the electron transport chain late in the chain are expected to produce a more reduced electron transport chain. Since the majority of O_2^- is thought to be generated by reaction of ubisemiquinone radical with molecular oxygen, it would be predicted that inhibition of the electron transport chain should lead to increased O_2^- generation. We observe, like others, that the MERRF mutation results in decreased oxygen consumption (76,77). Using the reduction of cytochrome c to measure the production of O_2^- from submitochondrial particles, we have measured the production of O_2^- from submitochondrial particles prepared from mitochondria bearing the MERRF mutation versus the normal controls in the identical nuclear background. We also observe that the MERRF mutation results in about an 11% increase in mean mitochondrial O_2^- production (Hutchin, Attardi, and Cortopassi, unpublished results) (Fig. 5). Whether this 11% increase in O_2^- production is relevant for MERRF pathophysiology is currently unknown.

X. PURE PATHOGENIC MERRF, MELAS, AND LHON mtDNA MUTATIONS CONFER SENSITIVITY TO OXIDANT STRESS

Although it is widely assumed that the phenotypic effects of inheritance of pathogenic mtDNA mutations result from deficits in bioenergetic functions, some alter-

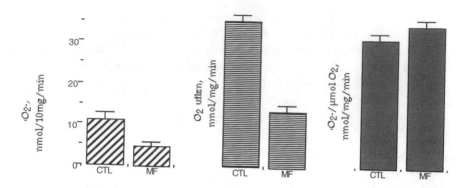

Figure 5 Lower O_2 consumption and higher O_2^- production from mutant MERRF than control mitochondria (T. P. Hutchin, G. Attardi, and G. A. Cortopassi, unpublished data.)

native possibilities exist. For example, in MERRF it has been observed that resting (ATP) levels are not any lower than in controls, but that cytoplasmic levels of CA^{2+} are higher in mutant versus control cells (78). Thus alternative models for the pathophysiological mechanism of mitochondrial genetic diseases might be entertained, including one in which mitochondrial deficits sensitized cells to oxidant and/or Ca^{2+} stress. Thus as a potential model of the mechanism of pathogenesis in germ-line mtDNA disease, and also potentially in aging of the mtDNA, we have tested the sensitivity of cells bearing the pathogenic mitochondrial mutations LHON, MERRF, and MELAS to an oxidative stress, exposure to H_2O_2 (79). The MERRF, MELAS, and LHON mutations each conferred significant sensitivity to oxidant stress (Fig. 6), which was rescued by depletion of extracellular Ca^{2+}, by depletion of Ca^{2+} from the tissue culture medium, and by depletion of intracellular Ca^{2+} by treatment with the intracellular Ca^{2+} chelator BAPTA-AM. Thus it is possible that oxidant stress results in an increase in intracellular Ca^{2+} that is required for toxicity. A potential downstream target of oxidant stress is the mitochondrial permeability transition (see below), which is sensitive to Ca^{2+} and reactive oxygen species and is inhibited by CsA (80). It was observed that CsA protected MERRF, MELAS, and LHON fibroblasts from oxidant stress, and specifically more than control cells.

XI. IMPLICATING MITOCHONDRIA IN CELL DEATH

Multiple lines of evidence have implicated mitochondria in the control of cell death, including the localization of the antiapoptotic protein Bcl-2 to the outer mitochondrial membrane (81), and the requirement for mitochondria (82) and

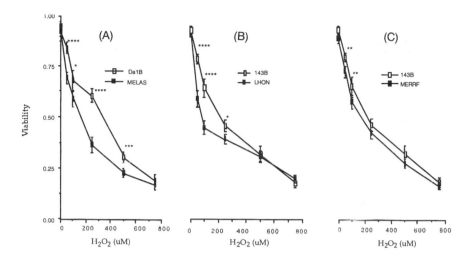

Figure 6 Pathogenic mitochondrial mutations confer cellular sensitivity to oxidant stress. Redrawn from Ref. 79.

specifically mitochondrial cytochrome c in cell free systems of apoptosis (83), the early loss of mitochondrial membrane potential in some forms of cell death (84,85), the regulation of mitochondrial Ca^{2+} by Bcl-2 (86), and the inhibition of release of cytochrome c by Bcl-2 (87,88). Currently the mechanism by which cytochrome c is released from mitochondria and by which Bcl-2 potentially inhibits the release of cytochrome c from mitochondria are of significant interest.

XII. MITOCHONDRIAL PERMEABILITY TRANSITION ASSOCIATED WITH CELLULAR DYSFUNCTION AND DEATH

The mitochondrial permeability transition (MPT) is an abrupt rise in mitochondrial permeability that is induced by Ca^{2+}, reactive oxygen species, and several other agents (80,89,90). Cyclosporin A inhibits MPT, presumably by inhibiting the peptidyl-prolyl-isomerase of the mitochondrial cyclophilin to which it binds (91). Induction of the MPT has recently been associated with cell death (85,92). Cytochrome c has recently been demonstrated to be required for induction of the biochemical hallmarks of apoptosis in some cell-free systems (83,87,93). One potential issue is by what mechanism is cytochrome c released from mitochondria to initiate apoptosis. Given the known sensitivity of the MPT to several known

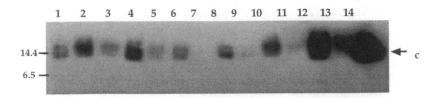

Figure 7 Induction of the mitochondrial permeability transition causes release of apoptogenic cytochrome c, redrawn from Ref. 93. Cytochrome c is detected in supernatants from mitochondria induced for MPT by western blot. Lane 1, control (uninduced). Lanes 2–12 are six pairs; the first sample of each pair is inducer alone, the second sample contains identical concentration of inducer plus 1 mM cyclosporin A. The six lanes and inducers were 2–3, 100 mM Ca^{2+}; 4–5, 5 mM atractyloside; 6–7, 100 mM tBOOH; 8–9, 5 mM MPP^+; 10–11, 5 mM 1 mM H_2O_2; 12–13, 0.5 mM dATP. Lane 14, 0.3 mg of cytochrome c.

toxins, a straightforward question is to ask whether the induction of the MPT results in the release of cytochrome c. We (93) and others (94) have observed that the induction of the mitochondrial permeability transition is associated with the release of cytochrome c. In Figure 7, several known inducers of the MPT in purified mitochondria were demonstrated to induce the release of cytochrome c by western blot. The inhibition of the MPT by cyclosporin A strongly inhibited the release of cytochrome c. However, it is likely that the bulk of this release is through mitochondrial bursting (95,96).

XIII. INCREASED MPT INDUCIBILITY IN AGED MOUSE HEPATOCYTES

Given the association of the MPT with cellular pathophysiology, and the association between deficits of mitochondrial dysfunction with age, one potential issue is whether the MPT is more inducible in aging, which might provide a biochemical and molecular underpinning for cellular-age-related susceptibility, that is, frailty. In a recent comparison of MPT in young (1 month) versus old (36 month) liver mitochondria, lower state 3 respiration, higher state 4 respiration, and more inducible MPT were observed in the old mitochondria (72; Fig. 8), which provides the first experimental support for the concept that MPT may be more inducible in aged animals. Further research will demonstrate whether or not these observations are general in other strains and species. In Figure 8, the inducibility of MPT was measured as the difference in the rate of induction of swelling, which is observed as an increased light scattering of the swollen mitochondria. MPT was induced faster in old mitochondria than young mitochondria on aver-

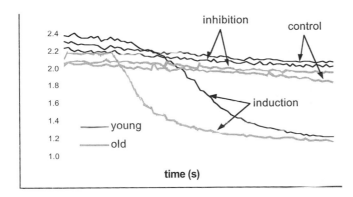

Figure 8 Sample spectrophotometric recording showing the design used for a comparison of rate of mitochondrial permeability transition in young versus old mice, redrawn from Ref. 72. Numbers on y-axis are absorbance, the induction of the mitochondrial permeability transition results in increased light scattering. Induction mitochondria received Ca^{2+} stimulus, control mitochondria received no Ca^{2+} stimulus, and inhibition samples received Ca^{2+} stimulus and cyclosporin A, a specific inhibitor of the mitochondrial permeability transition.

age, and the difference in MPT was dependent on Ca^{2+} (a known inducer), and inhibitable by cyclosporin A, a known MPT inhibitor.

XIV. FUTURE PROSPECTS

Several lines of evidence indicate that there is a decline in mitochondrial function with age at the genetic, biochemical, and bioenergetic level. However, to what extent these deficits of mitochondrial function underlie phenotypic changes that occur with aging is still an area of very active research. Although it is clear that mitochondrial deletion mutations increase exponentially with human age, and there is substantial support for an association between cox-negative fibers and mtDNA deletions, it is also clear that there is a substantial number, perhaps 60% of cox-negative fibers in muscle that do not have detectable mtDNA deletions. Although possible, it is not yet clear whether mitochondrial and or mtDNA damage is a cause of muscle atrophy, or atrophy and death of other cell types with age. Analytical techniques which globally and quantitatively assay mtDNA mutagenesis in a quantitative way are needed to resolve this issue, as well as the ability to assay such damage in situ in aging human tissue.

REFERENCES

1. Medawar PB, ed. The Uniqueness of the Individual. London: Methuen, 1957:44–70.
2. Kirkwood TB. Evolution of ageing. Nature 1977; 270:301–304.
3. Kirkwood TB, Cremer T. Cytogerontology since 1881: a reappraisal of August Weismann and a review of modern progress. Hum Genet 1982; 2:101–121.
4. Rose MR, Finch CE. The Janiform genetics of aging. Genetica 1993; 91:3–10.
5. Holliday R. Chance and longevity. Bioessays 1995; 17:465–467.
6. Cortopassi G, Liu Y. Genotypic selection of mitochondrial and oncogenic mutations in human tissue suggests mechanisms of age-related pathophysiology. Mutat Res 1995; 338:151–159.
7. Cortopassi GA, Wang E. There is substantial agreement among interspecies estimates of DNA repair activity. Mech Ageing Dev 1996; 91:211–218.
8. Cann RL, Stoneking M, Wilson AC. Mitochondrial DNA and human evolution. Nature 1987; 325:31–36.
9. Wallace DC. Mitochondrial DNA mutations in diseases of energy metabolism. J Bioenerg Biomembr 1994; 26:241–250.
10. Martin GM, Austad SN, Johnson TE. Genetic analysis of ageing: role of oxidative damage and environmental stresses. Nat Genet 1996; 13:25–34.
11. Morley AA. The somatic mutation theory of ageing. Mutat Res 1995; 338:19–23.
12. Liu Y, Hernandez AM, Shibata D, Cortopassi GA. Bcl-2 translocation frequency rises with age in humans. Proc Natl Acad Sci USA 1994; 91:8910–8914.
13. Nakazawa H, English D, Randell PL et al. UV and skin cancer: specific p53 gene mutation in normal skin as a biologically relevant exposure measurement. Proc Natl Acad Sci USA 1994; 91:360–364.
14. Brennan JA, Boyle JO, Koch WM et al. Association between cigarette smoking and mutation of the p53 gene in squamous-cell carcinoma of the head and neck. N Engl J Med 1995; 332:712–717.
15. Boerrigter ME, Vijg J. Sources of variability in mutant frequency determinations in different organs of lacZ plasmid-based transgenic mice: experimental features and statistical analysis. Environ Mol Mutagen 1997; 29:221–229.
16. Harman D. The biologic clock: the mitochondria? J Am Geriatr Soc 1972; 20:145–147.
17. Fleming JE, Miquel J, Bensch KG. Age dependent changes in mitochondria. Basic Life Sci 1985; 35:143–156.
18. Linnane AW, Marzuki S, Ozawa T, Tanaka M. Mitochondrial DNA mutations as an important contributor to ageing and degenerative diseases. Lancet 1989; 1:642–645.
19. Byrne E, Trounce I, Dennett X. Mitochondrial theory of senescence: respiratory chain protein studies in human skeletal muscle. Mech Ageing Dev 1991; 60:295–302.
20. Cortopassi GA, Arnheim N. Detection of a specific mitochondrial DNA deletion in tissues of older humans. Nucl Acid Res 1990; 18:6927–6933.
21. Corral-Debrinski M, Horton T, Lott MT, et al. Mitochondrial DNA deletions in

human brain: regional variability and increase with advanced age. Nat Genet 1992; 2:324–329.

22. Wallace DC. Mitochondrial genetics: a paradigm for aging and degencrative diseases? Science 1992; 256:628–632.

23. Melov S, Hinerfeld D, Esposito L, Wallace DC. Multi-organ characterization of mitochondrial genomic rearrangements in ad libitum and caloric restricted mice show striking somatic mitochondrial DNA rearrangements with age. Nucl Acid Res 1997; 25:974–982.

24. Cortopassi GA, Shibata D, Soong N-W, Arnheim N. A pattern of accumulation of a somatic deletion of mitochondrial DNA in aging human tissues. Proc Natl Acad Sci USA 1992; 89:7370–7374.

25. Ikebe S, Tanaka M, Ohno K, et al. Increase of deleted mitchndrial DNA in the striatum in Parkinson's disease and senescence. Biochem Biophys Res Commun 1990; 170:1044–1048.

26. Linnane AW, Baumer A, Maxwell RJ, et al. Mitochondrial gene mutation: the ageing process and degencrative diseases. Biochem Int 1990; 22:1067–1076.

27. Corral-Debrinski M, Horton T, Lott MT, et al. Mitochondrial DNA deletions in human brain: regional variability and increase with advanced age. Nat Genet 1992; 2:324–329.

28. Simonetti S, Chen X, DiMauro S, Schon EA. Accumulation of deletions in human mitochondrial DNA during normal aging: analysis by quantitative PCR. Biochim Biophys Acta 1992; 1180:113–122.

29. Lee CM, Chung SS, Kaczkowski JM, Weindruch R, Aiken JM. Multiple mitochondrial DNA deletions associated with age in skeletal muscle of rhesus monkeys. J Gerontol 1993; 48:B201–B205.

30. Gadaleta MN, Petruzzella V, Daddabbo L, et al. Mitochondrial DNA transcription and translation in aged rat. Effect of acetyl-L-carnitine. Ann NY Acad Sci 1994; 717:150–160.

31. Edris W, Burgett B, Stine OC, Filburn CR. Detection and quantitation by competitive PCR of an age-associated increase in a 4.8-kb deletion in rat mitochondrial DNA. Mutat Res 1994; 316:69–78.

32. Brossas J-Y, Barreau E, Courtois Y, Treton J. Multiple deletions in mitochondrial DNA are present in senescent mouse brain. Biochem Biophys Res Commun 1994; 202:654–659.

33. Chung SS, Weindruch R, Schwarze SR, Mckenzie DI, Aiken JM. Multiple age-associated mitochondrial DNA deletions in skeletal muscle of mice. Aging 1994; 6:193–199.

34. Tanhauser SM, Laipis PJ. Multiple deletions are detectable in mitochondrial DNA of aging mice. J Biol Chem 1995; 270:24769–24775.

35. Melov S, Hertz GZ, Stormo GD, Johnson TE. Detection of deletions in the mitochondrial genome of *Caenorhabditis elegans*. Nucl Acid Res 1994; 22:1075–1078.

36. Soong N-W, Hinton DR, Cortopassi G, Arnheim N. Mosaicism for a specific somatic mitochondrial DNA mutation in adult human brain. Nat Genet 1992; 2:318–323.

37. Moraes CT, Dimauro S, Zeviani M, et al. Mitochondrial DNA deletions in progressive external ophthalmoplegia and Kearns-Sayre syndrome. N Engl J Med 1989; 320:1293–1299.

38. Holt IJ, Harding AE, Morgan-Hughes JA, Deletions of muscle mitochondrial DNA in patients with mitochondrial myopathies. Nature 1988; 331:717–719.

39. Mita S, Schmidt B, Schon EA, DiMauro S, Bonilla E. Detection of "deleted" mitochondrial genomes in cytochrome-c oxidase-deficient muscle fibers of a patient with Kearns-Sayre syndrome. Proc Natl Acad Sci USA 1989; 86:9509–9513.

40. Shoffner JM, Lott MT, Voljavec AS, et al. Spontaneous Kearns-Sayre/chronic external ophthalmoplegia plus syndrome associated with a mitochondrial DNA deletion: a slip-replication model and metabolic therapy. Proc Natl Acad Sci USA 1989; 86: 7952–7956.

41. Corral-Debrinski M, Stepien G, Shoffner JM, et al. Hypoxemia is associated with mitochondrial DNA damage and gene induction. Implications for cardiac disease. JAMA 1991; 266:1812–1816.

42. Hattori K, Tanaka M, Sugiyama S, et al. Age-dependent increase in deleted mitochondrial DNA in the human heart: possible contributory factor to presbycardia. Am Heart J 1991; 121:1735–1742.

43. Kao SH, Liu CS, Wang SY, Wei YH. Ageing-associated large-scale deletions of mitochondrial DNA in human hair follicles. Biochem Mol Biol Int 1997; 42:285–298.

44. Wei YW, Kao SH, Lee HC. Simultaneous increase of mitochondrial DNA deletions and lipid peroxidation in human aging. Ann NY Acad Sci 1996; 786:24–43.

45. Kadenbach B, Munscher C, Frank V, Muller-Hocker J, Napiwotzki J. Human aging is associated with stochastic somatic mutations of mitochondrial DNA. Mutat Res 1995; 338:161–172.

46. Brierly EJ, Johnson MA, Bowman A, et al. Mitochondrial function in muscle from elderly athletes. Ann Neurol 1997; 41:114–116.

47. Zhang C, Liu VW, Nagley P. Gross mosaic pattern of mitochondrial DNA deletions in skeletal muscle tissues of an individual adult human subject. Biochem Biophys Res Commun 1997; 233:56–60.

48. Wang E, Wong A, Cortopassi G. The rate of mitochondrial mutagenesis is faster in mice than humans. Mutat Res 1997; 377:157–166.

49. Lee CM, Weindruch R, Aiken JM. Age-associated alterations of the mitochondrial genome. Free Radic Biol Med 1997; 22:1259–1269.

50. Liu VW, Zhang C, Linnane AW, Nagley P. Quantitative allele-specific PCR: demonstration of age-associated accumulation in human tissues of the A → G mutation at nucleotide 3243 in mitochondrial DNA. Hum Mutat 1997; 9:265–271.

51. Zhang C, Linnane AW, Nagley P. Occurrence of a particular base substitution (3243 A to G) in mitochondrial DNA of tissues of ageing humans. Biochem Biophys Res Commun 1993; 195:1104–1110.

52. Munscher C, Muller-Hocker J, Kadenbach B. Human aging is associated with various point mutations in tRNA genes of mitochondrial DNA. Biol Chem Hoppe Seyler 1993; 374:1099–1104.

53. Munscher C, Rieger T, Muller-Hocker J, Kadenbach B. The point mutation of mitochondrial DNA characteristic for MERRF disease is found also in healthy people of different ages. FEBS Lett 1993; 31:27–30.

54. Pallotti F, Chen X, Bonilla E, Schon EA, Evidence that specific mtDNA point muta-

tions may not accumulate in skeletal muscle during normal human aging. Am J Hum Genet 1996; 59:591–602.

55. Byrne E, Dennett X. Respiratory chain failure in adult muscle fibres: relationship with ageing and possible implications for the neuronal pool. Mutat Res 1992; 275: 125–131.

56. Muller-Hocken J. Cytochrome-c-oxidase deficient cardiomyocytes in the human heart—an age-related phenomenon. A histochemical ultracytochemical study. Am J Pathol 1989; 134:1167–1173.

57. Muller-Hocker J, Seibel P, Schneiderbanger K, Kadenbach B. Different in situ hybridization patterns of mitochondrial DNA in cytochrome c oxidase-deficient extraocular muscle fibres in the elderly. Virchows Arch 1993; 422:7–15.

58. Brierley EJ, Johnson MA, James OF, Turnbull DM. Effects of physical activity and age on mitochondrial function. Q J Med 1996; 89:251–258.

59. Weinbach EC, Garbus J. Age and oxidative phosphorylation in rat liver and brain. Nature 1956; 178: 1225–1226.

60. Weinbach EC, Garbus J. Oxidative phosphorylation in mitochondria from aged rats. J Biol Chem 1959; 234:412–417.

61. Weindruch RH, Cheung MK, Varity MA, Walford RL. Modification of mitochondrial respiration by aging and dietary restriction. Mech Ageing Dev 1980; 12:375–392.

62. Trounce I, Byrne E, Marzuki S. Decline in skeletal muscle mitochondrial respiratory chain function: possible factor in ageing. Lancet 1989; 1:637–639.

63. Papa S, Scacco S, Schliebs M, Trappe J, Seibel P. Mitochondrial diseases and aging. Mol Aspects Med 1996; 17:513–563.

64. Lezza AM, Boffoli D, Scacco S, Cantatore P, Gadaleta NM. Correlation between mitochondrial DNA 4977-bp deletion and respiratory chain enzyme activities in aging human skeletal muscles. Biochem Biophys Res Commun 1994; 205:772–779.

65. Hsieh RH, Hou JH, Hsu HS, Wei YH. Age-dependent respiratory function decline and DNA deletions in human muscle mitochondria. Biochem Mol Biol Int 1994; 32:1009–1022.

66. Filburn CR, Edris W, Tamatani N, et al. Mitochondrial electron transport chain activities and DNA deletions in regions of the rat brain. Mech Ageing Dev 1996; 87: 35–46.

67. Gold PH, Gee MV, Strehler BL. Effect of age on oxidative phosphorylation in the rat. J Gerontol 1968; 23:509–512.

68. Barrientos A, Casademont J, Rotig A, et al. Absence of relationship between the level of electron transport chain activities and aging in human skeletal muscle. Biochem Biophys Res Commun 1996; 229:536–539.

69. Linnane AW, Degli Esposti M, Generowicz M, Luff AR, Nagley P. The universality of bioenergetic disease and amelioration with redox therapy. Biochim Biophys Acta 1995; 1271:191–194.

70. Hagen TM, Yowe DL, Bartholomew JC, et al. Mitochondrial decay in hepatocytes from old rats: membrane potential declines, heterogeneity and oxidants increase. Proc Natl Acad Sci USA 1997; 94:3064–3069.

71. Pieri C, Recchioni R, Moroni F. Age-dependent modifications of mitochondrial

transmembrane potential and mass in rat splenic lymphocytes during proliferation. Mech Ageing Dev 1993; 70:201–212.

72. Goodell SC, Cortopassi GA. Analysis of oxygen consumption and mitochondrial permeability with age in mice. Mech Ageing Dev 1998; 101:245–256.

73. Sohal RS, Weindruck R. Oxidative stress, caloric restriction, and aging. Science 1996; 273:59–63.

74. Li Y, Huang T-T, Carlson EJ, et al. Dilated cardiomyopathy and neonatal lethality in mutant mice lacking manganese superoxide dismutase. Nat Genet 1995; 11:376–381.

75. Lebovitz RM, Zhang H, Vogel H, et al. Neurodegeneration, myocardial injury, and perinatal death in mitochondrial superoxide dismutase-deficient mice. Proc Natl Acad Sci USA 1996; 93:9782–9787.

76. Wallace DC, Zheng XX, Lott MT, et al. Familial mitochondrial encephalomyopathy (MERRF): genetic, pathophysiological, and biochemical characterization of a mitochondrial DNA disease. Cell 1988; 55:601–610.

77. Larsson NG, Tulinius MH, Holme E, et al. Segregation and manifestations of the mtDNA tRNA(Lys) A→G(8344) mutation of myoclonus epilepsy and ragged-red fibers (MERRF) syndrome. Am J Hum Genet 1992; 51:1202–1212.

78. Moudy AM, Handra SD, Goldberg MP, et al. Abnormal calcium homeostasis and mitochondrial polarization in a human encephalomyopathy. Proc Natl Acad Sci USA 1995; 92:729–733.

79. Wong A, Cortopassi G. mtDNA mutations confer cellular sensitivity to oxidant stress that is partially rescued by calcium depletion and cyclosporin A. Biochem Biophys Res Commun 1997; 239:139–145.

80. Bernardi P. Modulation of the mitochondrial cyclosporin A-sensitive permeability transition pore by the proton electrochemical gradient. Evidence that the pore can be opened by membrane depolarization. J Biol Chem 1992; 267:8334–8339.

81. Hockenbery D, Nunez G, Milliman C, Schreiber RD, Korsmeyer SJ. Bcl-2 is an inner mitochondrial membrane protein that blocks programmed cell death. Nature 1990; 348:334–336.

82. Newmeyer DD, Farschon DM, Reed JC. Cell-free apoptosis in *Xenopus* egg extracts: inhibition by Bcl-2 and requirement for an organelle fraction enriched in mitochondria. Cell 1994; 79:353–364.

83. Liu X, Kim CN, Yang J, Jemmerson R, Wang X. Induction of apoptotic program in cell-free extracts: requirement for dATP and cytochrome c. Cell 1997; 86:147–157.

84. Zamzami N, Marchetti P, Castedo M, et al. Sequential reduction of mitochondrial transmembrane potential and generation of reactive oxygen species in early programmed cell death. J Exp Med 1995; 182:367–377.

85. Marchetti P, Castedo M, Susin SA, et al. Mitochondrial permeability transition is a central coordinating event of apoptosis. J Exp Med 1996; 184:1155–1160.

86. Murphy AN, Bredesen DE, Cortopassi G, Wang E, Fiskum G. Bcl-2 potentiates the maximal calcium uptake capacity of neural cell mitochondria. Proc Natl Acad Sci USA 1996; 93:9893–9898.

87. Kluck RM, Bossy-Wetzel E, Green DR, Newmeyer DD. The release of cytochrome c from mitochondria: a primary site for Bcl-2 regulation of apoptosis. Science 1997; 275:1132–1136.

88. Yang J, Liu X, Bhalla K, et al. Prevention of apoptosis by Bcl-2: release of cytochrome c from mitochondria blocked. Science 1997; 275:1129–1132.
89. Hunter DR, Haworth RA. The Ca^{2+}-induced membrane transition in mitochondria. I. The protective mechanisms. Arch Biochem Biophys 1979; 195:453–459.
90. Gunter TE, Pfeiffen DR. Mechanisms by which mitochondria transport calcium. Am J Physiol 1990; 258:C755–C786.
91. Griffiths EJ, Halestrap AP. Further evidence that cyclosporin A protects mitochondria from calcium overload by inhibiting a matrix peptidyl-prolyl cis-trans isomerase. Implications for the immunosuppressive and toxic effects of cyclosporin. Biochem J 1991; 274:611–614.
92. Kroemer G, Petit PX, Zamzami N, Vayssiere J-L, Mignotte B. The biochemistry of programmed cell death. Am Soc Exp Biol J 1995; 9:1277–1287.
93. Yang JC, Cortopassi GA. Induction of the mitochondrial permeability transition causes release of the apoptogenic factor cytochrome c. Free Radic Biol Med 1998; 24:624–631.
94. Kantrow SP, Piantadosi CA. Release of cytochrome c from liver mitochondria during permeability transition. Biochem Biophys Res Commun 1997; 232:669–671.
95. Yang JC, Cortopassi GA. dATP causes specific release of cytochrome C from mitochondria. Biochem Biophys Res Commun 1998; 250:454–457.
96. Scarlett JL, Murphy MP. Release of apoptogenic proteins from the mitochondrial intermembrane space during the mitochondrial permeability transition. FEBS Lett 1998; 418:282–296.
97. Sohal RS. Aging, cytochrome oxidase activity, and hydrogen peroxide release by mitochondria. Free Radic Biol Med 4:583–588.
98. Fries JF, Crapo LM. Vitality and Aging. San Francisco: W.H. Freeman, 1981.

16

Mitochondrial Alteration in Aging and Inflammation: A Possible Site of Action of Nitrone-Based Free Radical Traps

Kenneth Hensley, Lindsay Maidt, C. A. Stewart, Quentin N. Pye, K. A. Robinson, and Robert A. Floyd
Oklahoma Medical Research Foundation, Oklahoma City, Oklahoma

I. INTRODUCTION

The mitochondrial organelle is an extremely efficient engine capable of enzymatically oxidizing various fuel molecules (amino acids, fatty acids, and carbohydrates) and harnessing the energy released from these exergonic reactions. Despite the extraordinary efficiency of mitochondrial enzyme complexes in catalyzing the release of energy from respiratory substrates, inevitable side reactions occur which result in a loss of efficiency and the release of oxidizing agents as side products. Since mitochondrial bioenergetics ultimately depends on the controlled four-electron reduction of oxygen to water, the side products most associated with mitochondrial inefficiency are the one- and two-electron reduction products of oxygen, superoxide (O_2^-) and hydrogen peroxide (H_2O_2), respectively. Both O_2^- and H_2O_2 are reactive oxidizing agents capable of attacking and lesioning membranes, enzymes, and genomic components to the detriment of the cell and the organism. The approximate yield of O_2^- plus H_2O_2 (collectively referred to as reactive oxygen species, or ROS) is generally estimated at 1–2% of total mitochondrial oxygen consumption (1,2).

   Because mitochondria are the major cellular site of constitutively active redox biochemistry, many attempts have been made to link mitochondrial dysfunction with oxidative damage to aging tissue, with various degrees of success. A consensus has emerged that much of the oxidative damage that occurs during aging, and which may exacerbate the aging process, originates from mitochondrial electron leakage. Unfortunately, no convincing model has yet been promulgated that adequately explains the chemistry of mitochondrial ROS production, or its dysregulation under pathophysiologic conditions known to induce oxidative damage to tissue. Two likely sources of ROS leakage in mitochondria, redox-cycling ubiquinone and redox-cycling flavin dehydrogenases, are discussed in this review. Additionally, recent data are discussed which implicates mitochondrial electron leakage as a contributing factor in inflammatory gene induction. We hypothesize that blockade of the mitochondrial electron leak may explain certain of the anti-inflammatory properties of specific antioxidant compounds, particularly nitrone-based free radical traps (NRTs).

II. BIOCHEMICAL MECHANISMS FOR ROS PRODUCTION BY MITOCHONDRIA

While ROS production by respiring mitochondria has been documented as a phenomenon for almost half a century (3), no complete model has emerged to explain the mechanism of this process. Seminal observations made in the laboratory of Britton Chance prior to 1975, and subsequently elaborated upon by other researchers, remain the crux of knowledge regarding mitochondrial ROS efflux.

Prior to 1973, Boveris and Chance (2) documented H_2O_2 production during state 4 mitochondrial respiration in pigeon and rat heart, while negligible peroxide was generated during state 3 respiration. H_2O_2 generation could be stimulated by complex I or complex II substrates (malate + glutamate or succinate, respectively), and could be markedly increased by antimycin blockade of the electron transport chain at the level of complex III (2). In the same work, rotenone inhibited NAD-linked peroxide efflux but had no effect on succinate-stimulated H_2O_2 generation. Ubiqunone depletion abolished peroxide production, while repletion restored peroxide-generating capacity (1,2). Interestingly, however, the concentration of rotenone required to inhibit H_2O_2 production in heart mitochondria by 50% were five-fold greater than the concentration needed to produce greater than 95% inhibition of NADH oxidation (2). Uncouplers decreased the rate of H_2O_2 flux in the absence of antimycin but increased the rate in the presence of the complex III inhibitor antimycin, while increasing pH or pO_2 increased the yield of peroxide.

Boveris and Chance (2) interpreted their results to indicate that the primary source of ROS production by mitochondria were "a variable potential component" upstream from complex II as "a source or a regulatory factor of maximal importance." However, Boveris and Chance acknowledged that their data did not allow discrimination among ubiquinone, complex I and II flavins, and Fe-S centers as ultimate sources for ROS leakage. Latter studies using extracted flavoproteins estimated that at least 50% of ROS production by submitochondrial particles arose from autoxidation of reduced ubiquinol or semiubiquinone (Fig. 1) with the remainder leaking from flavoproteins (complex I and II; Fig. 1) (4–6).

The mechanism of ROS production became clouded by subsequent studies which clearly showed that lower concentrations of rotenone markedly increased NADH-supported O_2^- and H_2O_2 production in bovine heart mitochondria, while higher rotenone concentrations inhibited the effect (5). Moreover, inhibition of electron flow downstream from complex I (i.e., at complex III or cytochrome c) exerts a stimulatory effect on complex I–dependent O_2^- yield (6–8). It can be concluded that much remains to be discerned regarding the mechanism of ROS production by mitochondria under conditions mimicking those likely to occur in vivo during normo- or pathophysiologic conditions. Particularly, the mechanism of oxygen reduction by reduced flavin dehydrogenases remains to be elucidated, as does the relative contribution to ROS flux of flavin enzymes versus reduced ubiquinone species.

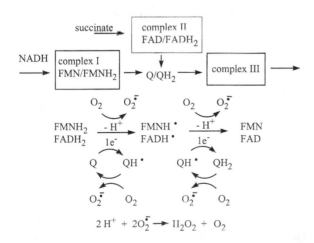

Figure 1 Pathway of electron transit through the initial portions of the mitochondrial electron transfer chain and illustration of putative electron leakage reactions. FMN = flavin mononucleotide; FAD = flavin adenine dinucleotide; Q = ubiquinone.

III. EVIDENCE FOR A MITOCHONDRIAL ROLE IN THE AGING PROCESS

Reduction in energy metabolism with age is well documented. As early as 1955, Shock and Yiengst (9) observed a decreased basal heat production with age in mammals, while Lassen et al. (10) noted an age-related decline in cerebral glucose utilization in humans. The intuitive and empirically supported notion that old organisms have less bioavailable "energy" than young organisms with which to meet physiologic demands led early investigators to study mitochondria as they searched for primary causes of senescence. Consequently, histologists of the 1950s and 1960s noted age-related declines in mitochondrial number regardless of tissue examined (11,12). Subsequent examination of mitochondrial ultrastructure by electron microscopy revealed age-related changes in mitochondrial morphology including a transition from tube-like to sphere-like mitochondria with matrix vacuolization and shortening of christae (13). Functional correlates of mitochondrial function, or dysfunction, with age proved more subtle and difficult to document. Weinbach and Garbus (14) noted that liver mitochondria from old rats were inefficient at utilizing β-hydroxybutyrate as a respiratory substrate, while no age-related change was observed with succinate or NAD-linked substrates. Subsequent studies of age-related changes in mitochondrial function were contradictory, with results depending on tissue type used for mitochondrial isolation and exact experimental protocol. Decline in activity of specific oxidative enzymes, particularly β-hydroxybutyrate dehydrogenase and NAD-isocitrate dehydrogenase, were repeatedly documented (15), while age-related alterations in the cytochrome system were not so thoroughly studied. A consensus gradually emerged that state 3 (ADP unlimited) respiration decreases with age by 25–75% depending on animal, tissue, and substrate, while state 4 (ADP limited) respiration is much less age dependent (16,17).

While documentation of age-related mitochondrial dysfunction is consistent with the mitochondrial hypothesis of aging, a complete mechanism explaining mitochondrial demise has never been adequately presented. The discovery that mitochondria leak ROS (2) provides some causal basis for mitochondrial perturbation in senescence. Progressive accumulation of oxidant-induced damage to mitochondrial DNA (mtDNA) might result in accrual of less-efficient mitochondria with even greater propensity for electron leakage (18,19). Thirteen polypeptide components of the electron transport chain are encoded by the mammalian mitochondrial genome, seven of which comprise portions of complex I (20). Interestingly, a 5 kb "common deletion" occurs in the region of the mitochondrial genome encoding for the rotenone-binding subunit of complex I in rat liver (21) and brain (22,23). This 5 kb deletion seems to correlate with a loss of NADH dehydrogenase activity and a decrease in sensitivity to rotenone inhibition, phenomena which have been proposed as biomarkers of aging (22–24).

It remains to be shown that mtDNA deletions are indeed caused by ROS, or that such mutations in mtDNA impart increased ROS-producing ability to aging mitochondria. Nonetheless, such a hypothesis is circumstantially supported by several recent studies. Based on quantitation of the oxidized nucleoside 8-hydroxy-2′-deoxyguanosine (8-OH-dG) in mitochondria and nuclei isolated from postmortem human brain tissue, Mecocci et al. (25) found a 20- to 50-fold greater rate of DNA oxidation with aging in mtDNA relative to nuclear DNA (nDNA). In this study, cortical DNA (both mtDNA and nDNA) was more vulnerable than cerebellar DNA. In a complementary study, Yakes and van Houten (26) have shown that application of exogenous H_2O_2 to cultured fibroblasts induces preferential oxidation of mtDNA over nDNA oxidation. Furthermore, skin fibroblast mitochondria taken from patients with disparate complex I deficiencies show evidence of enhanced reductive capacity and possibly enhanced O_2^- leakage (27). Regarding the issue of age-related ROS leakage from mitochondria, several studies have claimed to document age-dependent increases in H_2O_2 efflux from mitochondria isolated from insects as well as mammals (28,29). Finally, a statistically significant inverse correlation has been made between maximum life-span potential of mammalian species and rates of state 4 mitochondrial ROS generation (30). A similar correlation has been shown among species of fly (31). Taken together, these data suggest the possibility that interaction between mitochondrial ROS production and mtDNA damage may drive the progressive decline in bioenergetic capacity observed with senescence.

IV. MITOCHONDRIAL ROS PRODUCTION CONTRIBUTES TO ISCHEMIA-REPERFUSION INJURY

During a transient hypoxic episode, oxidative phosphorylation becomes impossible, and consequently the reduced forms of mitochondrial electron carriers increase as succinate and NADH-linked substrates are metabolized. Electron paramagnetic resonance has been used to directly monitor the accumulation of reduced flavin, hydroubiquinone, and Fe-S centers during cardiac ischemia (32). As tissue becomes reoxygenated, and oxidative phosphorylation resumes, controlled electron transit through the cytochrome chain is presumed to compete ineffectively with autoxidation of reduced electron carrier intermediates. The result is a "burst" of O_2^- and H_2O_2 during the reperfusion event. In carefully controlled experiments using the bilateral carotid artery occlusion model of cerebral ischemic injury (stroke), it has been shown that the reperfusion event rather than the ischemia per se induces brain protein oxidation and metabolic compromise (33–35). Other research points to mitochondria as the primary source of oxidizing equivalents generated during the immediate postischemic reperfusion event in brain and other tissue. Most strikingly, Piantadosi and Zhang (36) have used microdialysis techniques to administer respiratory substrates and inhibitors to rats subjected to transient global cerebral ischemia while monitoring oxidant production using salicylate as a trap for O_2^- or peroxide-derived hydroxyl radicals. Salicylate hydroxylation increased fivefold following reperfusion in the rat brain, and was blocked by rotenone. Moreover, the rotenone blockade could be overcome by coadministration of succinate, indicating substrate-linked redox cycling of ubiquinone and possibly flavin pools (36).

V. MITOCHONDRIAL ROS PRODUCTION AS AN EXACERBATING PHENOMENON OF CHRONIC INFLAMMATION

Although traditionally thought of as inadvertent and deleterious side products of oxidative metabolism, an emerging paradigm invokes H_2O_2 as a second messenger which acts, in a poorly understood manner, to amplify protein kinase cascade modules linked to cytokine-stimulated inflammatory gene induction (reviewed in Refs. 37–43). The basic pattern in cytokine-stimulated gene induction involves recruitment of specific kinases which phosphorylate and activate cytosol-resident transcription factors, most notably members of the NFκB/Rel family. These transcription factors, when phosphorylated, translocate to the nucleus and promote expression of various stress-response proteins as well as inflammation-associated cytokines and potentially cytotoxic enzymes (e.g., inducible nitric oxide synthase and cyclooxygenase) (38). Thus while activation of NFκB and induction of some proinflammatory gene products can be modeled in vitro by application of exogenous H_2O_2 mitochondria are likely to be major sources of oxidizing agents in actual inflamed organ tissue.

Hepatic mitochondria exhibit morphologic and functional alterations during severe acute septic episodes (i.e., sepsis), including markedly enhanced rates of nonenzymatic salicylate hydroxylation (44,45). Similarly, treatment of cultured L929 fibrosarcoid cell culture with the sepsis-associated inflammatory cytokine TNF-α produces rotenone-inhibitable induction of lucigenin chemiluminescence, presumably due to mitochondrial O_2^- efflux (46). Interestingly, partial blockade of electron transport in L929 cells by rotenone or barbituates (i.e., complex I inhibition) or thenoyltrifluoroacetone (i.e., complex II inhibition) reportedly protects the cells from TNF-α cytotoxicity (47,48). Antimycin A, which tends to stimulate mitochondrial ROS leakage (discussed above), exacerbates TNF-α toxicity in the same cell line (48). Strikingly, chemical antagonism of electron transport using established electron transport inhibitors strongly suppresses TNF-α-induced interleukin-6 in L929 cells, apparently at the level of inhibiting NFκB transcription factor activation (48). These recent observations concerning the role of mitochondria in TNF-α-mediated gene induction and cytotoxicity nicely complement, and partially explain, long-standing data that TNF-α induces mitochondrial Mn-SOD expression (49).

VI. NITRONE-BASED FREE RADICAL TRAPS AS MODULATORS OF OXIDANT-SENSITIVE SIGNAL TRANSDUCTION: A POSSIBLE EXPLANATION FOR ANTIOXIDANT, ANTISTROKE, AND ANTI-INFLAMMATORY ACTION

Nitrone-based free radical traps (NFTs) were designed as an analytical tool for the stabilization, detection, and study of transient free radicals (49). Free radical attack upon the nitrone functionality yields a nitroxide that is more stable than the original free radical species, and which can be detected by electron paramagnetic resonance spectroscopy (Fig. 2). Despite the utility of nitrones as analytical tools, most nitrones are no more efficient, and in many cases less efficient, at scavenging free radicals than aromatic compounds which lack the nitrone functionality (50; personal observations). In fact, in direct comparisons of ability as chain-breaking antioxidants, the archetypal nitrone, phenyl-*tert*-butyl-nitrone (PBN), is approximately 100 times less efficient at inhibiting lipid peroxidation than are sterically hindered phenols (51). Similarly, we have found that PBN is less effective as a substrate for nonenzymatic hydroxylation by iron/H_2O_2 than is salicylate (data not shown).

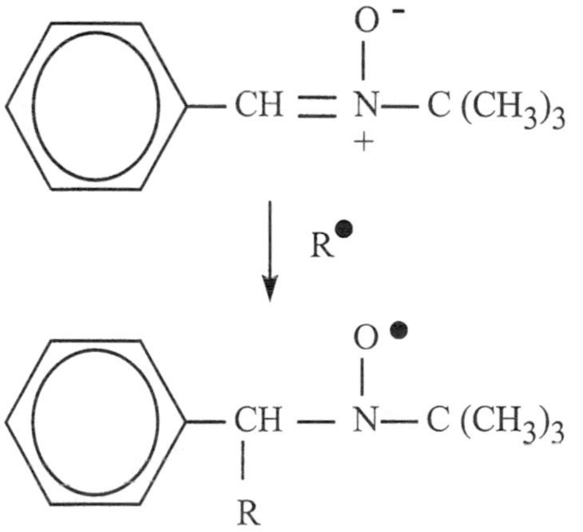

Figure 2 Structure of the archetypical nitrone α-phenyl-*tert*-butylnitrone (PBN) and a typical free radical trapping reaction.

Despite their relative inefficiency as chain-breaking antioxidants in vitro, nitrones have repeatedly shown themselves to suppress oxidative stress in vivo (reviewed in Ref. 52). Most notably, nitrones have been well-studied as protectants against cerebral ischemia-reperfusion injury (stroke). While attempting to use nitrones as a means of detecting hydroxyl radical production in ischemic gerbil brain, Floyd and Carney (33,34,53,54) discovered that nitrone treatment could dramatically increase survivability in this model of stroke. Decrease in postischemic mortality was accompanied by diminished protein oxidation and enzyme dysfunction (33,34,54). Other researchers confirmed these early observations (55–57). The ability of nitrones to protect mitochondria from ischemia-reperfusion damage was alluded to in a 1995 study by Folbergrova et al. (57) who showed that PBN accelerates postischemic decline of lactate acidosis and concomitant ATP rebound, two parameters indicating competence of oxidative phosphorylation.

In addition to being neuroprotective agents in models of ischemia-reperfusion injury, certain nitrones having potent anti-inflammatory properties are capable of rescuing animals from endotoxemia (58–61). Finally, chronic nitrone administration dramatically reverses age-related protein oxidation, enzyme dysfunction, and cognitive impairment in rodents (62,63) and extends the life span of senescence-accelerated mice (64). Significantly, both mean and maximum life span are extended by more than 30% in mice chronically treated with PBN (64), while protein oxidation and oxidation-associated alteration in membrane protein ultrastructure are attenuated with nitrone administration (65). Extension of maximum life span indicates that the drug therapy alters a fundamental limiting process in aging rather than altering mortality-causing correlates of senescence.

Explanations for the diverse neuroprotective, anti-inflammatory, antiaging, and nonotropic action of nitrones generally invoke free radical scavenging paradigms, though there is little direct evidence to support such a hypothesis. The maximum achievable concentration in liver or brain after intraperitoneal injection of PBN is approximately 500 µM, and the half-life for PBN in these organs is less than 120 min (66). It is unlikely that such low concentrations of nitrone significantly augment endogenous antioxidant scavenging and detoxification systems.

More recent models for nitrone action incorporate the observations that PBN and derivatives suppress inflammatory gene induction in systems where the nitrones are well documented to exert protective effects (reviewed in reference 52). For instance, PBN strongly suppresses TNF-α transcription after systemic administration of endotoxin (67). Similarly, in the postischemic gerbil brain and in ischemic liver, PBN suppresses immediate early gene expression (68). Miyajima and Kotake (61) report that transcription of inducible NO synthase is suppressed in PBN-treated septic rodents. A complete understanding of nitrone action in aging, ischemia, and inflamed tissue therefore requires a consideration of the relationship between oxidant production and inflammatory gene induction.

Recent studies conducted in our laboratory and by other researchers indicate that endogenous or exogenously supplied oxidants, particularly H_2O_2, can activate protein kinase cascades linked to the recruitment of transcription factors responsible for specific gene induction. For instance, H_2O_2 or various inflammatory cytokines such as interleukin-1β (IL-1β) and TNF-α activate NFκB as well as c-Jun amino terminal kinases (JNK) and extracellularly regulated, mitogen-activated protein kinases p42/p44 (ERK) (38–43,69,70).

We have begun investigating the role of PBN in antagonizing IL-1β-mediated activation of the p38 stress-response protein kinase pathway. This particular kinase cascade module is particularly cogent to inflammatory phenomenon because the phosphorylated form of p38 serves as an activator for transcription factors necessarily involved in cytokine-mediated gene expression (71,72). Moreover, p38 phosphorylation can be stimulated by exposure to interleukin and TNF-α as well as exogenous H_2O_2 (73–75). Treatment of cultured rat primary glial cells with 10 ng/ml recombinant IL-1β or 0.5 mM H_2O_2 induces p38 phosphorylation within 5 min, while these processes are almost completely blocked by pretreatment with 1 mM PBN (Fig. 3). Our results agree with previous findings regarding the oxidant sensitivity of p38 phosphorylation (74,75). It should be noted that PBN reacts at a negligible rate with H_2O_2 near physiologic pH and metal concentrations, so that bulk scavenging of peroxide is an unlikely explanation for the effects shown in Figure 3. Thus antagonism of oxidant-sensitive signal transduction pathways might explain much of the neuroprotective and anti-inflammatory action inherent to nitrones without recourse to invoke efficient free-radical scavenging prowess of such compounds.

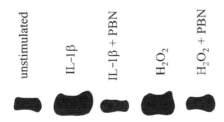

Figure 3 Western blot analysis of phosphorylated p38 in cultured rat primary glial cells. Phosphorylation of p38 is induced by exposure to IL-1β (10 ng/mL, 5 min) and H_2O_2 (0.5 mM, 5 min), both of which phenomena are inhibited by prior treatment with PBN (1 mM). Primary antibody was obtained from New England Biolabs (Cambridge, MA) and blots were visualized by enhanced chemiluminescence (Amersham, Buckinghamshire, UK).

VII. NITRONES MAY ANTAGONIZE CYTOKINE-STIMULATED SIGNAL TRANSDUCTION BY INHIBITING MITOCHONDRIAL OXIDANT GENERATION

As previously alluded to, exogenous application of the inflammatory cytokine TNF-α to L929 cells induces mitochondrial oxidant generation. Studies are under way in our laboratory to determine if a similar process is initiated by glial exposure to IL-1β. However, the observation that exogenous H_2O_2 mimics the IL-1β-induced p38 activation supports the possible involvement of H_2O_2 and other oxidants in IL-1β-mediated signal transduction. If this is the case, then mitochondrial ROS generation during cytokine stimulation could modulate the magnitude of inflammatory gene induction in a manner amenable to manipulation by appropriate mitochondrially active drugs.

In support of a mitochondrial target of nitrone action, we have measured H_2O_2 flux from isolated rat cortical mitochondria under conditions of complex I–stimulated state 4 respiration. Hydrogen peroxide efflux from these preparations is small relative to reported values obtained in heart and liver mitochondrial systems (16), yet represents 1–2% of total oxygen consumption in agreement with previous estimates of yield (2) (Fig. 4). Approximately 50% of the NADH-stimulated peroxide generation can be inhibited by PBN at concentrations roughly commensurate with those necessary to inhibit IL-Iβ and H_2O_2 activation of p38, with the remaining 50% being insensitive to PBN (Fig. 4). Although these data do not conclusively prove that PBN acts as an antioxidant and anti-inflammatory drug by suppressing mitochondrial ROS production, the data are strongly consistent with such a hypothesis.

VIII. CONCLUSIONS

Mitochondrial generation of ROS is a well-documented phenomenon that is being increasingly implicated as a causative factor of age-related pathophysiologies. Currently models are lacking to precisely describe ROS production mechanisms by mitochondria, especially under conditions of ischemia-reperfusion and inflammatory stimulation. A growing body of evidence suggests that mitochondrial ROS production contributes to inflammatory tissue damage, in part by stimulating recruitment of specific transcription factors which are activated by oxidant-sensitive protein kinase cascades. Data from our laboratory and others argue that nitrone-based compounds may be particularly beneficial compounds for the treatment of stroke as well as pathophysiologies involving a component of dysregulated inflammation. We hypothesize that one mechanism of nitrone action in such disorders is the mitigation of mitochondrial ROS efflux with consequent downregulation of oxidant-stimulated gene transcription.

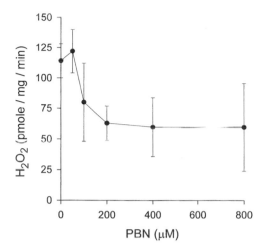

Figure 4 Inhibition by PBN of H_2O_2 production by NADH-stimulated mitochondria. Mitochondria were isolated as described by Mecocci et al. (25). Reaction mixture consisted of 1.5 mg/mL mitochondrial protein, 1 mM tetramethylbezidine (TMB), 0.1 mg/mL horseradish peroxidase (HRP), 0.4 mM NADH, and indicated concentrations of PBN in 20 mM HEPES, 5 mM KH_2PO_4/K_2HPO_4, and 80 mM KCL pH 7.3. H_2O_2-dependent, HRP-catalyzed TMB oxidation was monitored kinetically at 660 nm and 37°C and rates were compared to those of standard reaction mixtures containing HRP, TMB, and serial dilutions of H_2O_2.

ACKNOWLEDGMENTS

This work was supported in part by grants from the National Institutes of Health (N535747, N530457), the Oklahoma Center for the Advancement of Science and Technology (HR97-067), and a contract from Centaur Pharmaceuticals.

REFERENCES

1. Boveris A, Cadenas E, Stoppani AOM. Role of ubiquinone in the mitochondrial generation of hydrogen peroxide. Biochem J 1976; 156:435–444.
2. Boveris A, Chance B. The mitochondrial generation of hydrogen peroxide: General properties and effect of hyperbaric oxygen. Biochem J 1973; 134:707–716.
3. Chance B, Williams GR. Respiratory enzymes in oxidative phosphorylation. J Biol Chem 1955; 217:395–407.
4. Cadenas E, Boveris A, Ragan CL, Stoppani SOM. Production of superoxide radicals and hydrogen peroxide by NADH-ubiquinone reductase and ubiquinol-cytochrome

c reductase from beef-heart mitochondria. Arch Biochem Biophys 1977; 180:248–257.

5. Turrens JF, Boveris A. Generation of superoxide anion by the NADH dehydrogenase of bovine heart mitochondria. Biochem J 1980; 191:421–427.

6. Turrens JF, Freeman BA, Crapo JD. The effect of hyperoxia on superoxide production by lung submitochondrial particles. Arch Biochem Biophys 1982; 217:401–411.

7. Turrens JF, Alexandre A, Lehninger AL. Ubisemiquinone is the electron donor for superoxide formation by complex III of heart mitochondria. Arch Biochem Biophys 1985; 237:408–414.

8. Turrens JF. Superoxide production by the mitochondrial respiratory chain. Biosci Rep 1997; 17:3–8.

9. Shock NW, Yiengst MJ. Age changes in basal respiratory measurements and metabolism in males. J Gerontol 1955; 10:31–40.

10. Lassen NA, Feinberg I, Lane MH. Bilateral studies of cerebral oxygen uptake in young and aged normal subjects and in patients with organic dementia. J Clin Invest 1960; 39:491–500.

12. Tauchi H, Sato T. Age changes in size and number of mitochondria of human hepatic cells. J Gerontol 1968; 23:454.

13. Wilson PD, Franks LM. The effect of age on mitochondrial ultrastructure. Gerontologia 1975; 21:81–94.

15. Patel MS. Age-dependent changes in the oxidative metabolism in rat brain. J Gerontol 1977; 32:643–646.

16. Chiu YJD, Richardson A. Effect of age on the function of mitochondria isolated from brain and heart tissue. Exp Gerontol 1980; 15:511–517.

17. Horton AA, Spencer JA. Decline in respiratory control ratio of rat liver mitochondria in old age. Mech Ageing Dev 1981; 17:253–259.

18. Harman D. The biological clock: the mitochondia? J Am Geriatr Soc 1972; 20:145–147.

19. Miquel J, Economos AC, Fleming J, Johnson JE. Mitochondrial role in cell aging. Exp Gerontol 1980; 15:575–591.

20. Tzagoloff A, Myers AM. Genetics of mitochondrial biogenesis. Annu Rev Biochem 1986; 55:249–285.

21. Edris W, Burgett B, Colin-Stine O, Filburn CR. Detection and quantitation by competitive PCR of an age-associated increase in a 4.8 kb deletion in rat mitochondrial DNA. Mutat Res 1994; 316:69–78.

22. Genova ML, Castelluccio C, Fato R, et al. Major changes in complex I activity in mitochondria from aged rats may not be detected by direct assay of NADH: coenzyme Q reductase. Biochem J 1995; 311:105–109.

23. Genova ML, Bovina C, Marchetti M, et al. Decrease of rotenone inhibition is a sensitive parameter of complex I damage in brain non-synaptic mitochondria of aged rats. FEBS Lett 1997; 410:467–469.

24. Pich MM, Bovina C, Formiggini G, et al. Inhibitor sensitivity of respiratory complex I in human platelets: a possible biomarker of aging. FEBS Lett 1996; 380:176–178.

25. Mecocci P, MacGarvey U, Kaufman AE, et al. Oxidative damage to mitochondrial

DNA shows marked age-dependent increases in human brain. Ann Neurol 1993; 34:609–616.

26. Yakes FM, Van Houten B. Mitochondrial DNA damage is more extensive and persists longer than nuclear DNA damage in human cells following oxidative stress. Proc Natl Acad Sci USA 1997; 94:514–519.

27. Pitkanen S, Robinson BH. Mitochondrial complex I deficiency leads to increased production of superoxide radicals and induction of superoxide dismutase. J Clin Invest 1996; 98:345–351.

28. Sohal RS, Sohal BH. Hydrogen peroxide release by mitochondria increases during aging. Mech Ageing Dev 1991; 57:187–202.

29. Sohal RS, Arnold LA, Sohal BH. Age-related changes in antioxidant enzymes and prooxidant generation in tissues of the rat with special reference to parameters in two insect species. Free Rad Biol Med 1990; 9:495–500.

30. Ku H-H, Brunke UT, Sohal RS. Relationship between mitochondrial superoxide and hydrogen peroxide production and longevity of mammalian species. Free Radic Biol Med 1993; 15:621 627.

31. Sohal RS, Sohal BH, Orr WC. Mitochondrial superoxide and hydrogen peroxide generation, protein oxidative damage and longevity in different species of flies. Free Radic Biol Med 1995; 19:499–504.

32. Ruuge EK, Ledenev AN, Lakomkin VL, Konstantinov AA, Ksenzenko MY. Free radical metabolites in myocardium during ischemia and reperfusion. Am J Physiol 1991; 261:81–86.

33. Oliver CN, Starke-Reed PE, Stadtman ER, et al. Oxidative damage to brain proteins, loss of glutamine synthetase activity, and production of free radicals during ischemia/reperfusion-induced injury to gerbil brain. Proc Natl Acad Sci USA 1990; 87:5144–5147.

34. Floyd RA, Carney JM. Age influence on oxidative events during ischemia/reperfusion. Arch Gerontol Geriatr 1991; 12:155–177.

35. Funahashi T, Floyd RA, Carney JM. Age effect on brain pH during ischemia/reperfusion and pH influence on peroxidation. Neurobiol Aging 1994; 15:161–167.

36. Piantadosi CA, Zhang J. Mitochondrial generation of reactive oxygen species after brain ischemia in the rat. Stroke 1996; 27:327–332.

37. Schreck R, Baeuerle P. A role for oxygen radicals as second messengers. Trends Cell Biol 1991; 1:39–42.

38. Remackle J, Raes M, Toussaint O, Renard P, Rao G. Low levels of reactive oxygen species as modulators of cell function. Mutat Res 1995; 316:103–122.

39. Kyriakis JM, Avruch J. Sounding the alarm: protein kinase cascades activated by stress and inflammation. J Biol Chem 1996; 271:24313–24316.

40. Lo YYC, Wong JMS, Cruz TF. Reactive oxygen species mediate cytokine activation of c-Jun NH_2-terminal kinases. J Biol Chem. 1996; 271:15703–15707.

41. Sen CK, Packer L. Antioxidant and redox regulation of gene transcription. FASEB J 1996; 10:709–720.

42. Scherle PA, Pratta MA, Feeser WS, Tancula EJ, Arner EC. The effects of IL-1 on mitogen-activated protein kinases in rabbit articular chondrocytes. Biochem Biophys Res Commun 1997; 230:573–577.

43. Suzuki YJ, Forman HJ, Sevanian A. Oxidants as stimulators of signal transduction. Free Radic Biol Med 1997; 22:269–285.

44. Taylor DE, Ghio AJ, Piantadosi CA. Reactive oxygen species produced by liver mitochondria of rats in sepsi. Arch Biochem Biophys 1995; 316:70–76.

45. Kantrow SP, Taylor DE, Carraway MS, Piantadosi CA. Oxidative metabolism in rat hepatocytes and mitochondria during sepsis. Arch Biochem Biophys 1997; 345: 278–288.

46. Hennet T, Richter C, Peterhans E. Tumor necrosis factor-α induces superoxide anion generation in mitochondria of L929 cells. Biochem J 1993; 289:587–592.

47. Schulze-Osthoff K, Bakker AC, Vanhaesebroeck B, et al. Cytotoxic activity of tumor necrosis factor is mediated by early damage of mitochondrial functions: evidence for the involvement of mitochondrial radical generation. J Biol Chem 1992; 267: 5317–5323.

48. Schulze-Osthoff K, Beyaert R, Vandervoorde V, Haegeman G, Fiers W. Depletion of the mitochondrial electron transport abrogates the cytotoxic and gene inductive effects of TNF. EMBO J 1993; 12:3095–3104.

49. Wong GHW, Goeddel DV. Induction of manganous superoxide dismutase by tumor necrosis factor: possible protective mechanism. Science 1988; 242:941–944.

50. Janzen EG, Blackburn BJ. Detection and identification of short-lived free radicals by an electron spin resonance trapping technique. J Am Chem Soc 1968;90:5909.

51. Janzen EG, West MS, Poyer JL. Comparison of antioxidant activity of PBN with hindered phenols in initiated rat liver microsomal peroxidation. In: Asada K, Yoshikawa T, eds. Frontiers of Reactive Oxygen Species in Biology and Medicine. Amsterdam: Elsevier Science, 1994:431–434.

52. Hensley K, Carney JM, Stewart CA, et al. Nitrone-based free radical traps as neuroprotective agents in cerebral ischemia and other pathologies. In: Green AR, Cross AJ, eds. Neuroprotective Agents in Cerebral Ischemia. San Diego: Academic Press, 1997:299–313.

54. Carney JM, Floyd RA. Protection against oxidative damage to CNS by α-phenyl-tert-butylnitrone (PBN) and other spin-trapping agents: a novel series of nonlipid free radical scavengers. J Mol Neurosci 1991; 3:47–57.

55. Clough-Helfman C, Phillis JW. The free radical trapping agent N-tert-butyl-α-phenylnitrone (PBN) attenuates cerebral ischemic injury in gerbils. Free Radic Res Commun 1991; 15:177–186.

57. Folbergrova J, Zhao Q, Katsura KL, Seisjo BK. N-tert-butyl-α-phenylnitrone improves recovery of brain energy state in rats following transient focal ischemia. Proc Natl Acad Sci USA 1995; 91:5057–5061.

58. Hamburger SA, McCay PB. Endotoxin-induced mortality in rats is reduced by nitrones. Circ Shock 1989; 29:329–334.

60. French JF, Thomas CE, Downs TR, et al. Protective effects of a cyclic nitrone antioxidant in animal models of endotoxic shock and chronic bacteremia. Circ Shock 1994; 43:130–136.

61. Miyajima T, Kotake Y. Spin trapping agent, phenyl-N-tert-butyl nitrone, inhibits induction of nitric oxide synthase in endotoxin-induced shock in mice. Biochem Biophys Res Commun 1995; 215:114–121.

62. Carney JM, Starke-Reed PE, Oliver CN, et al. Reversal of age-related increase in

brain protein oxidation, decrease in enzyme activity, and loss in temporal and spatial memory by chronic administration of the spin trapping compound N-*tert*-butyl-α-phenylnitrone. Proc Natl Acad Sci USA 1991; 88:3633–3636.

63. Socci DJ, Crandall BM, Arendash GW. Chronic antioxidant treatment improves the cognitive performance of aged rats. Brain Res 1995; 693:88–94.

64. Edamatsu R, Mori A, Packer L. The spin trap N-*tert*-α-phenylnitrone prolongs the life span of the senescence accelerated mouse. Biochem Biophys Res Commun 1995; 211:847–849.

67. Pogrebniak HW, Merino MJ, Hahn SM, Mitchell JB, Pass HI. Spin trap salvage from endotoxemia: the role of cytokine down-regulation. Surgery 1992; 112:130–139.

68. Carney JM, Kindy MS, Smith CD, et al. Gene expression and functional changes after acute ischemia: age-related differences in outcome and mechanisms. In: Hartman A, Yatsu F, Kuschinsky W, eds. Cerebral Ischemia and Basic Mechanisms. Berlin: Springer-Verlag, 1994:301–311.

69. Sundaresan M, Yu Z-X, Ferrans VJ, Irani K, Finkel T. Requirement for H_2O_2 for platelet-derived growth factor signal transduction. Science 1995; 270:296–299.

70. Wilmer WA, Tan LC, Dickerson JA, Danne M, Rovin BH. Interleukin-1β induction of mitogen-activated protein kinases in human mesangial cells: role of oxidation. J Biol Chem 1997; 272:10877–10881.

71. Shapiro L, Dinarello CA. Osmotic regulation of cytokine synthesis in vitro. Proc Natl Acad Sci USA 1995; 92:12230–12234.

73. Raingeaud J, Gupta S, Rogers JS, et al. Proinflammatory cytokines and environmental stress cause p38 mitogen-activated protein kinase activation by dual phosphorylation of tyrosine and threonine. J Biol Chem 1995; 270:7420–7426.

74. Moriguchi T, Toyashima F, Gotoh Y, et al. Purification and identification of a major activator for p38 from osmotically shocked cells. Activation of mitogen-activated protein kinase kinase 6 by osmotic shock, tumor necrosis factor-alpha, and H_2O_2. J Biol Chem 1996; 271:26981–26988.

75. Guyton KZ, Liu Y, Gorospe M, Xu Q, Holgrook NJ. Activation of mitogen-activated protein kinase by H_2O_2. Role in cell survival following oxidant injury. J Biol Chem 1990; 271:4138–4142.

17

Muscle Mitochondrial Transmembrane Processes in Mitochondriocytopathies

Marjan Huizing, J. M. F. Trijbels, J. A. M. Smeitink, L. P. van den Heuvel, and Wim Ruitenbeek
University Hospital, Nijmegen, The Netherlands

Fernando Palmieri
University of Bari, Bari, Italy

I. INTRODUCTION

Several hundred patients with disturbances in the mitochondrial energy metabolism have been recognized in the last decades (1,2). Defects have been reported in the oxidation of fatty acids (3), the pyruvate dehydrogenase complex, the citric acid cycle, or in one or more of the multisubunit complexes of the respiratory chain (1,4). Abnormalities in the mitochondrial genome have also been reported (1,2). Mitochondria are present in nearly all types of cells and tissues, but the symptomatology of patients with mitochondrial disorders is very heterogeneous, although organs and tissues with a high energy demand are most often involved in the pathologic process.

Biochemical investigations in subjects suspected to suffer from a mitochondrial disorder on clinical, morphological, and clinical-chemical grounds are commonly performed on skeletal muscle. They may encompass measurements of oxidation of substrates, oxygen consumption, and/or ATP production by intact mitochondria in order to obtain information about the flux through the various parts of the mitochondrial energy metabolism. In case of impaired flux, various enzyme complexes of the respiratory chain and the pyruvate dehydrogenase complex are measured, possibly followed by immunoblotting or 2D-electrophoresis to precisely define the defect at the enzyme and protein level (5).

Despite progress in the diagnostic approach, we experienced that, in almost one-quarter of the patients with evidently reduced substrate oxidation rates in muscle mitochondria, these abnormalities could not be ascribed to a known defect in one of the mitochondrial enzymes. From this observation we concluded that other defects must account for the established defect in the mitochondrial energy generation in muscle. Earlier we focused our attention on the possible deficiency of mitochondrial creatine kinase as a cause of the disturbed energy production. However, investigation of more than 100 muscle samples with such a disturbed oxidative capacity of mitochondria did not reveal a patient with a mitochondrial creatine kinase deficiency (6). Another so-called postrespiratory chain enzyme which might be involved in the pathogenesis of mitochondrial myopathies concerns mitochondrial ATPase (complex V). At present only a few patients with an ATPase deficiency have been described (1). These observations prompted us to investigate other possible causes of disturbed mitochondrial energy generation. We speculated that malfunctioning of transporting systems in the inner or outer mitochondrial membrane might be the primary cause of the biochemical aberrations in some of these patients (7,8).

Particularly the adenine nucleotide translocator (ANT) and phosphate carrier (PiC), which are directly involved in the process of oxidative phosphorylation (9), seem to be good candidates. Also transport systems taking care of transmembrane ion transport likely can induce imperfect energy metabolism due to osmotic disturbances in the mitochondrial matrix. Hence it seems advantageous to estimate the contents of the K^+, Na^+, Mg^{2+}, H^+, and Ca^{2+} transporting proteins (10,11). The voltage-dependent anion channel (VDAC) probably plays a role in both ion transport and ATP/ADP exchange (12,13), which has to be taken into consideration, too. Other mitochondrial transmembrane carriers that can influence mitochondrial energy metabolism are transporters responsible for the transport of different substrates like pyruvate, oxoglutarate, citrate, succinate, carnitine, glutamate, and aspartate (9). Also mitochondrial transmembrane processes like protein import and O_2 transport should be considered. These carriers and channels can be malfunctioning due to mutations in their genes, but might also be defective as a consequence of membrane impairment, for example, by increased oxygen radical production.

We started screening suspected patients for defects in some of these transport systems. Consequently, subjects lacking ANT or VDAC were diagnosed applying immunochemical techniques with specific antibodies raised against these proteins (14–16). Few other subjects with a defect in the mitochondrial transport system have been reported, among which are defects in the protein import machinery (17), defects in the aspartate-malate shuttle (18,19), deficiencies of the pyruvate carrier (20), defects in the carnitine-acylcarnitine carrier (21), and defects in the ornithine carrier (22).

In this chapter we review features of the mitochondrial carrier systems in eukaryotes and if known in humans. We discuss the possible consequences of malfunctioning of these carriers for pathology. We will restrict the review to transport systems involved in energy metabolism of muscle mitochondria. Other carriers present in mitochondria, involved in the urea cycle, lipogenesis, and gluconeogenesis, are briefly mentioned.

II. PROTON TRANSLOCATION AND ENERGY CONSERVATION

Protons play a very important role in physiological processes. The biological activity of the majority of enzymes is pH dependent. Maximum activity can only be reached at the optimal pH of an individual enzyme. Most intramitochondrial enzymes have their optimal pH between 7.2 and 7.8. The affinity of enzymes, transport proteins, and channels toward substrates and cofactors is often pH dependent. Several enzyme reactions are associated with a change in the proton concentration, but the pH of the mitochondrial matrix must be maintained within rather narrow limits to guarantee adequate metabolic fluxes. A correct pH is also required for the maintenance of integrity of cellular and subcellular structures like membranes. Low pH values induce a reduced rate of energy production (23). Loss of membrane integrity can easily result in proton leak. Proton and/or hydroxyl ion transport is part of many cation translocating systems in the inner mitochondrial membrane (see Sec. IV). Disturbance in one or more of these proton translocating processes could be the cause of mild or severe reduction in substrate oxidation rates in patients' muscle mitochondria in situ and in vitro. It is not clear whether the matrix pH per se can form the trigger for activation of the cation transporting systems.

By far the most important mitochondrial process in which proton translocation plays a role is oxidative phosphorylation. Electron flux through complex I, III, and IV of the respiratory chain leads to extrusion of protons. The energy from the redox reactions performed by the three enzyme complexes is used for generation of a proton gradient over the inner mitochondrial membrane. The abundance of protons can be transported back to the relatively alkalic matrix by complex V or Mg^{2+}-ATPase. The F_0 part of complex V forms a channel for protons, while the energy of the proton gradient, that is, the proton motive force, can be applied for phosphorylation of intramitochondrial ADP by the F_1 part of the complex. The mechanism of this complicated process of oxidative phosphorylation is only partly known. The hypothesis of the chemiosmotic model proposed by Mitchell and Moyle (24) is now generally accepted. The flux of electrons through the respiratory chain only leads to an efficient rate of ADP phosphorylation if the so-called coupling state of the mitochondria is intact. If mitochondria are well coupled the ADP concentration in the matrix is rate limiting for respiration. In uncoupled mitochondria, no proton gradient can be built because the inner membrane has become freely permeable for these ions. No ATP is formed, although substrate oxidation introduces adequate electron flux through the respiratory chain. Loosely coupled mitochondria generate ATP in the presence of ADP, while the electron flux through their respiratory chain is maximal in the absence of ADP, like in uncoupled mitochondria. In mitochondria with a less well-coupled state, the permeability of the inner membrane is increased for protons. Proton permeability depends on various conditions and can be influenced by, for example, hormones (25). Pathological intracellular concentrations of free fatty acids can readily induce uncoupling (26), thus affecting the proton gradient.

The energy generated by respiring mitochondria is not exclusively used for ATP production, but can also be used for uptake of calcium ions by mitochondria or for heat generation. Well-coupled mitochondria always produce some heat because of the nonoptimal efficacy of the oxidative phosphorylation process. In stress susceptible pigs, used as a malignant hyperthermia model, even the pathologically high temperature after challenge by dantrolene (27) can be ascribed to heat production by maximal functioning and well-coupled muscle mitochondria (28). Perhaps there is a mechanism that constantly weights the contribution of energy employment to ATP production against heat generation. A special form of mitochondrial heat production in nature is regulated by the uncoupling protein (UCP). One protein (UCP1), expressed exclusively in brown adipose tissue, takes care of body temperature control, for example, in hibernating animals. The role of UCP in other animals is less obvious, although energy balance checking may be a general function. UCP has partly the same amino acid sequence as other mitochondrial transmembrane proteins, and hence belongs to the so-called family of transporter proteins (29,30). Recently, two other forms of UCP have been detected (31,32). UCP2 is ubiquitously distributed in human tissues, and may have a role not only in thermoregulation but also in body weight regulation and inflammatory processes. UCP3 is almost exclusively expressed in skeletal muscle. It is not yet known if UCP2 and/or UCP3 are involved in (regulation of) basal heat generation in mitochondria.

Actually, the first described patient with a mitochondrial cytopathy was suffering from a defective coupling state: Luft disease (33). The woman's mitochondria generated in vivo a massive amount of heat. In vitro their respiration rate was high in the absence of ADP, and normal in its presence. Apart from a few patients with more or less comparable features (34,35), no other patients with uncoupled mitochondria have been described during the last three decades. Several subsequent functional, biochemical, and histochemical studies have been performed on these patients. The biochemical abnormality in the patient's skeletal muscle mitochondria has not yet been explained at the molecular level. It has been supposed that in Luft disease lack of the ATPase inhibitor PMI (Pullman–Monroy inhibitor) plays a crucial role in the pathogenetic mechanism (36), but this hypothesis seems not to be the ultimate explanation. Overexpression of the lately discovered UCP3 protein can theoretically cause Luft disease.

We found in some patients suspected to suffer from a mitochondriocytopathy whose muscle mitochondria showed in vitro a decreased ADP dependency for the pyruvate oxidation (37). Possibly this kinetic aspect points to a less well-coupling state of the oxidative phosphorylation, either only in vitro or also in vivo. The accumulation of lactic acid in the patients reflects a hampered flux through the mitochondrial respiratory chain rather than an unrestricted respiration rate. Müller-Höcker et al. (38) concluded from enzyme-histochemical investigations that loose coupling is related to proliferation of mitochondria.

Some patients have been reported with a deficiency of the ATPase activity of complex V. Disturbed proton uptake by complex V has never been described. Several patients have been found with an encephalomyopathy of the Leigh or NARP type, in whom a point mutation in the coding region of subunit 6 of ATPase (position 8993) was found (2). This subunit is participating in the proton translocating channel of complex V. It is likely that the mutation has its consequences for the clinical phenotype and lactic acidosis observed in these patients, although in only part of the patients with this point mutation the in vitro activity of the ATPase is decreased.

Calcium uptake by mitochondria is replacing ATP synthesis under various (patho-) physiological conditions. Increased intramitochondrial Ca^{2+} concentration can lead in a concentration-dependent way to either activation or inactivation of mitochondrial energy metabolism (see Sec. IV). Consequently, these circumstances can form the origin of not-localized and not-well defined dysfunctioning of mitochondria.

III. ADP, ATP, AND P$_i$ TRANSPORT

Adenine nucleotide phosphorylation and transportation are a crucial part of mitochondrial energy metabolism. Carriers involved in this process are the adenine nucleotide translocator (ANT) and the phosphate carrier (PiC) in the inner mitochondrial membrane, and the voltage-dependent anion channel (VDAC) in the outer mitochondrial membrane. Defects in one of these carriers may affect mitochondrial energy metabolism.

A. Adenine Nucleotide Translocator

The adenine nucleotide translocator (ANT) is an integral inner mitochondrial membrane protein. It catalyzes the transmembrane exchange of cytosolic ADP^{3-} and matrix ATP^{4-} (in the process of oxidative phosphorylation) (39,40). ANT has a low turnover rate, but a high concentration, so the capacity for ADP and ATP transport across the inner mitochondrial membrane can easily cope with the energy requirement of the different cell types (41,42). It has been shown that ANT contributes to the control of respiration (43,44). Two different conformations of ANT have been demonstrated on the basis of interactions with specific inhibitors, carboxyatractyloside (CATR) and bongkrekic acid (BA). The two conformations, referred to as CATR and BA conformations, are interconvertible, provided that ADP or ATP are present. The functional ANT is probably organized as a homodimer (45) or as a tetramer (40,46). In the presence of CATR or BA the tetramer is split into two dimers combined with either of the two inhibitors (40,46).

ANT is a member of the "mitochondrial carrier protein family," a group of mitochondrial membrane transporters with similar structures (29,30). The amino acid sequence of the carriers in this family consists of three tandemly repeated homologous domains, each about 100 amino acids in length. Each domain forms two membrane-spanning α-helical segments, linked by a hydrophilic loop, forming a structure with six transmembrane α-helices. There is evidence that the central part of the second domain of ANT is involved in substrate binding (47,48).

Three human ANT isoforms (ANT1, ANT2, and ANT3) have been identified and sequenced. They are expressed in a tissue-specific manner, which seems to be related to their function (41,42,49–51). ANT1 is encoded by a gene located on chromosome 4 (52), and may permit rapid exchange of ADP and ATP to accommodate the high energy demand associated with contraction of striated muscle fibers (50). ANT1 is highly expressed in heart and skeletal muscle (53). The ANT2 gene is located on the X chromosome (54). ANT2 is mainly associated with smooth muscle cells (53) and it is induced in rapidly dividing cells such as fibroblasts, human leukemic cells (41), and myoblasts (53). The ANT3 isoform is ubiquitous in humans, expressed in all tissues, mainly in kidney (55). The ANT3 gene is described by Cozens et al. (51), but its chromosomal localization is not described yet. The ANT genes have different promotor and intronic regions, supporting the hypothesis that the expression of ANT genes is under distinct regulatory controls (53).

In 1993, the first deficiency of ANT in muscle was demonstrated by immunochemical techniques (14). This patient showed myopathy with lactic acidosis. A 2- to 20-fold increase in activity of all measured mitochondrial enzymes except for cytochrome c oxidase was found in muscle tissue. Remarkably, the decreased pyruvate oxidation was strongly increased by the uncoupler CCCP and only mildly stimulated by ADP. Immunostaining of Western blots, using polyclonal antiserum against ANT, revealed a fourfold decrease in muscle ANT content. The defect was not present in fibroblasts and lymphocytes, pointing to a muscle-specific deficiency of ANT, likely of ANT1. No mutation in the patient's ANT1 cDNA was found (H. D. Bakker, personal communication).

Schultheiss et al. (56) demonstrated in heart muscle tissue of patients suffering from myocarditis and dilated cardiomyopathy a decreased ANT transport capacity accompanied by an elevation in total ANT protein content. The alteration in ANT protein amount was due to an ANT isoform shift: an increase in ANT1, a decrease in ANT2, the ANT3 content being unchanged. The isoform shift in these patients is not a progressive process, but occurs in the early stages of illness and becomes permanent. The underlying mechanism and pathognostic significance are still obscure (56).

Recently we examined muscle from patients with "Sengers syndrome" (McKusick 212350) (57,58), clinically presenting with congenital cataract, hypertrophic cardiomyopathy, mitochondrial myopathy, and lactic acidosis. In four muscle specimens a lack of ANT was demonstrated by western blot. Other patients with this syndrome are studied in order to establish if an ANT deficiency is a predominant cause (59).

Mice with a null mutation in the heart and skeletal muscle ANT (ANT1) gene have been raised. They show cardiomyopathy, proliferation of muscle mitochondria, ragged red fibers, lactic acidosis, and exercise intolerance (60).

B. Phosphate Carrier

The supply of intramitochondrial inorganic phosphate (P_i) required for oxidative phosphorylation is maintained by two systems which transport P_i across the inner mitochondrial membrane. The P_i/dicarboxylate carrier mediates electroneutral P_i:dicarboxylate or P_i:P_i exchange, whereas the phosphate carrier (PiC) mediates the P_i/H^+ symport (or P_i/OH^- antiport) (61–63). Because the supply of inorganic phosphate for oxidative phosphorylation depends mainly on PiC activity, PiC plays a role in regulation of the mitochondrial energy metabolism (64).

PiC has been isolated and purified from different mammalian mitochondria and its activity reconstituted in liposomes (65–67). Lipids, in particular cardiolipin, may have an important role in the regulation of PiC in vivo, as demonstrated by the specific requirement of this phospholipid for the reconstitution of the isolated PiC activity in liposomes (68).

The human PiC amino acid sequence has been determined (69), its molecular weight is 32 kDa, and PiC is, like ANT, a member of the mitochondrial carrier protein family: a tripartite structure with three repeated units of about 100 amino acids, and the presence of some highly conserved residues (29,30). A full-length cDNA encoding the precursor of the human heart PiC has been synthesized and characterized (69). The gene encoding human PiC is located on chromosome 12q23 (70) and is spread over 7.9 kb of DNA, containing nine exons (71). Two alternatively spliced mRNAs for PiC (IIIA and IIIB) have been reported, both being present in a human heart cDNA library. The alternative splicing mechanism introduces a different region of 13 amino acids into the human carrier protein, the functional consequences of which are not yet understood (71). Expression studies on the bovine heart PiC gene, which is 94% identical to human PiC, showed a ubiquitous expression of the IIIB isoform, whereas IIIA is only expressed in heart, skeletal muscle, and pancreas (72).

We investigated immunochemically more than 100 muscle specimens from patients with a mitochondriocytopathy. So far no deficiency in PiC has been detected.

C. Voltage-Dependent Anion Channel

A substantial part of the molecular traffic across the outer mitochondrial membrane is mediated by an abundant pore-forming protein, the voltage-dependent anion channel (VDAC; or mitochondrial porin). At low transmembrane voltage the VDAC pore is open for anions such as phosphate, chloride, and adenine nucleotides. At higher transmembrane voltage or in the presence of a VDAC modulating protein, VDAC functions as a selective channel for cations and uncharged molecules. These features make it likely that VDAC plays a regulatory role in mitochondrial energy metabolism (12,13,73). The NADH concentration (74) and colloidal osmotic pressure (75) may also function as a VDAC regulator by shifting the membrane potential required for opening and closing of the pores. In addition to the outer mitochondrial membrane, VDAC may be localized in the plasmalemma of certain cells (76), although this extramitochondrial localization is still under discussion (77).

With respect to human VDAC (HVDAC), five genes encoding different isoforms have been reported. The cDNAs of HVDAC1 (on chromosome X) and HVDAC2 (chromosome 21) are completely sequenced (78), as well as a HVDAC2' sequence (chromosome 21), very closely related to HVDAC2 (79). HVDAC3 (chromosome 12) and HVDAC4 (chromosome 1) are only partial sequenced (80), and may be pseudogenes. Until now only HVDAC1 and HVDAC2 have been shown to be expressed at the protein level, HVDAC1 being the most abundantly expressed (78,81).

In 1994 we reported the first patient with a VDAC deficiency (15,16). The patient presented with high birth weight, macrocephaly, and dysmorphic features. Biochemical studies on muscle mitochondria showed impaired rates of pyruvate oxidation and ATP production. No deficiency of one of the respiratory chain complexes or of PDHc was found. Western blotting experiments indicated an almost complete deficiency of VDAC protein in skeletal muscle, but not in fibroblasts. The occurrence of isoforms might be an explanation for this tissue-specific VDAC disturbance. The monoclonal antiserum used in the immunochemical experiments recognizes only the HVDAC1 isoform, and more specifically the acetylated N-terminal region (81). So most probably the HVDAC1 protein in the muscle tissue of our patient is almost completely absent, mutated, or nonacetylated. Studies are in progress to detect the defect at the molecular level. The VDAC deficiency might cause abnormal composition and/or altered osmolarity of the mitochondrial matrix, which could lead to impairment of the mitochondrial energy metabolism in the patient.

Up to now, we investigated a group of 200 suspected patients for a defect in HVDAC. Apart from the already described patient, we have indications for two other HVDAC deficiencies. The clinical picture of these patients is different from the first detected patient. Further studies for more detailed characterization are in progress.

IV. CATION TRANSLOCATING PROTEINS

The maintenance of volume and ion composition of the mitochondrial matrix within narrow ranges is crucial for proper mitochondrial energy metabolism. Permeability changes of the inner or outer mitochondrial membrane may influence the respiration rate (82,83). H^+, K^+, Na^+, Ca^{2+}, and Mg^{2+} ions have their specific role in mitochondrial physiology and are transported across the inner mitochondrial membrane by one or more specific carriers. Few cation carriers have been isolated from mammalian tissues and identified, but none have been cloned and sequenced yet (10,84,85). So far no studies on human cation transporters have been reported.

The mitochondrial volume is mainly controlled by the K^+ cycle, which includes the K^+/H^+ antiporter (82 kDa), the K_{ATP} channel (54 kDa), and a K^+ leak induced by high membrane potential (84). However, Mg^{2+} is also involved in volume maintenance, besides stimulation of numerous enzymes and transporters, and ligation of different metabolites. Whereas the influx of Mg^{2+} occurs by a nonspecific diffusion in response to elevated membrane potentials, the efflux of Mg^{2+} may occur in exchange for H^+ (85). The transport of protons is discussed in Section II.

Mitochondrial Ca^{2+} plays a special role in muscle metabolism. Besides its uniform role in activating the citric acid cycle (by activating various dehydrogenases) (86) it contributes to the regulation of the cytosolic Ca^{2+} concentration in muscle, thereby influencing the contraction-relaxation process (87). Electrophoretic Ca^{2+} uptake across the inner mitochondrial membrane proceeds via a Ca^{2+} uniporter, whereas the electroneutral Ca^{2+} release proceeds by a Na^+/Ca^{2+} antiporter (110 kDa). This process is balanced by a Na^+/H^+ antiporter (59 kDa) (10,84,87).

The role of matrix Ca^{2+} in mitochondrial energy metabolism was described by McCormack et al. (88). The primary or secondary role of Ca^{2+} in human myopathies may be underestimated. In diseases such as Duchenne muscular dystrophy, myotonic dystrophy, and polymyositis, an increased Ca^{2+} concentration in muscle fibers has been reported (89). In some patients with disorders in mitochondrial energy metabolism calcification of the basal ganglia is reported (88,90). Accumulation of calcium in cytoplasm and mitochondria can lead to cell death (90,91). Ca^{2+} ions can bind to many macromolecules and metabolites. They activate or inactivate many enzymes in a direct way, but also play an indirect regulating role by induction of a phosphorylation or dephosphorylation process after being effected by a hormone signal.

V.   MITOCHONDRIAL SUBSTRATE CARRIERS

Several mitochondrial carriers specific for different substrates are described. They are directly or indirectly involved in mitochondrial energy metabolism. This group includes carriers for mitochondrial substrates such as pyruvate, oxoglutarate, citrate, succinate, carnitine, glutamate, and aspartate. In this section we give an overview of the present knowledge (if possible about human mitochondria) of the mitochondrial substrate carriers. Other mitochondrial transport systems (not directly involved in energy metabolism) involved in the urea cycle, gluconeogenesis, or fatty acid oxidation are discussed.

A.   Pyruvate Carrier

The pyruvate carrier (PyC), or monocarboxylate carrier, mediates the exchange of pyruvate and OH^- across the inner mitochondrial membrane. Pyruvate formed in the cytosol from glucose must be transported into mitochondria to supply the pyruvate dehydrogenase complex and pyruvate carboxylase located in the matrix space with their substrate. Gluconeogenesis and pyruvate oxidation depend on proper functioning of PyC.

The molecular mass for rat PyC is 34 kDa (92). In rat, PyC is mainly expressed in the brain (92). The chromosomal localization of the PyC gene in humans and the cDNA of this carrier are unknown at this moment.

So far, no proven case of a deficiency of this carrier has been reported. Recently, Selak et al. (20) described four patients presenting with hypotonia, developmental delay, seizures, severe headaches, and ophthalmological abnormalities. They examined isolated skeletal muscle mitochondria from 150 children. Four of these children had a high respiratory control index for all substrates tested except for pyruvate. In three successive cycles of ADP-stimulated oxygen consumption there was a progressive decrease of oxygen consumption with pyruvate plus malate as substrates, being 17%, 28%, and 58% below control values, respectively. Respiratory rates with glutamate, succinate, α-glycerolphosphate, acetyl-, octanoyl-, and palmitoyl-carnitine were normal. Activity of all respiratory chain enzymes and of the pyruvate dehydrogenase complex were also in the normal range. From these investigations the authors concluded that the rate-limiting step in pyruvate oxidation in these four patients appeared to reflect a decreased entry of pyruvate into the matrix. A defect in the pyruvate carrier may account for the decreased pyruvate oxidation in these patients.

We did not have indications for the presence of a deficiency in the pyruvate carrier among several hundreds of patients from whom a muscle specimen has been investigated biochemically on clinical and clinical-chemical grounds. This conclusion has been drawn indirectly from our observation that in all investigated patients in whom only the pyruvate oxidation rate was diminished, but not that of malate and succinate, the impairment could be ascribed to a defect in the pyruvate dehydrogenase complex.

B. Dicarboxylate Carrier

The mitochondrial dicarboxylate carrier (DIC) catalyzes an electroneutral exchange across the inner mitochondrial membrane of dicarboxylates (e.g., malonate, malate, succinate) for inorganic phosphate and certain sulfur-containing compounds (e.g., sulfite, sulfate, thiosulfate) (61,93). The carrier plays an important role in hepatic gluconeogenesis, urea synthesis, and sulfur metabolism (94). The activity of the dicarboxylate carrier is found to be elevated in type 1 diabetes and can be normalized with insulin therapy, thereby suggesting a role for insulin in regulation of the carrier (95,96).

The only nucleotide and amino acid sequence of DIC known up to now is that of yeast (97,98), which appeared to be a member of the mitochondrial transmembrane carrier protein family (29,30). No deficiencies in the dicarboxylate carrier have been reported.

C. Oxoglutarate Carrier

The oxoglutarate carrier (OGC) translocates 2-oxoglutarate across the inner mitochondrial membrane in exchange for malate or other dicarboxylic acids. This carrier forms part of the aspartate-malate shuttle (see Fig. 1) (99,100). There is evidence that the isolated and reconstituted OGC functions as a homodimer (101). The amino acid sequence of the OGC from different mammals has been determined by cDNA sequencing (102–104). The OGC cDNA encodes a protein of 314 amino acids, which belongs to the mitochondrial carrier family (29,30). Like

Cytosol   Mitochondrion

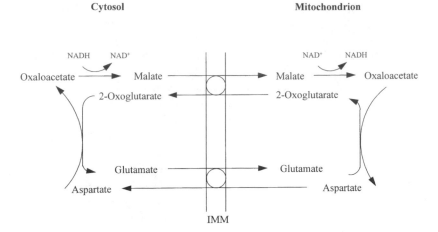

Figure 1   The aspartate-malate shuttle.

many other members of the family, the OGC has no processed import presequence. The human OGC gene has six exons and five introns (103). Its chromosomal localization is unknown.

We examined 12 patients with a 2-oxoglutaric aciduria and/or acidemia,
not due to a 2-oxoglutarate dehydrogenase deficiency, for a possible deficiency
of the OGC. So far we have not identified a patient with a deficiency of this
carrier using an immunochemical method, but it has to be stressed that real identification of such a deficiency is hampered by the wide range of immunochemical
responses observed in control muscle samples. Additional studies are necessary
in order to draw a definite conclusion concerning a possible deficiency of the 2-
oxoglutarate carrier in the aforementioned patients.

D.   Aspartate-Malate Shuttle

The inner mitochondrial membrane is impermeable for NADH. NADH generated
in the cytosol must therefore be transported across the inner mitochondrial membrane in another form before it can be oxidized by the mitochondrial respiratory
chain. This process is mediated by shuttles, the aspartate-malate shuttle
(100,105,106) being the most important one. The α-glycerolphosphate shuttle
(107), exhibiting an analogous function as the aspartate-malate shuttle, is of little
importance in mammalian muscle.

Two carrier proteins participate in the aspartate-malate shuttle: the 2-oxo-
glutarate carrier (OGC), exchanging 2-oxoglutarate and malate, and the aspartate-

glutamate carrier (AGC), exchanging aspartate and glutamate across the inner mitochondrial membrane (see Fig. 1) (100).

A defect in one of these shuttles, especially in the aspartate-malate shuttle, might be associated with a disturbance in mitochondrial energy generation. So far two patients with a defect in the aspartate-malate shuttle have been reported. The first patient was described by Hayes et al. (18). This patient, a 27-year-old man, showed muscle pain and myoglobinuria during severe exercise. There were no abnormally structured mitochondria. In vivo NMR studies during exercise showed an overutilization of PCr. In the resting state blood lactate and pyruvate concentrations were normal (700 μM and 46 μM, respectively). In an exercise protocol these values increased pathologically. The lactate:pyruvate ratio in the patient rose from 15 (at rest) to 60 (during exercise). The oxygen consumption by isolated muscle mitochondria was normal with several substrates. The oxidation rate of exogenous NADH by isolated mitochondria in the patient was only 20% of that in controls. From these data the authors concluded that the aspartate-malate shuttle was deficient.

Recently a second patient with an aspartate-malate shuttle defect was reported (19). This boy was well until 2 months of age, when suckling difficulties occurred. Examination revealed failure to thrive, axial hypotonia, mild hepatomegaly, and metabolic acidosis (pH 7.3). Investigations of the redox status in plasma showed permanent hyperlactatemia (3–6 mM) and increased lactate:pyruvate ratios (15–30). By MRI a hypersignal in the putamen nucleus was detected. The patient died suddenly at 3 months. In mitochondria isolated from skeletal muscle, fibroblasts, and lymphocytes, low oxidation rates were observed for pyruvate, whereas succinate oxidation was normal. Activities of respiratory chain complexes were all normal. The activity of the aspartate-malate shuttle was tested in digitonin permeabilized fibroblasts using a reconstituted system: the shuttle activity was significantly decreased in the patient's fibroblasts (35% of control values).

It should be emphasized that it is difficult to obtain indications for an aspartate-malate shuttle defect from routine screening, and the defect is difficult to establish by studies on tissues or cells.

E. Carnitine-Acylcarnitine Carrier

The carnitine-acylcarnitine carrier (CAC) shuttles acylcarnitine esters in exchange for free carnitine across the inner mitochondrial membrane (108,109). This transport is an essential step in the process of long-chain fatty acid oxidation (3,110,111). The oxidation of fatty acids in mitochondria plays an important role in energy production. During fasting, fatty acids are used for hepatic ketone body synthesis. Furthermore, fatty acids are an important source of energy for heart and also for skeletal muscle during exercise, while ketone bodies are excellent substrates for brain (3,110).

The cloning and sequencing of rat CAC cDNA has recently been described (112). Very recently we determined the nucleotide sequence of the human CAC cDNA and the corresponding amino acid sequence, as well as the distribution of CAC mRNA in human tissues (113). Also the CAC protein is a member of the mitochondrial carrier protein family (29,30). The gene encoding human CAC is located on chromosome 3p21.31 (114).

So far several cases with deficient CAC activity have been reported (108,111,113,115–120). The main features in these severely affected patients with onset in the neonatal period are hypoketotic hypoglycemia, mild hyperammonemia, variable dicarboxylic aciduria, hepatomegaly with abnormal liver functions, hypertrophic cardiomyopathies with or without dysrhythmias, and skeletal muscle weakness. In all cases CAC activity in cultured skin fibroblasts is below detectable levels. We recently found the first mutation in the cDNA of a CAC deficient patient (113).

F.   Oxygen Transport

A crucial substrate for mitochondria is oxygen. Hemoglobin in circulating erythrocytes takes care of oxygen transport to the tissues. Uptake of oxygen by cells takes place in capillaries. In some cell types, that is, muscle fibers and cardiomyocytes, intracellular transport is facilitated by a specific protein, myoglobin. In most other cell types the small oxygen molecules are transported by diffusion and convection. The apolar oxygen molecule crosses both the cellular and mitochondrial membranes by diffusion. It can easily bridge the short intramembranous distance to the oxygen reduction site of cytochrome-c oxidase in the inner mitochondrial membrane. A recent study on crystalline cytochrome-c oxidase makes plausible that several different subunits of cytochrome-c oxidase with hydrophobic domains can form one or more oxygen channels in the embedded enzyme (121).

Cytochrome-c oxidase has a very high affinity toward oxygen, some orders of magnitude higher than myoglobin does (122). Maximum oxygen consumption by intact rat cardiomyocytes can be reached at 8 μM oxygen (123). Even under hypoxic conditions, that is, at oxygen pressures as low as 10% of the normal cytosolic value, red muscle can maintain maximal oxygen consumption (124). In other kinds of tissue, for example, in brain, the oxidative phosphorylating capacity may be more susceptible to decreased oxygen pressure. It is the (unfacilitated) oxygen transport through the cytosolic compartment which is rate limiting rather than the transmembrane transport. There is, even at high respiration rate, almost no gradient of oxygen over the inner mitochondrial membrane: less than 1% of the intracellular oxygen pressure (122).

A disturbance in the oxygen uptake by mitochondria can theoretically induce a defective mitochondrial energy metabolism and so a mitochondrial encephalomyopathy. So far, however, no indications have been gathered for such a defect. The above-mentioned features of oxygen transport render the expression of a functional defect in it less presumably. A disturbance in the oxygen transport from the capillaries to muscle mitochondria has been hypothesized by us to occur in a specific group of patients (McKusick 212350), suffering from hypertrophic cardiomyopathy, cataract, and mitochondrial myopathy (57). The severe increase in blood lactate concentration, even after minimal exercise, is very suggestive of dysfunction of skeletal muscle mitochondria. However, the mitochondria showed an undisturbed oxidative phosphorylation in vitro (58,125). The hypothesis of abnormal myoglobin molecules in these patients could not be confirmed (58).

The presence of so-called ragged-red fibers, often used as a hallmark of mitochondrial myopathy (1,2), is possibly caused by stress of the mitochondria. Such accumulations of subsarcolemmally located mitochondria can be induced in rat by occlusion of the leg muscle artery (A. M. Stadhouders, personal communication). Skeletal muscle reacts to the hypoxic circumstances by proliferation of mitochondria. Their subsarcolemmal location guarantees the most efficient manner of oxygen uptake.

G. Other Mitochondrial Carriers

The main single carriers and channels, which can be of importance for physiological and pathophysiological functioning of muscle mitochondria, have been discussed in previous sections. Here we mention other mitochondrial carrier systems, not (directly) involved in mitochondrial energy metabolism. These carriers function in processes such as ureagenesis, lipogenesis, and gluconeogenesis.

In ureagenesis, urea is formed from ammonia. This process takes place in the liver (126). One of the involved carriers is the aspartate-glutamate carrier, which translocates aspartate out of the mitochondria in cotransport with glutamate. This glutamate can in turn be transported out of the matrix by the glutamate carrier (127). An essential component of the urea cycle is the ornithine carrier, which catalyses the entry of ornithine into the mitochondrial matrix in exchange for citrulline (128–130). The only nucleotide and amino acid sequence known so far of the ornithine carrier is that of yeast (131), which also belongs to the mitochondrial carrier family.

In lipogenesis, fatty acids are synthesized from glucose in the cytosol, thereby using acetyl-CoA formed in the matrix of mitochondria. The acetyl units are transferred across the inner mitochondrial membrane to the cytosol in the form of citrate by the citrate carrier (CIC; 132). In the cytosol, citrate is cleaved by citrate lyase into acetyl-CoA and oxaloacetate. The mitochondrial CIC, also called tricarboxylate carrier, plays an important role in fatty acid and sterol synthesis, gluconeogenesis, and the transfer of reducing equivalents across the membrane. This carrier is present in liver, but absent or very low in heart, skeletal muscle, and brain (133). The human CIC gene has been completely sequenced, and appears to be a member of the mitochondrial transporter family (134). Another shuttle system, the malate-citrate shuttle, is also described to play a role in the transport of citrate (133).

For gluconeogenesis in the cytosol, malate, aspartate, citrate, glutamate, and 2-oxoglutarate are used as precursors. These metabolites are transported out of the mitochondrial matrix by different carrier systems: CIC, DIC (Sec. V.B), OGC (Sec. V.C), and the aspartate-malate shuttle (Sec. V.D). Furthermore, gluconeogenesis depends on proper functioning of the PyC (Sec. V.A) (135,136).

The discovery of the ATP-Mg/P$_i$ carrier in the inner membrane of liver mitochondria (137,138) was prompted by the need to explain how matrix adenine nucleotide content could change in vivo as observed under many physiological situations. In liver, changes in matrix adenine nucleotide concentrations that are brought about by the ATP-Mg/P$_i$ carrier can affect the activity of adenine nucleotide-dependent enzymes which are located in the mitochondrial compartment. These enzymes in turn contribute to the overall regulation of bioenergetic function, the flux through the gluconeogenesis and urea synthesis pathways, and organelle biosynthesis. It is not clear if this carrier is also involved in adenine nucleotide distribution in other tissues than liver (137,138).

Other translocating processes in mitochondrial membranes that should be mentioned are the transport of amino acids, vitamins, and cofactors. The neutral amino acid carrier present in the inner mitochondrial membrane transports a broad range of neutral amino acids (139), but for the neutral amino acid proline a separate carrier system may be present (139). Transport of other amino acids occurs via more specific carrier systems (e.g., glutamate and glutamine carriers) (105). The knowledge about the import of vitamins and cofactors is very restricted. Different mechanisms have already been proposed, including diffusion, diffusion followed by binding to certain mitochondrial proteins, and carrier-mediated transport (140). Vitamins and cofactors described to be transported by carrier mechanisms are riboflavin (141), coenzyme A (142), thiamine (143), and folates (144). The knowledge of these transporter proteins and their genes in men is very restricted. Up to now no deficiencies have been published.

VI. INNER–OUTER MEMBRANE INTERACTIONS

Mitochondrial transmembrane carriers do not only act as individually functioning transmembrane carriers, but can also be part of complicated structures in which several types of transport proteins and other components are cooperating. The protein import machinery and some megachannels belong to these structures.

A. Mitochondrial Megachannel

The presence of VDAC, ANT, hexokinase, mitochondrial creatine kinase, glycerol kinase, and the benzodiazepine receptor have been described in structures such as contact sites and megachannels (12,83,145,146). The interactions between the various components are tissue specific and depend on the developmental and metabolic stages of the cells. Their role in human mitochondrial energy metabolism is speculative at this time.

There is evidence that hexokinase and glycerol kinase interact with VDAC in a tissue-specific fashion, and that their binding is altered in certain disease states. The most prominent example of the involvement of the VDAC-kinase interaction in disease concerns tumor cell transformation. A number of tumor cells have been characterized and appear to have an increased proportion of hexokinase bound to VDAC on the outer mitochondrial membrane. In brain, it has been shown that bound hexokinase activity increases under conditions of ischemia.

B. Mitochondrial Protein Import and Biogenesis

Over the last decade many studies have been performed in order to elucidate the protein import mechanism by mitochondria. Some hundreds of nuclear-encoded proteins have to be imported and guided to reach their ultimate location in the matrix, inner or outer membrane, or intermembrane space. It has become obvious that a large number of different components are involved in the machinery responsible for protein import across both mitochondrial membranes (147–149).

At least nine transmembrane proteins have been found which are involved in the protein import process. Some of them have receptor properties and recognize the leader sequences on the N-terminus of the proteins to be imported. Perhaps mRNAs of these precursor proteins also play a role in targeting (150). The proteins are, after being unfolded by cytoplasmic chaperones like heat shock protein 70 (Hsp70), dragged into membrane channels. The unfolded proteins can be translocated via the so-called contact sites, where the mitochondrial outer and inner membrane collaborate to form a channel: the mitochondrial megachannel (MMC; see Sec. VI.A). In the outer membrane the PSC (peptide-sensitive channel), and in the inner membrane the MCC (megaconductance channel) play a central role. The relation of these two channels to the MMC is not clear. In the matrix the leader sequences are removed by protease activity and again heat shock proteins and other chaperones take care of refolding of the mature proteins. Likely the assemblage and insertion of the proteins into the inner membrane are also guided by Hsp's. Some steps of the import process are driven by the transmembrane potential or energetic state.

Although the rough mechanism of protein import in purified systems is unraveled, its consequences for mitochondrial protein import in vivo are still obscure. The specificity of the importing systems toward various proteins must still be established. Which property of the precursor protein determines whether the mature protein will be embedded in the inner membrane or in the outer membrane, or suspended in the matrix? In this context it is interesting that the VDAC molecule as an outer membrane protein (see Sec. III.C) has no leader sequence which is removed in a later stage (151). Remarkably, the C-terminus of the VDAC precursor seems to be more important for recognition by mitochondria than the usual N-terminus (152). Also ANT and the yeast PiC contain an uncleavable targeting sequence (153). The substrate carriers of the inner mitochondrial membrane are typically synthesized as mature-sized proteins on cytosolic polysomes. In only a few cases, such as the phosphate carrier of mammalian mitochondria (154), the carrier proteins are synthesized with cleavable N-terminal extensions. Studies on the role of the presequence of the mammalian phosphate carrier have shown that the major import information resides in the mature part of the phosphate carrier, as in the case of the presequence-deficient members of the mitochondrial carrier family (155). Knowledge about the protein import process in men is very scarce (156).

Only a couple of patients has been described with a dysfunctioning of one of the above-mentioned mechanisms. Schapira et al. (17) describe a girl with generalized muscle weakness and exercise intolerance from birth on. Her blood lactate concentration increased pathologically after standardized exercise. In isolated skeletal muscle mitochondria many parameters of the energy metabolism were decreased, succinate dehydrogenase being the most deficient enzyme. The presence in the total homogenate of the iron-sulfur containing "Rieske" protein of complex III, in combination with its absence in isolated mitochondria, strongly suggests a disturbance in the import machinery of the Rieske protein.

A patient with much more severe clinical symptoms appeared to suffer from a lack of Hsp60 (157,158). Severe hypotonia, facial dysmorphism, frontal bossing, abundant body hair, hepatomegaly, and lactic acidosis were observed. The baby died at her second day of life. Previously another child of this consanguineous couple died at the third day of life. The patient's fibroblasts showed deficiencies of all measured mitochondrial enzymes, while the peroxisomal, lysosomal, and cytosolic enzymes had normal activities. An immunochemical study revealed a strongly decreased content of Hsp60 and a normal content of Hsp70 in the fibroblasts. A second patient with a lack of Hsp60 has been reported (159). Symptoms in this girl were hypotonia, facial dysmorphisms, feeding problems, and failure to thrive. Besides lactic acidosis many organic acids were excreted in urine in a pathological range. The child died at age 4.5 years. Also in this case, all mitochondrial enzyme activities were reduced. The Hsp60 content was decreased to about 30% of control values. In the last category of patients the nuclear-encoded proteins could apparently not be imported or not be assembled after having been imported, leading to an early breakdown.

VII. DISCUSSION

A. Mitochondrial Transmembrane Processes

The inner mitochondrial membrane contains several transport systems for metabolites and reduced equivalents which are necessary for important metabolic pathways such as oxidative phosphorylation, fatty acid oxidation, gluconeogenesis, and the urea cycle. Defects in transport systems which play a role in muscle tissue can cause mitochondrial myopathies. Several of these carriers are cloned and sequenced. It appears that they have similarities in their amino acid sequences, particularly in the membrane-spanning regions, indicating that they are encoded by a gene family and therefore they have evolved from a common ancestor. Their common characteristics are that their molecular weights are within the same range (28–34 kDa) and their polypeptide chains consist of three tandemly repeated related sequences of approximately 100 amino acids each. Each element is folded into two transmembrane α-helices linked by an extensive polar region, forming a structure with six transmembrane α-helices (29,30).

    In spite of the enormous progress in the research on mitochondrial transport systems in the last decade, many details about functioning under physiological and pathophysiological conditions are still lacking. Knowledge of the transport systems in men is even more restricted.

    The various types of carriers do not have the same absolute and mutual concentrations in mitochondria of different tissues. Most carriers are expressed according to the metabolic function of the tissue. For example, the DIC activity is high in liver and kidney, where it is involved in gluconeogenesis, and the CIC activity is high in tissues, such as liver and adipose tissue, where transport of citrate out of mitochondria is required for extramitochondrial fatty acid synthesis. In heart muscle, where these biosynthetic processes do not occur to any significant extent, the activity of these carriers is low or absent. ANT, PiC, and PyC are ubiquitously expressed with higher concentrations in heart and skeletal muscle.

B. Human Mitochondrial Carrier Deficiencies

It is well known that the mitochondrial ADP and P_i concentration, the ATP: ADP ratio, as well as the membrane potential may significantly contribute to the control of mitochondrial respiration (43,44). As a consequence defects of carriers such as ANT and PiC must be considered as a cause of mitochondriopathy. This also holds for other mitochondrial carrier proteins or ion channels which contribute to maintaining the osmotic homeostasis of the mitochondrial matrix (10,160).

So far only a few subjects with a specific defect in mitochondrial transmembrane transport have been reported, among which are ANT and VDAC deficiency (14–16), a defect in the protein import machinery (17), a disturbed aspartate-malate shuttle (18,19), and several CAC deficiencies (108,111,113,115–120). An abnormality in the level of cDNA has been documented in only one patient (113). More systematic studies on the different transport proteins will likely reveal additional disturbances causing mitochondriopathies.

Since some of these carriers occur in tissue-specific isoforms it can be imagined that a defect in one of these carriers may cause a disorder with a tissue-specific nature, a phenomenon not uncommon in mitochondriopathies (1,14–16,161). The presence of isoforms also has consequences for selection of tissue for diagnostic examinations.

C. Reactive Oxygen Species

The number of patients in whom a carrier defect has been established is low, hitherto only for one case in which a mutation in the corresponding cDNA has been established (113). Therefore a secondary cause of carrier defects must be considered. Recently, increased concentrations of oxygen radicals have been thought to cause disturbances in intermediary metabolism. Particularly in membrane-bound enzymatic and transport processes, membranes contain very susceptible unsaturated fatty acid moieties.

During oxidative phosphorylation most oxygen is converted into water via tetravalent reduction. However, even under physiological conditions, about 1–2% of oxygen metabolized during the process of oxidative phosphorylation is converted into highly reactive substances, the reactive oxygen species (ROS) (162). The physiological significance of ROS under conditions as regulation and stimulation of signal transduction are becoming better understood (163).

Univalent reduction of molecular oxygen results in the formation of the O_2^- radical. Dismutation of O_2^- causes H_2O_2 production. Furthermore, O_2^- can interact with its dismutation product, H_2O_2 resulting in the generation of the hydroxyl radical (164). All oxygen species formed are highly reactive, the hydroxyl radical being the most potent one, and react with many biological components in their direct surrounding; in the case of mitochondria, mtDNA, enzymes, membrane proteins, free iron, and phospholipids (165).

To counteract ROS activity, cells have been provided with enzymatic and nonenzymatic antioxidant defense systems like superoxide dismutases (SoDs), glutathione peroxidases, the lipophilic antioxidant α-tocopherol and β-carotene (164).

In vitro experiments have clearly shown that in the presence of inhibitors of distinct parts of the mitochondrial respiratory chain ROS generation can increase several fold (166). Sites of O_2^- production have been proposed in complex I and III of the respiratory chain (167). Conceivably, genetically determined dysfunction of the mitochondrial respiratory chain may give rise to in vivo ROS overproduction, thereby disturbing the balance between ROS and its defense mechanisms, leading to a sequela of harmful cell biological events like glutathione depletion, lipid peroxidation, an increase in the cytosolic Ca^{2+} concentration and oxidation of protein thiol groups (168).

Pitkänen and Robinson (169) were the first to show that mitochondrial complex I deficiency in skin fibroblasts may lead to increased production of O_2^- radicals and induction of Mn-SoD, an intramitochondrial ROS scavenging enzyme.

Besides respiratory chain complexes numerous other proteins are located in or attached to the inner mitochondrial membrane, among which are most of the mitochondrial channels and carriers. Due to the relatively close proximity of the various inner mitochondrial membrane proteins, it seems theoretically possible that a disturbance in one of these proteins has an influence on one or more of the other inner mitochondrial membrane proteins. An increase in the level of products of the one-electron reduction of oxygen is known to open the mitochondrial permeability transition pores (170). So far no combined deficiency of one of the respiratory chain complexes and proteins of the mitochondrial transport system has been reported.

In an attempt to reduce possibly enhanced levels of free radicals and/or to promote membrane stabilization, the described ANT-deficient boy (14) was treated with vitamin E. This produced an obvious improvement of the patient's myopathic complaints. The clinical improvement was confirmed by ^{31}P-NMR spectroscopy 10 weeks after starting the therapy (171). It is very unlikely that vitamin E has a direct effect on the functioning and/or biogenesis of ANT. An explanation for the beneficial vitamin E therapy might be that because of the elevated activity of almost all enzymes of the respiratory chain in the patient, enhanced production of O_2^- radicals and H_2O_2 might occur. If this is not accompanied by increased activity of enzymes involved in scavenging these highly reactive compounds, reactive hydroxyl radicals might accumulate, as seen in normal ageing (172,173). These high levels of free radicals might result in enhanced mitochondrial phospholipid peroxidation. Vitamin E may (partly) counteract this effect by scavenging free radicals. Since most mitochondrial inner-membrane enzymes depend on membrane phospholipids for optimal activity, it is possible that the improvement in the patient is caused by a better functioning of the partially deficient ANT (171).

VIII. CONCLUDING REMARKS

Several newly discovered translocating systems in mitochondria have been reported in eucaryotes over the last three decades, and knowledge about already known carriers and channels has evidently increased. Application of new sensitive techniques has played an important role in this development. Electrophysiological studies have produced many results. Biochemical properties of transport proteins have been settled in more detail as the DNA sequences and amino acid composition of several carrier proteins have been elucidated. Besides, the potential of diagnostic programs for mitochondrial diseases has also obviously been expanded. The number of detected patients has increased, and certainly also the heterogeneity and complexity of the observed biochemical pathological findings. But unfortunately the progress in both areas has not been associated with firm interactions between the fundamental and applied fields of science. Lack of knowledge about other people's potentials is without doubt one of the reasons. Another reason is the fact that the availability of appropriate material from patients is often too restricted to study transport processes. The functional capacity of many transport systems, discussed in this chapter, can only be estimated in intact mitochondria originating from fresh tissues. The coupling state and aspartate-malate shuttle are good illustrations. Robinson et al. (174) described in 1983 two patients with neurological symptoms and lactic acidosis whose fibroblasts showed a hampered transport of reducing equivalents across the mitochondrial membranes. But up to now no responsible protein has been suggested. The molecular origin of the disturbed coupling state of muscle mitochondria in Luft disease is also still obscure. At least in one case a defect at the molecular level was established (113).

Examination of the functioning of many channel or carrier systems is difficult and can only be done in specialized laboratories. Drawing conclusions about the consequences for in vivo metabolism and for the patient is even harder. The impact of substrate carrier deficiencies is more predictable than the impact of a deficiency of an ion channel. Defects in ion translocating systems probably can induce less specific and hardly recognizable abnormalities in the mitochondrial energy metabolism, because the involved ions like Ca^{2+}, protons, and others influence many metabolic steps, as well as the osmotic value of a compartment (see also Ref. 16). Nevertheless it seems worthwhile to investigate the most obvious transport systems in those cases in which no enzyme deficiency has been found to explain a disturbed flux from pyruvate to ATP. Such translocation defects may be caused by a mutation in their encoding gene (113,120) or transcription process (8), but also by pathological circumstances like accumulation of oxygen radicals, free fatty acids, acyl-CoA esters, Ca^{2+}, nitric oxide, and so on (26,90,113,175, 176).

Routine diagnostic programs for mitochondrial disorders (1,2,4) have not been developed for detection of membrane-bound transport processes. Also, it will be difficult to find patients with such a defect in a muscle biopsy specimen, for instance, a defective protein import component. If the defect does not occur in a tissue-specific way the suspected pathway can be established in cultured skin fibroblasts of the patient. Fibroblast investigations can be helpful in distinguishing primary and secondary mitochondrial abnormalities. For example, channel or enzyme malfunctioning caused by membrane alteration as a consequence of oxygen radical attack in vivo is likely absent under cultivation conditions.

Studies on a possible relationship between defects in the regulation of mitochondrial membrane permeability and mitochondriocytopathies have not yet been performed. The insight into permeability and its potential importance for several mitochondrial interactions is drastically increasing (177). Mitochondrial permeability is involved in the process of programmed cell death—apoptosis. Many of the systems and compounds discussed here are involved in apoptosis: Ca^{2+} accumulation, lack of, ATP nitric oxide, decreased membrane potential, oxidative stress with oxygen radicals, increased membrane permeability, benzodiazepine receptor (170,178,179). In necrosis, the so-called mitochondrial permeability transition pore in the inner membrane plays a role. This channel is influenced by Ca^{2+} and proton concentrations, by oxidative stress, and by the mitochondrial membrane potential (180–182). Cyclosporin A and the oncoprotein Bcl-2, both being able to interact with membranes, have a protective effect in cell death and in apoptosis, respectively. The role of mitochondrial membrane channels in apoptosis, in necrosis, in the likely related process of ageing, and in mitochondrial disorders will undoubtedly draw more attention in the near future.

ACKNOWLEDGMENTS

The Prinses Beatrix Fonds (grants 93-018 and 95-0501 to M.H.) and Telethon-Italy (grant 985 to F.P.) are gratefully acknowledged for financial support.

REFERENCES

1. DiMauro S. Mitochondrial encephalomyopathies. In: Rosenburg RN, Prusiner SB, DiMauro S, Barchi LR, Kunkel LM, eds. The Molecular and Genetic Basis of Neurological Disease. Boston: Butterworth-Heinemann, 1993:665–674.
2. Shoffner JM, Wallace DC. Oxidative phosphorylation diseases and mitochondrial DNA mutation: diagnosis and treatment. Annu Rev Nutr 1994; 14:535–568.
3. Coates PM, Tanaka K. Molecular basis of mitochondrial fatty acid oxidation defects. J Lipid Res 1992; 33:1099–1110.
4. Ruitenbeek W, Wendel U, Trijbels F, Sengers R. Mitochondrial energy metabolism. In: Physician's Guide to the Laboratory Diagnosis of Metabolic Diseases. Blau N, Duran M, Blaskovics ME, eds. London: Chapman & Hall, 1996:391–406.
5. Trijbels JMF, Scholte HR, Ruitenbeek W, et al. Problems with the biochemical diagnosis in mitochondrial (encephalo-)myopathies. Eur J Pediatr 1993; 152:178–184.
6. Smeitink J, Ruitenbeek W, Sengers R, et al. Mitochondrial creatine kinase activity

in patients with disturbed energy generation in muscle mitochondria. J Inherit Metab Dis 1994; 17:67–73.

7. Ruitenbeek W, Huizing M, DePinto V, et al. Defects of mitochondrial membrane-bound transport proteins in human mitochondriopathies: a biochemical approach. In: Palmieri F, ed. Progress in Cell Research, vol. 5 Amsterdam: Elsevier Science, 1995:225–229.

8. Huizing M, DePinto V, Ruitenbeek W. et al. Importance of mitochondrial trans-membrane processes in human mitochondriopathies. J Bioenerg Biomembr 1996; 28:109–114.

9. Tylor DD, Sutton CM. Mitochondrial transporting systems. In: Bittar EE, ed. Membrane Structure and Function, vol. V. New York: John Wiley 1984:181–270.

10. Li W, Shariat-Madar Z, Powers M, et al. Reconstitution, identification, purification, and immunological characterization of the 110 kDa Na^+/Ca^{2+} antiporter from beef heart mitochondria. J Biol Chem 1992; 267:17983–17989.

11. Beavis AD, Lu Y, Garlid KD. On the regulation of K^+ uniport in intact mitochondria by adenine nucleotides and analogs. J Biol Chem 1993; 268:997–1004.

12. Benz R. Permeation of hydrophilic solutes through mitochondrial outer membranes: review on mitochondrial porins. Biochim Biophys Acta 1994; 1197:167–196.

13. Mannella CA. The 'ins' and 'outs' of mitochondrial membrane channels. Trends Biochem Sci 1992; 17:315–320.

14. Bakker HD, Scholte HR, Van den Bogert C, et al. Deficiency of the adenine nucleotide translocator in muscle of a patient with myopathy and lactic acidosis: a new mitochondrial defect. Pediatr Res 1993; 33:412–417.

15. Huizing M, Ruitenbeek W, Thinnes FP, DePinto V. Lack of voltage-dependent anion channel in human mitochondrial myopathies. Lancet 1994; 344:762.

16. Huizing M, Ruitenbeek W, Thinnes FP, et al. Deficiency of the voltage-dependent anion channel (VDAC): a novel cause of mitochondriopathy. Pediatr Res 1996; 39:760–765.

17. Schapira AHV, Cooper JM, Morgan-Hughes JA, Landon DN, Clark JB. Mitochondrial myopathy with a defect of mitochondrial protein transport. N Engl J Med 1990; 323:37–42.

18. Hayes DJ, Taylor DJ, Bore PJ, et al. An unusual metabolic myopathy: a malate-aspartate shuttle defect. J Neurol Sci 1987; 82:27–39.

19. Brivet M, Slama A, Rustin P, et al. A mitochondrial encephalomyopathy with a presumptive defect at the level of aspartate/malate shuttle [abstract P69]. Abstractbook 7th International Congress of Inborn Errors of Metabolism, Vienna, 1997.

20. Selak MA, Grover WD, Foley CM, Miles DK, Salganicoff L. Possible defect in pyruvate transport in skeletal muscle mitochondria from four children with encephalomyopathies and myopathies [abstract 59]. International Conference on Mitochondrial Diseases, "Challenges in the Study of Encephalomyopathies of Mitochondrial Origin," Philadelphia, 1997.

21. Pande SV, Murthy MSR. Carnitine-acylcarnitine translocase deficiency: implications in human pathology. Biochim Biophys Acta 1994; 1226:269–276.

22. Inoue I, Saheki T, Kayanuma K, et al. Biochemical analysis of decreased ornithine transport activity in the liver mitochondria from patients with hyperornithinemia,

hyperammonemia and homocitrullinuria. Biochim Biophys Acta 1988; 964:90–95.

23. Hak JB, Van Beek JHGM, Westerhof N. Acidosis slows the response of oxidative phosphorylation to metabolic demand in isolated rabbit heart. Pflügers Arch 1993; 423:324–329.

24. Mitchell P, Moyle J. Estimation of membrane potential and pH difference across the cristae membrane of rat liver mitochondria. Eur J Biochem 1969; 7:471–484.

25. Brand MD, Chien LF, Ainscow EK, Rolfe DF, Porter RK. The causes and functions of mitochondrial proton leak. Biochim Biophys Acta 1994; 1187:132–139.

26. Wojtczak L, Schonfeld P. Effect of fatty acids on energy coupling processes in mitochondria. Biochim. Biophys. Acta 1993; 1183:41–57.

27. Verburg MP, Oerlemans FTJJ, Van Bennekom CA, et al. *In vivo* induced malignant hyperthermia in pigs. I. Physiological and biochemical changes and the influence of dantrolene sodium. Acta Anaesthesiol Scand 1984; 28:1–8.

28. Ruitenbeek W, Verburg MP, Janssen AJM, Stadhouders AM, Sengers RCA. *In vivo* induced malignant hyperthermia in pigs. II. Metabolism of skeletal muscle mitochondria. Acta Anaesthesiol Scand 1984; 28:9–13.

29. Walker JE, Runswick MJ. The mitochondrial transport protein super-family. J Bioenerg Biomembr 1993; 25:435–446.

30. Palmieri F. Mitochondrial carrier proteins. FEBS Lett 1994; 346:48–54.

31. Fleury C, Neverova M, Collins S, et al. Uncoupling protein-2: a novel gene linked to obesity and hyperinsulinemia. Nat Genet 1997; 15:223–224.

32. Boss O, Samec S, Paoloni-Giacobino A, et al. Uncoupling protein-3: a new member of the mitochondrial carrier family with tissue-specific expression. FEBS Lett 1997; 408:39–42.

33. Luft R, Ikkos D, Palmieri G, Ernster L, Afzelius B. A case of severe hypermetabolism of nonthyroid origin with a defect in the maintainance of mitochondrial respiratory control: a correlated clinical, biochemical, and morphological study. J Clin Invest 1962; 41:1776–1804.

34. Van Wijngaarden GK, Bethlem J, Meijer AE, Hulsmann WC, Feltkamp CA. Skeletal muscle disease with abnormal mitochondria. Brain 1967; 90:577–592.

35. Haydar NA, Conn HL, Afifi A, et al. Severe hypermetabolism with primary abnormality of skeletal muscle mitochondria. Ann Intern Med 1971; 74:548–558.

36. Yamada EW, Huzel NJ. Distribution of the ATPase inhibitor proteins of mitochondria in mammalian tissues including fibroblasts from a patient with Luft's disease. Biochim Biophys Acta 1992; 1139:143–147.

37. Ruitenbeek W, Joosten E, Janssen A, Trijbels F, Sengers R. A defective *in vitro* coupling of mitochondria in myopathies. J Neurol Sci 1990; 98(suppl.):306.

38. Müller-Höcker J, Pongratz D, Hubner G. Activation of mitochondrial ATPase as evidence of loosely coupled oxidative phosphorylation in various skeletal muscle disorders. J Neurol Sci 1986; 74:199–213.

39. Klingenberg M. Membrane protein oligomeric structure and transport function. Nature 1981; 290:449–454.

40. Brandolin G, Le Saux A, Trezeguet V, Lauquin GJM, Vignais PV. Chemical, immunological, enzymatic, and genetic approaches to studying the arrangement of the

peptide chain of the ADP/ATP carrier in the mitochondrial membrane. J Bioenerg Biomembr 1993; 25:459–472.

41. Battini R, Ferrari S, Kaczmarek L, et al. Molecular cloning of a cDNA for a human ADP/ATP carrier which is growth-regulated. J Biol Chem 1987; 9:4355–4359.

42. Houldsworth J, Attardi G. Two distinct genes for ADP/ATP translocase are expressed at the mRNA level in adult human liver. Proc Natl Acad Sci USA 1988; 85:377–381.

43. Tager JM, Wanders RJA, Groen AK, et al. Control of mitochondrial respiration. FEBS Lett 1983; 151:1–9.

44. Letellier T, Malgat M, Mazat JP. Control of oxidative phosphorylation in rat muscle mitochondria: implications for mitochondrial myopathies. Biochim Biophys Acta 1993; 1141:58–64.

45. Hackenberg H, Klingenberg M. Molecular weight and hydrodynamic parameters of the adenosine 5′-diphosphate-adenosine 5′-triphosphate carrier in Triton X-100. Biochemistry 1980; 19:548–555.

46. Vignais PV, Block MR, Boulay F, Brandolin G, Lauquin GJM. Molecular aspects of structure-function relationships in mitochondrial adenine nucleotide carrier. In: Benga G, ed. Structure and Properties of Cell Membranes, vol. II. Boca Raton, FL: CRC Press, 1985; 139–304.

47. Dalbon P, Brandolin G, Boulay F, Hoppe J, Vignais PV. Mapping of the nucleotide binding sites in the ADP/ATP carrier of beef heart mitochondria by photo-labeling with 2-azido[α-^{32}P]adenosine diphosphate. Biochemistry 1988; 27:5141–5149.

48. Mayinger P, Winkler E, Klingenberg M. The ADP/ATP carrier from yeast (AAC-2) is uniquely suited for the assignment of the binding center by photoaffinity labeling. FEBS Lett 1989; 244:421–426.

49. Walker JE, Cozens AL, Dyer MR, et al. DNA sequence of two expressed nuclear genes for human mitochondrial ADP/ATP translocase. Chem Crypta 1987; 27B: 97–105.

50. Neckelmann N, Li K, Wade RP, Shuster R, Wallace DC. cDNA sequence of a human skeletal muscle ADP/ATP translocator: lack of a leader peptide, divergence from a fibroblast translocator cDNA, and coevolution with mitochondrial DNA genes. Proc Natl Acad Sci USA 1987; 84:7580–7584.

51. Cozens AL, Runswick MJ, Walker JE. DNA sequences of two expressed nuclear genes for human mitochondrial ADP/ATP translocase. J Mol Biol 1989; 206:261–280.

52. Li K, Warner CK, Hodge JA, et al. A human muscle adenine nucleotide translocator gene has four exons, is located on chromosome 4, and is differentially expressed. J Biol Chem 1989; 264:13998–14004.

53. Stepien G, Torroni A, Chung AB, Hodge JA, Wallace DC. Differential expression of adenine nucleotide translocator isoforms in mammalian tissues and during muscle cell differentiation. J Biol Chem 1992; 267:14592–14597.

54. Ku DH, Kagan J, Chen ST, et al. The human fibroblast adenine nucleotide translocator gene. J Biol Chem 1990; 265:16060–16063.

55. Torroni A, Stepien G, Hodge JA, Wallace DC. Neoplastic transformation is associ-

ated with coordinate induction of nuclear and cytoplasmic oxidative phosphorylation genes. J Biol Chem 1990; 265:20589–20593.

56. Schultheiss HP, Schulze K, Dörner A. Significance of the adenine nucleotide translocator in the pathogenesis of viral heart disease. Mol Cell Biochem 1996; 163/164:319–327.

57. Sengers RCA, Trijbels JMF, Willems JL, Daniëls O, Stadhouders AM. Congenital cataract and mitochondrial myopathy of skeletal and heart muscle associated with lactic acidosis after exercise. J Pediatr 1975; 86:873–880.

58. Smeitink JAM, Sengers RCA, Trijbels JMF, et al. Fatal neonatal cardiomyopathy associated with cataract and mitochondrial myopathy. Eur J Pediatr 1989; 148:656–659.

59. Smeitink J, Huizing M, Ruitenbeck W, et al. Adenine nucleotide translocator deficiency in a patient with fatal congenital cardiomyopathy, cataract and mitochondrial myopathy [abstract 013] J Inherit Metab Dis 1997; 20(suppl. 1):7.

60. Graham BH, Waymire KG, Cottrell B, et al. A mouse model for mitochondrial myopathy and cardiomyopathy resulting from a deficiency in the heart/muscle isoform of the adenine nucleotide translocator. Nat Genet 1997; 16:226–234.

61. Palmieri F, Prezioso G, Quagliariello E, Klingenberg M. Kinetic study of the dicarboxylate carrier in rat liver mitochondria. Eur J Biochem 1971; 22:66–74.

62. Palmieri F, Quagliariello E, Klingenberg M. Quantitative correlation between the distribution of anions and the pH difference across the mitochondrial membrane. Eur J Biochem 1970; 17:230–238.

63. Stappen R, Krämer R. Kinetic mechanism of phosphate/phosphate and phosphate/OII⁻ antiports catalyzed by reconstituted phosphate carrier from beef heart mitochondria. J Biol Chem 1994; 269:11240–11246.

64. Ferreira GC, Pederson PL. Phosphate transport in mitochondria: past accomplishments, present problems and future challenges. J Bioenerg Biomembr 1993; 25:483–492.

65. DePinto V, Tommasino M, Palmieri F, Kadenbach B. Purification of the active mitochondrial phosphate carrier by affinity chromatography with an organomercurial agarose column. FEBS Lett 1982; 148:103–106.

66. Bisaccia F, Palmieri F. Specific elution from hydroxylapatite of the mitochondrial phosphate carrier by cardiolipin. Biochim Biophys Acta 1984; 766:386–394.

67. Kolbe HVJ, Costello D, Wong A, Lu RC, Wohlrab H. Mitochondrial phosphate transport. J Biol Chem 1993; 259:9115–9120.

68. Kadenbach B, Mende P, Kolbe HVJ, Stipani I, Palmieri F. The mitochondrial phosphate carrier has an essential requirement for cardiolipin. FEBS Lett 1982; 139:109–112.

69. Dolce V, Fiermonte G, Messina A, Palmieri F. Nucleotide sequence of a human heart cDNA encoding the mitochondrial phosphate carrier. DNA Seq 1991; 2:133–135.

70. Marsh S, Carter NP, Dolce V, Iacobazzi V, Palmieri F. Chromosomal localization of the mitochondrial phosphate carrier gene PHC to 12q23. Genomics 1995; 29:814–815.

71. Dolce V, Iacobazzi V, Palmieri F, Walker JE. The sequences of human and bovine genes of the phosphate carrier from mitochondria contain evidence of alternatively spliced forms. J Biol Chem 1994; 269:10451–10460.

72. Dolce V, Fiermonte G, Palmieri F. Tissue-specific expression of the two isoforms of the mitochondrial phosphate carrier in bovine tissues. FEBS Lett 1996; 399:95–98.

73. Liu MY, Torgrimson A, Colombini M. Characterization and partial purification of the VDAC-channel-modulating protein from calf liver mitochondria. Biochim Biophys Acta 1994; 1185:203–212.

74. Zizi M, Forte M, Blachly-Dyson E, Colombini M. NADH regulates the gating of VDAC, the mitochondrial outer membrane channel. J Biol Chem 1994; 269:1614–1616.

75. Zimmerberg J, Parsegian VA. Polymer inaccessible volume changes during opening and closing of voltage-dependent ionic channel. Nature 1986; 323:36–39.

76. Thinnes FP. Evidence for extra-mitochondrial localization of the VDAC/porin channel in eucaryotic cells. J Bioenerg Biomembr 1992; 24:71–75.

77. Yu WH, Wolfgang W, Forte M. Subcellular localization of human voltage-dependent anion channel isoforms. J Biol Chem 1995; 270:13998–14006.

78. Blachly-Dyson E, Zambronicz EB, Yu WH, et al. Cloning and functional expression in yeast of two human isoforms of the outer mitochondrial membrane channel, the voltage-dependent anion channel. J Biol Chem 1993; 268:1835–1841.

79. Ha H, Hajek P, Bedwell DM, Burrows PD. A mitochondrial porin cDNA predicts the existence of multiple human porins. J Biol Chem 1993; 268:12143–12149.

80. Blachly-Dyson E, Baldini A, Litt M, McCabe ERB, Forte M. Human genes encoding the voltage-dependent anion channel (VDAC) of the outer mitochondrial membrane: mapping and identification of two new isoforms. Genomics 1994; 20:62–67.

81. Winkelbach H, Walter G, Morys-Wortmann C, et al. Studies on human porin XII. Biochem Med Metab Biol 1994; 52:120–127.

82. Halestrap AP, Connern CP, Griffiths EJ. Mechanisms involved in the regulation of mitochondrial inner membrane permeability and their physiological and pathological significance. In: Quagliariello E, Palmieri F, eds. Molecular Mechanisms of Transport. Amsterdam: Elsevier Science, 1992:259–266.

83. Brdiczka D, Wallimann T. The importance of the outer mitochondrial compartment in regulation of energy metabolism. Mol Cell Biochem 1994; 133/134:69–83.

84. Garlid KD. Mitochondrial cation transport: a progress report. J Bioenerg Biomembr 1994; 26:537–542.

85. Jung DW, Brierley GP. Magnesium transport by mitochondria. J Bioenerg Biomembr 1994; 26:527–535.

86. Gunter TE. Cation transport by mitochondria. J Bioenerg Biomembr 1994; 26:465–469.

87. Carafoli E, Caroni P, Chiesi M, Famulski K. Ca^{2+} as a metabolic regulator: mechanisms for the control of its intracellular activity. In: Sies H, ed. Metabolic Compartmentation. London: Academic Press, 1982:521–547.

88. McCormack JG, Halestrap AP, Denton RM. Role of calcium ions in regulation of mammalian intramitochondrial metabolism. Physiol Rev 1990; 70:391–425.

89. Stadhouders AM. Cellular calcium homeostasis, mitochondrial and muscle cell disease. In: Busch HFM, Jennekens FGI, Scholte HR, eds. Mitochondria and Muscular Diseases. Beetsterzwaag, The Netherlands: Mefar B.V., 1981:77–88.

90. Samsom JF, Barth PG, De Vries JIP, et al. Familial mitochondrial encephalomyopathy with fatal ultrasonographic ventriculomegaly and intracerebral calcifications. Eur J Pediatr 1994; 153:510–516.

91. Engel WK. Dagen des oordeels. Pathokinetic mechanisms and molecular messengers (a dramatic view). Arch Neurol 1979; 36:329–339.

92. Nalecz KA, Kaminska J, Nalecz M, Azzi A. The activity of pyruvate carrier in a reconstituted system: substrate specificity and inhibitor sensitivity. Arch Biochem Biophys 1992; 297:162–168.

93. Crompton M, Palmieri F, Capano M, Quagliariello E. A kinetic study of sulphate transport in rat liver mitochondria. Biochem J 1975; 146:667–673.

94. Krämer R, Palmieri F. Metabolite carriers in mitochondria. In: Ernster L, ed. Molecular Mechanisms in Bioenergetics. Amsterdam: Elsevier Science, 1992:359–384.

95. Kaplan RS, Oliveira DL, Wilson GL. Streptozotocin-induced alterations in the levels of functional mitochondrial anion transport proteins. Arch Biochem Biophys 1990; 280:181–191.

96. Kaplan RS, Mayor JA, Blackwell R, Maughon RH, Wilson GL. The effect of insulin supplementation on diabetes-induced alterations in the extractable levels of functional mitochondrial anion transport proteins. Arch Biochem Biophys 1991; 287:305–311.

97. Palmieri L, Palmieri F, Runswick MJ, Walker JE. Identification by bacterial expression and functional reconstitution of the yeast genomic sequence encoding the mitochondrial dicarboxylate carrier protein. FEBS Lett 1996; 399:299–302.

98. Kakhniashvili D, Mayor JA, Gremse DA, Xu Y, Kaplan RS. Identification of a novel gene encoding the yeast mitochondrial dicarboxylate transport protein via overexpression, purification, and characterization of its protein product. J Biol Chem 1997; 272:4516–4521.

99. Bisaccia F, Indiveri C, Palmieri F. Purification of reconstitutively active alpha-oxoglutarate carrier from pig heart mitochondria. Biochim Biophys Acta 1985; 810:362–369.

100. Indiveri C, Krämer R, Palmieri F. Reconstitution of the malate/aspartate shuttle from mitochondria. J Biol Chem 1987; 262:15979–15983.

101. Bisaccia F, Zara V, Capobianco L, et al. The formation of a disulfide cross-link between the two subunits demonstrates the dimeric structure of the mitochondrial oxoglutarate carrier. Biochim Biophys Acta 1996; 1292:281–288.

102. Runswick MJ, Walker JE, Bisaccia F, Iacobazzi V, Palmieri F. Sequence of the bovine 2-oxoglutarate/malate carrier protein: structural relationship to other mitochondrial transport proteins. Biochemistry 1990; 29:11033–11040.

103. Iacobazzi V, Palmieri F, Runswick MJ, Walker JE. Sequences of the human and bovine genes for the mitochondrial 2-oxoglutarate carrier. DNA Seq 1992; 3:79–88.

104. Dolce V, Messina A, Cambria A, Palmieri F. Cloning and sequencing of the rat cDNA encoding the mitochondrial 2-oxoglutarate carrier protein. DNA Seq 1994; 5:103–109.

105. LaNoue KL, Schoolwerth AC. Metabolic transport in mitochondria. Annu Rev Biochem 1979; 48:871–922.

106. Meijer AJ, Van Dam K. The metabolic significance of anion transport in mitochondria. Biochim Biophys Acta 1974; 346:213–244.
107. McLeod DS, Lutty GA. Menadione-dependent alpha glycerolphosphate and succinate dehydrogenases in the developing canine retina. Curr Eye Res 1995; 14:819–826.
108. Pande SV, Brivet M, Slama A, et al. Carnitine-acylcarnitine translocase deficiency with severe hypoglycemia and auriculo ventricular block. J Clin Invest 1993; 91:1247–1252.
109. Ramsay RR, Tubbs PK. The mechanism of fatty acid uptake by heart mitochondria: an acylcarnitine-carnitine exchange. FEBS Lett 1975; 54:21–25.
110. Stanley CA. New genetic defects in mitochondrial fatty acid oxidation and carnitine deficiency. Adv Pediatr 1987; 34:59–88.
111. Stanley CA, Hale DE, Berry GT, et al. A deficiency of carnitine-acylcarnitine translocase in the inner mitochondrial membrane. N Engl J Med 1992; 327:19–22.
112. Indiveri C, Iacobazzi V, Giangregorio N, Palmieri F. The mitochondrial carnitine carrier protein: cDNA cloning, primary structure and comparison with other mitochondrial transport proteins. Biochem J 1997; 321:713–719.
113. Huizing M, Iacobazzi V, Ijlst L, et al. Cloning of the human carnitine-acylcarnitine carrier cDNA, and identification of the molecular defect in a patient. Am J Hum Genet 1997; 61:1239–1245.
114. Viggiano L, Iacobazzi V, Marzella R, et al. Assignment of the carnitine/acylcarnitine translocase (CACT) gene to human chromosomal band 3p21.31 by in situ hybridization. Cytogenet Cell Genet. In press.
115. Brivet M, Slama A, Ogier H, et al. Diagnosis of carnitine acylcarnitine translocase deficiency by complementation analysis. J Inherit Metab Dis 1994; 17:271–274.
116. Niezen-Koning KE, Van Spronsen FJ, Ijlst L, et al. A patient with lethal cardiomyopathy and a carnitine-acylcarnitine translocase deficiency. J Inherit Metab Dis 1995; 18:230–232.
117. Ogier de Baulney H, Slama A, Touati G, et al. Neonatal hyperammonemia caused by a defect of carnitine-acylcarnitine translocase. J Pediatr 1995;127:723–728.
118. Brivet M, Slama A, Millington DS, et al. Retrospective diagnosis of carnitine-acylcarnitine translocase deficiency by acylcarnitine analysis in the proband Guthrie card and enzymatic studies in the parents. J Inherit Metab Dis 1996; 19:181–184.
119. Chalmers RA, Stanley CA, English N, Wigglesworth JS. Mitochondrial carnitine-acylcarnitine translocase deficiency presenting as sudden neonatal death. J Pediatr 1997; 131:220–225.
120. Huizing M, Wendel U, Ruitenbeek W, et al. Carnitine-acylcarnitine carrier deficiency: identification of the molecular defect in a patient. J Inherit Metab Dis 1998; 21:262–267.
121. Tsukihara T, Aoyama H, Yamashita E, et al. The whole structure of the 13-subunit oxidized cytochrome c oxidase at 2.8 A. Science 1996; 272:1136–1144.
122. Clark A, Clark PAA, Connett RJ, Gayeski TEJ, Honig CR. How large is the drop in PO_2 between cytosol and mitochondrion? Am J Physiol 1987; 252:C583–C587.
123. Kennedy FG, Jones DP. Oxygen dependence of mitochondrial function in isolated rat cardiac myocytes. Am J Physiol 1986; 250:C374–C383.

124. Gayeski TEJ, Connett RJ, Honig CR. Minimum intracellular PO_2 for maximum cytochrome turnover in red muscle *in situ*. Am J Physiol 1987; 252:H906–H915.
125. Sengers RCA, Stadhouders AM, Van Lakwijk-Vondrovicova E, Kubat K, Ruitenbeek W. Hypertrophic cardiomyopathy associated with a mitochondrial myopathy of voluntary muscles and congenital cataract. Br Heart J 1985; 54:543–547.
126. Meijer AJ, Hensgens HESJ. Ureogenesis. In: Sies H, ed. Metabolic Compartmentation. New York: Academic Press, 1982:259–286.
127. Meijer AJ, Gimpel JA, De Leeuw G, et al. Interrelationships between gluconeogen esis and ureogenesis in isolated hepatocytes. J Biol Chem 1978; 253:2308–2320.
128. Indiveri C, Tonazzi A, Palmieri F. Identification and purification of the ornithine/citrulline carrier from rat liver mitochondria. Eur J Biochem 1992; 207:449–454.
129. Indiveri C, Palmieri L, Palmieri F. Kinetic characterization of the reconstituted ornithine carrier from rat liver mitochondria. Biochim Biophys Acta 1994; 1188:293–301.
130. Indiveri C, Tonazzi A, Palmieri F. The purified and reconstituted ornithine/citrulline carrier from rat liver mitochondria: electrical nature and coupling of the exchange reaction with H^+ translocation. Biochem J 1997; 327:349–356.
131. Palmieri L, De Marco V, Iacobazzi V, et al. Identification of the yeast ARG-11 gene as a mitochondrial ornithine carrier involved in arginine biosynthesis. FEBS Lett 1997; 410:447–451.
132. Kaplan RS, Mayor JA, Wood DO. The mitochondrial tricarboxylate transport protein. cDNA cloning, primary structure, and comparison with other mitochondrial transport proteins. J Biol Chem 1993; 268:13682–13690.
133. Meijer AJ, Van Dam K. The metabolic significance of anion transport in mitochondria. Biochim Biophys Acta 1974; 346:213–244.
134. Iacobazzi V, Lauria G, Palmieri F. Organization and sequence of the human gene for the mitochondrial citrate transport protein. DNA Seq 1997; 7:127–139.
135. Williamson JR. Role of anion transport in the regulation of metabolism. In: Hanson RW, Mehlman MA, eds. Gluconeogenesis. London: John Wiley, 1976:165–220.
136. Snell K. Muscle alanine synthesis and hepatic gluconeogenesis. Biochem Soc Trans 1980; 8:205–213.
137. Nosek MT, Dransfield DT, Aprille JR. Calcium stimulates ATP-Mg/P_i carrier activity in rat liver mitochondria. J Biol Chem 1990; 265:8444–8450.
138. Aprille JR. Mechanism and regulation of the mitochondrial ATP-Mg/Pi carrier. J Bioenerg Biomembr 1993; 25:473–481.
139. Cybulsky RL, Fisher RR. Mitochondrial neutral amino acid transport: evidence for a carrier mediated mechanism. Biochemistry 1977; 16:5116–5120.
140. Passarella S, Atlante A, Barile M. New aspects in mitochondrial transport and metabolism of metabolites and vitamin derivatives. In: Palmieri F, ed. Progress in Cell Research, vol. 5. Amsterdam: Elsevier Science, 1995:89–93.
141. Barile M, Passarella S, Bertoldi A, Quagliariello E. Flavin adenine dinucleotide synthesis in isolated rat liver mitochondria caused by imported flavin mononucleotide. Arch Biochem Biophys 1993; 305:442–447.
142. Tahilani AG. Evidence for net uptake and efflux of mitochondrial coenzyme A. Biochim Biophys Acta 1991; 1067:29–37.

143. Barile M, Passarella M, Quagliariello E. Thiamine pyrophosphate uptake into isolated rat liver mitochondria. Arch Biochim Biophys 1990; 280:352–357.

144. Horne DW, Holloway RS, Said HM. Uptake of 5-formyltetrahydro-folate in isolated rat liver mitochondria is carrier-mediated. J Nutr 1992; 122:2204–2209.

145. McEnery MW, Dawson TM, Verma A, et al. Mitochondrial voltage-dependent anion channel. Immunochemical and immunohistochemical characterization in rat brain. J Biol Chem 1993; 268:23289–23296.

146. Gellerich FN, Kapischke M, Kunz W, et al. The influence of the cytosolic oncotic pressure on the permeability of the mitochondrial outer membrane for ADP:implications for the kinetic properties of mitochondrial creatine kinase and for ADP channelling into the intermembrane space. Mol Cell Biochem 1994; 133/134:85–104.

147. Haucke V, Schatz G. Import of proteins into mitochondria and chloroplasts. Trends Cell Biol 1997; 7:103–106.

148. Neupert W. Protein import into mitochondria. Annu Rev Biochem 1997; 66:863–917.

149. Pfanner N, Craig EA, Meijer M. The protein import machinery of the mitochondrial inner membrane. Trends Biochem Sci 1994; 19:368–372.

150. Lithgow T, Cuezva JM, Silver PA. Highways for protein delivery to the mitochondria. Trends Biochem Sci 1997; 22:110–113.

151. Freitag H, Janes M, Neupert W. Biosynthesis of mitochondrial porin and insertion into the outer mitochondrial membrane of *Neurospora crassa*. Eur J Biochem 1982; 126:197–202.

152. Court DA, Kleene R, Neupert W, Lill R. Role of the N- and C-termini of porin in import into the outer membrane of *Neurospora* mitochondria. FEBS Lett 1996; 390:73–77.

153. Moczko M, Ehmann B, Gartner F, et al. Deletion of the receptor Mom19 strongly impairs import of cleavable preproteins into *Saccharomyces cerevisiae* mitochondria. J Biol Chem 1994; 269:9045–9051.

154. Zara V, Rassow J, Wachter E, et al. Biogenesis of the mitochondrial phosphate carrier. Eur J Biochem 1991; 198:405–410.

155. Zara V, Palmieri F, Mahlke K, Pfanner N. The cleavable presequence is not essential for import and assembly of the phosphate carrier of mammalian mitochondria, but enhances the specificity and efficiency of import. J Biol Chem 1992; 267:12077–12081.

156. Seki N, Moczko M, Nagase T, et al. A human homology of the mitochondrial protein import receptor Mom19 can assemble with the yeast mitochondrial receptor complex. FEBS Lett 1995; 375:307–310.

157. Agsteribbe E, Huckriede A, Veenhuis M, et al. A fatal, systemic mitochondrial disease with decreased mitochondrial enzyme activities, abnormal ultrastructure of the mitochondria and deficiency of heat shock protein 60. Biochem Biophys Res Commun 1993; 193:146–154.

158. Huckriede A Agsteribbe E. Decreased synthesis and inefficient mitochondrial import of Hsp60 in a patient with a mitochondrial encephalomyopathy. Biochim Biophys Acta 1994; 1227:200–206.

159. Briones P, Vilaseca MA, Ribes A, et al. A new case of multiple mitochondrial

enzyme deficiencies with decreased amount of heat shock protein 60. J Inherit Metab Dis 1997; 20:569–577.

160. Cox DA, Matlib MA. A role for the mitochondrial Na^+-Ca^{2+} exchanger in the regulation of oxidative phosphorylation in isolated heart mitochondria. J Biol Chem 1993; 268:938–947.

161. Trijbels JMF, Sengers RCA, Ruitenbeek W, et al. Disorders of the mitochondrial respiratory chain: clinical manifestations and diagnostic approach. Eur J Pediatr 1988; 148:92–97.

162. Chance B, Sies H, Boveris A. Hydroperoxide metabolism in mammalian organs. Physiol Rev 1979; 59:527–605.

163. Suzuki YJ, Forman HJ, Sevanian A. Oxidants as stimulators of signal transduction. Free Radic Biol Med 1997; 22:269–285.

164. Halliwell B. Reactive oxygen species in living systems: source, biochemistry and role in human disease. Am J Med 1991; 91:14–22.

165. Richter C. Reactive oxygen and DNA damage in mitochondria. Mutat Res 1992; 275:249–255.

166. Nohl H, De Silva D, Summer KH. 2,3,7,8-tetrachlorodibenzo-p-dioxin induces oxygen activation with cell respiration. Free Radic Biol Med 1989; 6:369 374.

167. Beyer RE. An analysis of the role of coenzyme Q in free radical generation and as an oxidant. Biochem Cell Biol 1992; 70:390–403.

168. Takeyama N, Matsuo N, Tanaka T. Oxidative damage to mitochondria is mediated by the Ca^{2+}-dependent inner membrane permeability transition. Biochem J 1993; 294:719–725.

169. Pitkänen S, Robinson BH. Mitochondrial complex I deficiency leads to increased production of superoxide radicals and induction of superoxide dismutase. J Clin Invest 1996; 98:345–351.

170. Marchetti P, Castedo M, Susin SA, et al. Mitochondrial permeability transition is a central coordinating event of apoptosis. J Exp Med 1996; 184:1155–1160.

171. Bakker HD, Scholte HR, Van den Bogert C, et al. Adenine nucleotide translocase deficiency in muscle: potential therapeutic value of vitamin E. J Inherit Metab Dis 1993; 16:548–552.

172. Nohl H, Hegner D. Do mitochondria produce oxygen radicals *in vitro*? Eur J Biochem 1978; 82:563–567.

173. Linnane AW, Marzuki S, Ozawa T, Tanaka M. Mitochondrial DNA mutations as an important contributor to ageing and degenerative disease. Lancet 1989; 8639: 642–645.

174. Robinson BH, Taylor J, Francois B, Beaudet AL, Peterson DF. Lactic adosis, neurological deterioration and compromised cellular pyruvate oxidation due to a defect in the reoxidation of cytoplasmically generated NADH. Eur J Pediatr 1983; 140: 98–101.

175. Tritschler HJ, Packer L, Mcdori R. Oxidative stress and mitochondrial dysfunction in neurodegeneration. Biochem Mol Biol Int 1994; 34:169–181.

176. Okada S, Takehara Y, Yabuki M, et al. Nitric oxide, a physiological modulator of mitochondrial function. Physiol Chem Phys Med NMR 1996; 28:69–82.

177. Zoratti M, Szabo I. The mitochondrial permeability transition. Biochim Biophys Acta 1995; 1241:139–176.

178. Zorov DB. Mitochondrial damage as a source of diseases and aging: a strategy of how to fight these. Biochim Biophys Acta 1996; 1275:10–15.

179. Richter C, Schweizer M, Cossarizza A, Franceschi C. Control of apoptosis by the cellular ATP level. FEBS Lett 1996; 378:107–110.

180. Bernardi P, Vassanelli S, Veronese P, et al. Modulation of the mitochondrial permeability transition pore. Effect of protons and divalent cations. J Biol Chem 1992; 267:2934–2939.

181. Bernardi P. The permeability transition pore. Control points of a cyclosporin A-sensitive mitochondrial channel involved in cell death. Biochim Biophys Acta 1996; 1275:5–9.

182. Chernyak BV, Bernardi P. The mitochondrial permeability transition pore is modulated by oxidative agents through both pyridine nucleotides and glutathione at two separate sites. Eur J Biochem 1996; 238:623–630.

Index